MINNESOTA STUDIES IN THE PHILOSOPHY OF SCIENCE

Minnesota Studies in the
PHILOSOPHY OF SCIENCE

HERBERT FEIGL, FOUNDING EDITOR

VOLUME X

Testing Scientific Theories

EDITED BY

JOHN EARMAN

UNIVERSITY OF MINNESOTA PRESS, MINNEAPOLIS

Library of Congress Cataloging in Publication Data

Main entry under title:

Testing scientific theories.

(Minnesota studies in the philosophy of science; v. 10)
Includes index.
1. Science—Philosophy.   2. Theory (Philosophy)   3. Evidence.
I. Earman, John.   II. Series.
Q175.M64 vol. 10 501s [501]   82-24816
ISBN 0-8166-1158-0
ISBN 0-8166-1159-9 (pbk.)

Preface

Progress in philosophy is often hard to detect—perhaps, the cynic will urge, because it is often nonexistent. However, I submit that the volumes of this series demonstrate a steady advance in our understanding of the structure, the function, and the testing and confirmation of scientific theories. The last mentioned topic was the subject of a Center research conference held in June of 1980; many of the papers of the present volume arose from that conference. The focus for the conference was provided by Clark Glymour's *Theory and Evidence*. The negative thesis of Glymour's book is that the two most widely discussed accounts of the methodology of theory testing—hypothetico-deductivism and Bayesianism—are flawed, the latter seriously, the former irremediably. Hempel's notion of instance confirmation comes closer to capturing the sorts of structural relations between evidence and hypothesis which, by Glymour's lights, lies at the heart of theory testing. But Hempel's original account was too narrow in not permitting hypotheses stated in the theoretical language to be confirmed by evidence stated in the observational language. Glymour proposed to remedy this defect with the ingenious idea of 'bootstrapping,' which, with some false modesty, he attributes to Reichenbach, Weyl, and others: the basic relation of confirmation is three-place (E confirms H relative to T) and auxiliary assumptions drawn from T may be used in deducing instances of H from E.

As the papers in the first section indicate, the bootstraps may have to be shortened; Edidin and van Fraassen, for example, argue that the hypothesis under test need not and should not be used as an auxiliary. Some Bayesians remain unfazed by Glymour's criticisms, while others have been led to abandon the assumption of logical omniscience, implicit in most Bayesian learning models, in order to overcome Glymour's 'problem of old evidence.' Glymour, while still rejecting the 'never-never land' approach of orthodox Bayesianism, has moved to consider how to combine bootstrapping relations with partial knowledge of probabilities; the appropriate tool turns out to be the belief functions developed by Glen Schafer. It is thus heartening to be able to report that the various opposing camps learned from each other. I like to think that the interactions initiated by our conference contributed to this learning process.

All is not bootstrapping and Bayesianism. The volume also contains still other accounts of the methodology of theory testing. In addition, there are some valuable historical case studies against which the theories of methodology can be tested. And there are some timely discussions of the problems of testing psychoanalytic hypotheses and hypotheses about the completeness of the fossil record. In short, enough new ideas are germinated in this collection that I am confident in predicting that philosophers of science will reap the harvest for years to come.

While this volume was in preparation, Grover Maxwell left us to struggle on with the problems of philosophy of science without the benefit of his always gentle and insightful guidance. The editor and authors join in affectionately dedicating the volume to Grover's memory.

John Earman

Contents

I. GLYMOUR'S BOOTSTRAPPING THEORY OF CONFIRMATION

Strap →

The Boot

On Testing and Evidence

1. Introduction

If we knew the probabilities of all things, good thinking would be apt calculation, and the proper subject of theories of good reasoning would be confined to the artful and efficient assembly of judgments in accordance with the calculus of probabilities. In fact, we know the probabilities of very few things, and to understand inductive competence we must construct theories that will explain how preferences among hypotheses may be established in the absence of such knowledge. More than that, we must also understand how the procedures for establishing preferences in the absence of knowledge of probabilities combine with fragmentary knowledge of probability and with procedures for inferring probabilities. And, still more, we need to understand how knowledge of probabilities is obtained from other sorts of knowledge, and how preferences among hypotheses, knowledge of probabilities, and knowledge of other kinds constrain or determine rational belief and action.

For at least two reasons, the first and second of these questions have been neglected for many years by almost everyone except groups of philosophers who have been virtually without influence on practitioners of "inductive logic." One reason, unfortunately, is that a long tradition of attempts to construct an account of hypothesis testing, confirmation, or comparison outside of the calculus of probabilities has failed to achieve a theory possessing both sufficient logical clarity and structure to merit serious interest, and sufficient realism to be applied in scientific and engineering contexts with plausible results. A second reason is the recent predominance of the personal or subjective interpretation of probability: probability is degree of rational belief, and rational inference consists in appropriate changes in the distribution of intensity of belief. Rationality at any moment requires only that belief intensities be so distributed as to satisfy the axioms of probability. Thus, even when we reason about things in the absence of knowledge of objective chances, we still retain degrees of belief, and hence if we are rational our reasoning is an application of the calculus of probabilities or of principles for changing probabilities. The community of philosophical methodologists has been captured by this vision, and in consequence many have come to think that there is properly no separate question regarding how hypotheses may be assessed in the absence of knowledge of probabilities. I continue to disagree.

Theory and Evidence aimed to show that (and in some measure, why) the tradition of work on questions of testing and comparing hypotheses outside of the calculus of probability had failed to produce a plausible and nontrivial theory of the matter; quite as much, I aimed to argue that the Bayesian assimilation of scientific reasoning carried on without knowledge of probabilities to a species of probable reasoning is both inaccurate and unrevealing about the procedures that, in the absence of knowledge of probabilities, make for inductive competence. To no very great surprise, arguments on these scores have failed to persuade either the advocates of hypothetico-deductivism or Bayesian confirmation theorists. On the positive side, *Theory and Evidence* described and illustrated an elementary strategy for testing and discriminating among hypotheses. Questions regarding how such a strategy may be combined with fragmentary knowledge of probabilities were addressed only briefly and in passing.

What follows is unpolemical. The characterization of the testing relation for systems of equations and inequations was left undeveloped in *Theory and Evidence* and was in effect given through a class of examples, and rather special examples at that, in Chapter 7 of the book. Here I have tried to give a more general characterization of the strategy, and to describe in detail how particular features may be varied. I have also tried to characterize some of the structural relations that the account of testing permits us to define, and to indicate some questions about these relations that deserve further exploration. *Theory and Evidence* contained an abundance of informal illustrations of applications of the testing strategy. In this essay a quite different illustration, suggested to me by William Wimsatt, is given: It is shown that for systems of Boolean equations, certain variants of bootstrap testing are equivalent to a formalized testing strategy used in the location of faults in computer logic circuits. Since the context is a very special one, this is an interesting but not a distinctive result. A hypothetico-deductive approach will yield the same equivalence, assuming a special axiomatization of the system of equations.

This essay begins to discuss how to combine certain of the features of bootstrap testing with fragmentary knowledge of probabilities. In reflecting on these questions, I have been driven to reliance on work by Arthur Dempster and its generalization and reconception by Glenn Shafer. Shafer's interesting work has not found an eager reception among methodologists, but in certain contexts it seems to me to be just the right thing. At the very least, I hope the reader will come to see from the discussion below how elements of Shafer's theory of belief functions arise in a natural way when one considers evidence in combination with fragmentary knowledge of probabilities. I have applied Shafer's theory in a "bootstrap" setting to the same engineering problem discussed earlier—fault detection in logic circuits.

The attempt in Chapter 5 of *Theory and Evidence* to characterize a confirmation or testing relation for formalized theories in a first-order language is palpably clumsy. With five complex clauses, the analysis reads, as one commentator noted, like the fine print in an auto-rental contract. It speaks for the intellectual good will of many commentators that they have nonetheless patiently investigated its consequences and examined its motivations. I am especially indebted to David Christensen, Aron Edidin, Kevin Kelly, Robert Rynasiewicz, and Bas van Fraassen. Edidin showed that the attempt to allow computations to entail the hypothesis to be tested was unsound, and could not be realized without permitting trivial confirmation relations. Van Fraassen saw the essential ideas were contained in the applications to systems of equations, reformulated the ideas more elegantly, and gave some intriguing yet simple examples. Rynasiewicz showed that the rental clauses had loopholes. Christensen showed that the bootstrapping idea has a fundamental difficulty in the context of the predicate calculus: any purely universal hypothesis containing only a single non-logical predicate can be "tacked on" to any testable theory in which that predicate does not occur, and the "tacked on" hypothesis will be bootstrap tested with respect to the extended theory. Although Kelly has shown that Christensen's difficulty does not arise when the account of testing is developed in the context of relevance logic, in what follows I have applied the testing strategy only to systems of equations and inequations.

2. Three Schemas for Bootstrap Testing of Equations and Inequations

Let A be any algebra with a natural partial or complete ordering. I shall assume it is understood what is meant by an assignment of values in A to variables *satisfying* an equation or inequality H, and what it means for H to be *valid*. For algebraic equations or inequalities, H and K, over A, we shall say that H *entails* K if every set of values satisfying H satisfies K. Analogously if H is a set of equations or inequations. A *subtheory* of a system of equations or inequalities T is any system that is satisfied whenever T is satisfied. Two systems are *equivalent* if each is a subtheory of the other. A variable is said to occur *essentially* in an equation or inequality H if that variable occurs in every system equivalent to H. *Var (H)* will denote the set of variables occurring essentially in H. The simplest scheme for bootstrap testing is then:

Schema I: Let H and T be as above and mutually consistent; let e* be a set of values for the variables e_j, consistent with H and T. For each x_i in Var(H) let T_i be a subtheory of T such that

(i) T_i determines x_i as a (possibly partial) function of the e_j, denoted $T_i(e_j)$;

(ii) the set of values for Var(H) given by $x_i^* = T_i(e^*)$ satisfies H;

(iii) if &T is the collection of T_i for all of the x_i in Var(H), &T does *not* entail that H is equivalent to any (in) equation K such that Var(K) $\subset$ Var(H);

(iv) for all i, H does not entail that T_i is equivalent to any (in) equation K such that Var(K) $\subset$ Var(T_i).

If these conditions are met, the e* are said to provide a *positive test* of H with respect to T. These are roughly the conditions given, for a special case, in Chapter 7 of *Theory and Evidence* and illustrated there in applications in the social sciences. The motivation for each condition in Schema I is reasonably straightforward. The requirement that a value be determined for each quantity occurring essentially in H reflects a common prejudice against theories containing quantities that cannot be determined from the evidence; more deeply, it reflects the sense that when values for the basic quantities occurring in H have not been determined from the evidence using some theory, then that evidence and the relevant fragment of that theory do not of themselves provide reason to believe that those basic quantities are related as H claims them to be.

The reason for the second requirement in Schema I is obvious. The third requirement is imposed because if it were violated there would be some quantity x occurring *essentially* in H such that the evidence and the method of computing quantities in Var(H) from the evidence would fail to *test* the constraints H imposes on x. Thus a significant component of what H says would go untested. The fourth condition is motivated by the fact, noted by Christensen, that the first three conditions alone permit a value of y and values of c and d to test the hypothesis $x = y$ with respect to the pair of equations $x = y$ and $c = d$: simply measure y and compute x using the consequence $x = y + (c-d)$ and measured values of c and d.

The third condition might even be strengthened plausibly as follows:

Schema II: Let H and T be as above and mutually consistent, and let e^* be a set of values for the variables e_j consistent with H and T. For each x_i in Var(H) let T_i be a subtheory of T such that

(i) T_i determines x_i as a (possibly partial) function of the e_j, denoted $T_i(e_j)$;

(ii) the set of values for Var(H) given by $x_i^* = T_i(e^*)$ satisfies H;

(iii) for each x_i in Var(H), there exists a set of values e_i^+ for the e_j such that for $k \neq i$, $T_k(e_i^+) = T_k(e^*)$ and the set of values for Var(H) given by $x_m^+ = T_m(e_i^+)$ does not satisfy H;

(iv) for all i, H does not entail that T_i is equivalent to any equation K such that $Var(K) \subset Var(T_i)$.

When the conditions of Schema II are met, the values e^* are again said to provide a positive test of H with respect to T.

Schemas I and II are not equivalent, and Schema II imposes stronger requirements on positive tests. For example, let H be $x_1 = x_2 \cdot x_3$. Suppose that the x_i are measured directly, so that $x_i = e_i$ in Schemas I and II. Let the evidence be that $x_1 = 0$, $x_2 = 0$, and $x_3 = 1$. Then H is positively tested according to Schema I but not according to Schema II.

The difference between these two schemas lies entirely in the third clause, for which there is a natural further alternative:

Schema III

(iii) If K is any sentence such that &T entails K and $Var(K) \subseteq Var(H)$, then K is valid.

Schema III is neither stronger nor weaker than Schema II. For the example just given, $x_1 = x_2 = 0$, and $x_3 = 1$ does count as a positive test according to III. On the other hand, if we have

H: $x > 3$

T: $x = e^2$

and evidence e = 2, where variables are real valued, then we have a positive test according to I and II but not according to III since T entails $x \geq 0$. Hence all three schemas are inequivalent. Further, Schema III satisfies a consequence condition: If H is positively tested with respect to T, then so is any consequence G of H, so long as $\mathrm{Var}(G) \subseteq \mathrm{Var}(H)$.

3. Variant Schemes

3.1 Set Values

Schemas I, II, and III are stated in such a way that the hypothesis tested, H, may state an equation or an inequality. The T_i in these schemas, however, must be equations. In situations in which the T_i include inequations the schemas must be modified so that the variables become, in effect, set valued. Rynasiewicz and Van Fraassen have both remarked that they see no reason to require that *unique* values of the computed quantities be determined from the evidence, and I quite agree—in fact, in the lamented Chapter 8 of *Theory and Evidence*, the testing scheme for equations is applied without the requirement of unique values.

A set valuation for a collection of variables is simply a mapping taking each variable in the collection to a subset of the algebra A. For any algebraic function f, the value of f(x) when the value of x is (in) $\triangle$ is {f(a) I a is in $\triangle$} , written f($\triangle$). If $x_i = T_i(e_j)$ is an equation, the value of the x_i determined by that equation for values of e_j in $\triangle_j$ is the value of $T_i(e_j)$ when the value of e_j is $\triangle_j$. If $x_i > T_i(e_j)$ is an inequation, the value of x_i determined by that inequation for values of e_j in $\triangle_j$, is {a in A : b is in $T_i(\triangle_j)$ implies a > b}. (There is an alternative definition {a in A I there exists b in $T_i(\triangle_j)$ such that b > a}). For a collection of inequations determining x_i , the set-value of x_i determined by the inequations jointly is the intersection of the set-values for x_i that they determine individually.

An equation or inequation $H(x_i)$ is *satisfied* by set values $\triangle_i$ for its essential variables x_i if there exists for each i a value x_i^* in $\triangle_i$ such that the collection of x_i^* satisfy H in the usual sense. The notions of equivalence and entailment are explained just as before. With these explications, Schemas I, II, and III make sense if the e* are set values or if the T assert inequalities. (Of course, in the schemas we must then understand a value of x to be any *proper* subset of the set of all possible values of x.)

3.2 Extending "Quantities"

The first condition of each schema requires that for H to be positively tested, a value for each quantity occurring essentially in H must be determined. That is not a condition we always impose when testing our theories. We do not nowadays impose it when testing, say, hypotheses of statistical mechanics, where the quantities that experiment determines are not, e.g., values of the momenta of individual molecules, but instead only averages of large numbers of such values. Yet at the turn of the century it was in considerable part the inability of statistical mechanicians to determine values for such basic quantities reliably that gave anti-atomists such as Ostwald reason to doubt kinetic theory. Ostwald's methodological sensibilities were not, in that particular, idiosyncratic, and even his atomist opponents, such as Perrin, realized something of what was required to disarm his criticisms. I view Schemas I, II, and III as providing accounts of when, in the absence of special knowledge about the quantities occurring in an hypothesis, a piece of evidence is determined by a theory to be positively relevant to the hypothesis. But sometimes—as with statistical mechanical hypotheses *nowadays*—we do have special knowledge about the quantities that occur in the hypothesis to be tested. We do not need some special evidence that, for example, the value of the quantity determined to be the average kinetic energy really is the average of the many distinct quantities, the kinetic energies of the individual molecules. Schemas I, II, and III can be altered to apply to situations in which we do have knowledge of this sort.

Let $f_1(x_i)$ be functions of the collection of variables x_i occurring essentially in H, and suppose the following hold:

(a) All f_1 occur in H.

(b) No one occurrence of any quantity x_i in H is within the scope of (i.e., is an argument of) two distinct occurrences of functions in the set of f_1.

(c) There exist values V for the f_1 such that every set of values for the x_i consistent with V satisfies H, and there exist values U for the f_1 such that every set of values for the x_i consistent with U fails to satisfy H.

Call a set of variables or functions satisfying (a), (b), and (c) *sufficient* for H. Values V of the functions f_1 are said to *satisfy* H if every set of values for the x_i in Var(H) consistent with V satisfies H. Analogously, values U for f_1 *contradict* H if every set of values for the x_i in Var(H) consistent with U fail to satisfy H. Now Schema I can be reformulated as follows:

Schema IV: Let H be an equation or inequation, let T be a system of equations or inequations consistent with H, and let Q be a set of variables or functions sufficient for H. Let e* be a set of values for the variables e_j consistent with H and T. For each f_l in Q let T_l be a subtheory of T such that:

   (i) T_l determines f_l as a (possibly partial) function of e_j , denoted $T_l(e_j)$.
   (ii) the set of values for elements of Q given by $f_l* = T_l(e*)$ satisfies H.
   (iii) the collection of T_l for f_l in Q does not entail that H is equivalent to any (in) equation E such that $Var(E) \subset Var(H)$.
   (iv) for all l, H does not entail that T_l is equivalent to any (in) equation K such that $Var(K) \subset Var(T_l)$.

Schema IV does not satisfy the equivalence condition; hence a hypothesis K must be regarded as positively tested according to Schema IV if and only if K is equivalent to a hypothesis H satisfying the Schema.

An analogous version of Schema II, call it Schema V, can readily be stated. I shall forbear from giving the whole of the thing, and instead only a version of the third clause.

Schema V:

   (iii) For each f_l in Q, there exists a set of values e^+, for e_j such that for $k \neq l$, $T_k(e^+) = T_k(e*)$, and the set of values for Q given by $f_m^+ = T_m(e^+)$ contradicts H.

The reformulation of Schema III is obvious.

4. Questions of Separability

I view the bootstrap testing of a theory as a matter of testing one equation in a theory against others, with an eye in theory construction towards establishing a set of equations each of which is positively tested with respect to the entire system. Clearly, in many systems of equations there are epistemological entanglements among hypotheses, and that fact constitutes one of the grains of truth in holism. Are there natural ways to describe or measure such entanglements?

Let E be a set of variables. I shall assume that all possible evidence consists in all possible assignments of values to variables in E. Explicit reference to E is sometimes suppressed in what follows, but it should be understood that all notions are relative to E. We consider some definite system of equations or inequations, T. All hypotheses considered are assumed to be among the consequences of T. Again, even where not made explicit, notions in this section are relative to T and E.

One natural way of characterizing the epistemic interdependence of hypotheses is through the following notions:

1. H is *E-dependent* on K if for every (possible) positive test of H, K is a consequence of the set of T_i used in the test,
2. H is *cyclic* if there is a sequence of hypotheses $H_1, \ldots H_n$, $n > 2$, such that H_i is inequivalent to H_{i+1} and H_i is E-dependent on H_{i+1}, for all i, and $H_1 = H_n = H$.
3. H *E-entails* K if every set of values of E that provides a positive test of H also provides a positive test of K.
4. H and K are *interdependent* if each is E-dependent on the other; *E-equivalent* if each E entails the other.

Notice that the notion of two equations being E-equivalent is not the same as that of each being E-dependent on the other. For example, among the consequences of the theory

$$a = e_1$$
$$b = e_2$$
$$a = b$$

with e_1 and e_2 directly measurable quantities, the equations $e_1 = b$ and $e_2 = b$ are E-dependent on one another, but not on $a = b$, whereas all three of these equations are E-equivalent. Each of the claims $e_1 = b$ and $e_2 = b$ is cyclic. E-entailment relations have no neat or natural relation to ordinary entailment when the notion of positive test is explicated according to Schema I or Schema II. For Schema III, however, I make the following conjecture:

If X is a collection of equations or inequations, A is an equation or inequation such that every variable occurring essentially in A occurs essentially in some member of X, and X entails A, then X E-entails A.

Thus we can understand E-entailment for Schema III as an extension of ordinary entailment over a restricted language.

From the point of view of methodology, E-dependence relations seem more important than E-entailment relations, even though they are formally less interesting.

5. Fault Detection in Logic Circuits

A standard problem in computer science is that of determining faulty circuit elements in a complex logic circuit from data as to circuit inputs and outputs. William Wimsatt has remarked that this is in effect a problem of localizing confirmation (or disconfirmation) within a complex theory, so bootstrap techniques should apply to it. Indeed, because such circuits can be represented by a system of equations with variables taking values in a two-element Boolean algebra, bootstrap testing generates a technique for the determination of faults. Computer scientists have developed methods for the solution of the problem, and although their methods may not look as if they are bootstrap tests, the two procedures are equivalent.

In the simplest way, bootstrap testing (in the sense of Schemas I or III), applies to logic circuits as follows: Consider a circuit such as the one illustrated below:

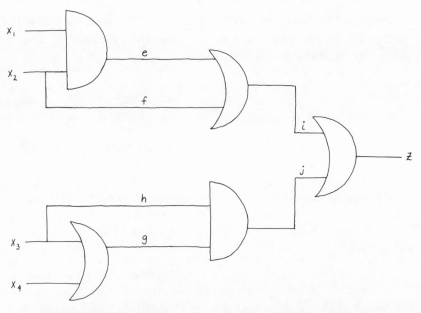

The input signals are the x_i, the output signal is z. Cusps signify an "or" or "+" gate, and dees signify an "and" or "·" gate. Each line of the circuit is an immediate function either of inputs or of lines to which it is connected by a gate. Thus expressing each line, and also the output z, as a function of those lines that lie to the left of it and are connected to it by a gate, or as a function of the input variables where no gate intervenes, results in a system of equations (e.g., $e = x_1 \cdot x_2$), which can be thought of as an axiomatization of a theory.

A bootstrap test of, for example,

(1) $e = x_1 \cdot x_2$

is carried out by finding computations of each of the quantities in the equation in terms of input and output variables. The computations for the two variables on the right hand side of (1) are trivial since they are themselves input variables. A computation of e is obtained as follows:

$$z = i + j$$
$$z = (e + f) + (g \cdot h)$$
$$z = (e + x_2) + (x_3 \cdot (x_3 + x_4))$$
$$x_2' \cdot x_3' \cdot z = x_2' \cdot x_3' \cdot e$$
$$e = \frac{x_2' \cdot x_3' \cdot z}{x_2' \cdot x_3'}$$

where $x' = 1 - x$. Each step in the computation of e is warranted either by some Boolean identity or by some consequence of the "theory" described above. The equation in measured variables that must be satisfied if these computed values are to confirm (1) is therefore:

(2) $\dfrac{x_2' \cdot x_3' \cdot z}{x_2' \cdot x_3'} = x_1 \cdot x_2,$

and this equation is called the *representative* of (1) for the computations given. Values of the input variables and of the output variables will constitute a test of hypothesis (1) if and only if:

(3) $x_2' \cdot x_3' = 1,$

since this condition is necessary and sufficient for the computation of e from the data to determine a value for e, and since, also, (2) is not a Boolean identity. Thus, assuming that condition (3) is met by the data, the representative (2) of the hypothesis becomes:

(4) $z = 0.$

Any of the following inputs will therefore test hypothesis (1):

$(x_1,$	$x_2,$	$x_3,$	$x_4)$
$(1,$	$0,$	$0,$	$0)$
$(0,$	$0,$	$0,$	$0)$
$(1,$	$0,$	$0,$	$1)$
$(0,$	$0,$	$0,$	$1)$

and the output z must in each case be 0 if the hypothesis is to pass the test, 1 if it is to fail the test.

Alternative hypotheses usually considered by computer scientists are that the e gate is stuck at 1 (s.a.1.) or that it is stuck at 0 (s.a.0.). The same computations apply to either of these hypotheses as to hypothesis (1). For the stuck at 1 fault the hypothesis to be tested is e = 1, and the representative of the hypothesis for the computations is:

$$(5) \quad \frac{x_2' \cdot x_3' \cdot z}{x_2' \cdot x_3'} = 1,$$

which is, again assuming the condition (3) for the determinacy of e,

$$(6) \quad z = 1.$$

As with the first case, we obtain tests of the hypothesis that e is s.a.1. if and only if the data meet condition (3), and the value of z required for the hypothesis to pass the test is in each case the complement of the value required for hypothesis (1) to pass the tests. Thus each of the inputs is a test of e = 1 if and only if it is a test of (1), and each such test *discriminates* between the two hypotheses, since the predicted outputs are always different, i.e., the representatives are incompatible.

For the hypothesis that e is s.a.0., everything is as before, but when condition (3) for the determinacy of e is met, the representative of the hypothesis becomes

$$(4) \quad z = 0.$$

Thus no bootstrap test will discriminate between a s.a.0. fault and the normal condition given by hypothesis (1).

A standard technique in the computer science literature for detecting faculty in logic circuits derives from the observation that, for example, the s.a.0. fault can be discriminated from the normal condition of a line, provided there is a combination of inputs that, in the normal condition, gives the line the value 1, and, for these inputs, the output variable z would be different from the value determined by the normal circuit if the line value were 0 instead of 1. This "Boolean derivative" method, applied to the case discussed above in which the normal hypothesis is (1) and the two alternative hypotheses are that e is s.a.0. and that e is s.a.1., yields the following necessary and sufficient condition for a set of input values (a_1, a_2, a_3, a_4) to be a test that discriminates the normal condition from the s.a.0. fault:

(7) $e(x_1, x_2) \cdot \dfrac{dz\ (e, x_2, x_3, x_4)}{de} \Bigg|_{(a_1 \ldots a_4)} = 1$

and likewise the necessary and sufficient condition for a discrimination between the normal and the s.a.l. fault is:

(8) $(1 - e(x_1 \cdot x_2)) \cdot \dfrac{dz(e, x_2, x_3, x_4)}{de} \Bigg|_{(a_1 \ldots a_4)} = 1.$

In equations (7) and (8) $\dfrac{dz}{de}$ is the Boolean derivative, i.e., the Boolean function whose value is 1 if and only if a change in the value of e changes the value of z when x_2, x_3, x_4 are kept constant.

For the case given it can easily be shown that:

$$\frac{dz(e, x_2, x_3, x_4)}{de} = x_2' \cdot x_3',$$

and since $e = x_1 \cdot x_2$, it follows that $e \cdot \dfrac{dz}{de} = 0$ identically, and therefore in computer parlance the s.a.0. fault is not testable, i.e., not discriminable from the normal circuit element. This is the same result obtained from bootstrap testing. For the s.a.l. fault, the Boolean derivative method yields as a condition necessary and sufficient for discriminating the fault:

$$(1-e) \cdot \frac{dz}{de} \Bigg|_{(a_1 \ldots a_4)} = 1,$$

hence

$$(1-(x_1 \cdot x_2)) \cdot (x_2' \cdot x_3') = 1$$
$$(x_1' \cdot x_2' \cdot x_3') + (x_2' \cdot x_3') = 1$$
$$x_2' \cdot x_3' = 1$$

Thus every input such that $x_2 = x_3 = 0$ will discriminate between the normal case and the s.a.l. fault. Again this is the same result the bootstrap method gives.

The equivalence of the methods is general: Provided a circuit contains no loops, a set of inputs discriminates between a fault and a normal circuit element according to the bootstrap method if and only if that set of inputs discriminates between that fault and the normal circuit element according to the Boolean derivative method. (See appendix.)

6. Probability and Error

Many philosophers view the analysis of confirmation and testing as a branch of the theory of probability or of statistics. I do not. I do, however, believe that applications of statistical methods in science and engineering are often dependent on the structural features of hypothesis testing described in previous sections, and I believe that bootstrap testing provides one natural setting for certain probabilistic conceptions of the bearing of evidence.

In many cases measurement results are treated as values for random variables. Following the customary theory of error, values for the random variable E are taken to be normally distributed about the true value of the (nonstochastic) variable e. The mean and variance of the sampling distribution of E are in practice used both for qualitative indications of the value of the evidence for some hypothesis about e, and also for more technical procedures—hypothesis testing, determination of confidence intervals, etc. The same is true when e and E are not single quantities but a set or vector of quantities and random variables. When some theory T relates the quantities $(e_1, \ldots, e_n)$ to other quantities $x_j = f_j(e_1, \ldots, e_n)$, there are corresponding statistics $X_j = f_j(E_1, \ldots, E_n)$ that may be used for testing hypotheses about $(e_1, \ldots, e_n)$ *based on T*. The very point of the sections on "propagation of error" contained in every statistical manual is to describe how the variances of the E_i determine the variances of the X_i, so as to permit the application of standard statistical testing and estimation techniques for hypotheses about the x_i. The conditions of the bootstrap testing schemas are no more than attempts to describe necessary conditions for the application of standard statistical hypothesis testing-techniques to hypotheses about unmeasured quantities.

7. Bayesian Bootstrapping

In scientific contexts, Bayesian probability not only rests on a different conception of probability from the one embraced by the more "orthodox" tradition in statistics, it also has different applications. Despite my distaste for Bayesian confirmation theory as a general theory of rational belief and rational change of belief—a theory that is, in Glenn Shafer's words, suited for "never-never land"—I think Bayesian and related techniques are practical and appealing in various scientific and engineering circumstances. In this and the next two sections, we focus on the first of the conditions in the three schemas for bootstrap testing, and its interaction with probability.

How can we associate a probability measure, closed under Boolean operations, with a set of equations or inequations in specified variables? Straightforwardly, by considering for n-variables an n-dimensional space S_n and associating with every set K of equations or inequations the solution set of K in S_n. The probability of K is then the value of a probability measure on S_n for the solution set of K.

Consider a set of variables x_i essential to a hypothesis H, and a (prior) probability distribution Prob over the space of the x_i. Let Prob(H) be the prior probability of H, i.e., of the solution set of H. Suppose values x_i* are computed for the variables x_i from the measured variables e_j using some theory T. Let T_i be the fragment of T used to compute x_i, and let (&T) be the set of T_i for all of the x_i occurring in H. We can associate T and &T with subspaces in the space of the variables x_i and e_j. Prob and the computed values for the variables x_i could be taken to determine

(1) $\text{Prob}(H/x_i = x^*) = \text{Prob}(H/e_j = e^* \ \& \ (\&T))$

where the e^* are the measured values of the e_j.

Where we are dealing with point values x_i and where the T_i are equations, (1) will typically require that we conditionalize on a set with probability measure zero, and so we would need either approximation techniques or an extended, nonstandard, notion of conditionalization.

Finally, one might take the values e^* for e_j to confirm H with respect to T if

(2) $\text{Prob}(H/x_i = x^*) = \text{Prob}(H/e_j = e^* \ \& \ (\&T)) > \text{Prob (H)}.$

This is essentially the proposal made by Aaron Edidin. For Schemas I and II, if we take this procedure strictly, then hypotheses involving the x_i that are not testable because they are entailed by &T will always get posterior probability one. On the other hand, if we avoid this difficulty as Edidin suggests by conditionalizing only when H is not entailed by &T, then $\text{Prob}(H/(\&T) \ \& \ e_j = e^*)$ will not necessarily be a probability distribution.

8. Jeffrey Conditionalization

Sometimes we do have established probabilities for a collection of hypotheses regarding a set of variables, and we regard a consistent body of such probabilized hypotheses as forming a theory that we wish to test or to use to test other hypotheses. Now the Bayesian approach previously described makes no use of the probabilities of the hypotheses used to compute values of unmeasured variables from values of measured variables. But it is very natural to let the probabilities of the T_i determine probabilities for the computed values of the x_i, and to use this information in judging both the posterior probability of any hypothesis $H(x_i)$, as well as the confirmation of any such hypothesis.

To begin with we shall confine attention to a definite theory T, a set of variables x_i, and for each x_i a *single* computation $x_i = T_i(e_j)$ derivable from T. For each such equation (or inequality) there is assumed a probability $\text{Prob}(x_i) \equiv \text{Prob}(x_i = T_i(e_j))$ or Prob(i) for brevity; the joint probabilities for these hypotheses may or may not be known. Suppose the e_j are measured and values $e*$ are found. From the computations, values $x_i^* = T_i(e*)$ are determined. What confidence do we have that $x_i = T_i(e*)$, assuming we have no doubts about the evidence? We might suppose that

$$\text{Prob}(x_i = x_i^*) = \text{Prob}(x_i = T_i(e_j)) = \text{Prob}(i).$$

For hypotheses about the x_i, how can the probabilistic conclusions, when independent, be combined with the prior probabilities of hypotheses to generate a new, posterior probability? Let E_i be set of mutually exclusive, jointly exhaustive propositions, Prob a prior probability distribution and $\text{Pos}(E_i)$ a new probability for each E_i. Richard Jeffrey's rule for obtaining a posterior distribution from this information is:

$$(3) \quad \text{Pos}(H) = \sum_i \text{Pos}(E_i) \cdot \text{Prob}(H/E_i).$$

Applied to the present context, Jeffrey's rule suggests that we have for any hypothesis $H(x_i)$

$$(4) \quad \text{Pos}(H) = \text{Prob}(\&T) \cdot \text{Prob}(H/x*) + (1-\text{Prob}(\&T)) \cdot \text{Prob}(H/\bar{x}*)$$
where $x* = T_i(e*)$ and $\bar{x}*$ is the set complementary to $x*$ and, e.g., $\text{Prob}(H/x*)$ is short for $\text{Prob}(H/x_i = x_i^*$ for all x_i in Var (H)).

Equation (4) is not a strict application of Jeffrey's rule (3), however, for (3) uses the *posterior probability* of the evidence, $\text{Pos}(E_i)$, whereas in (4) the prior probability of the computed values, $\text{Prob}(\&T)$, is used. Carnap, and more recently H. Field and others, have noted that Jeffrey's $\text{Pos}(E_i)$ depends not only on the impact of whatever new evidence there is for E_i, but also on the prior probability of E_i. In contrast, the $\text{Prob}(\&T)$ of (4) need not depend on the prior distribution of the x_i.

Rule (4) has the following properties:

(i) Evidence E can disconfirm hypothesis H with respect to &T even though &T entails H.

(ii) Evidence E cannot confirm hypothesis H with respect to &T if &T entails H and H entails $x_i = T_i(e^*)$ for all i.

(iii) Evidence E cannot confirm hypothesis H with respect to &T if H entails $x_i = T_i(e^*)$ for all i and the prior of H is greater than or equal to the prior of &T.

(iv) Evidence E *confirms* hypothesis H with respect to T if $\bar{x}^*$ entails H and $1 - \text{Prob}(\&T) > \text{Prob}(H)$.

"Confirmation" here means the posterior probability is greater than the prior probability.

Results (ii) and (iv) show that the notion of confirmation implicit in (4) is not very satisfactory. (iv) says, for instance, that even if H is inconsistent with the computed values of the x_i, H is confirmed provided the probability of the theory used is sufficiently low. The reason is that a low probability for the theory gives a low probability to x^*, hence a high probability to $\bar{x}^*$ and thus to hypotheses it entails; (iii), which implies (ii), seems much too strong a feature.

9. Shafer's Belief Functions

Reconsider the interpretation we have given to the relation between the probability of &T and the number, also equal to the probability of &T, associated with the values of the x_i computed from the evidence using &T. The evidence, and the hypotheses &T, and their probability, give us reason to believe that the true value of each x_i is x_i^*. Prob(&T) measures the strength of that reason, on the assumption that the evidence is certain. We have supposed that Prob(&T) is a probability and that it is numerically equal to the *degree of support* that e^* gives to the $x_i = x_i^*$ hypothesis with respect to &T. But why should we assume that the degree of support is a *probability* in the usual sense? On reflection, we should not. To illustrate, suppose &T is a body of hypotheses relating x_i to e_j, which has been articulated and investigated and has an established probability about equal to two-thirds. Further suppose that no other comparably detailed hypotheses relating x_i to e_j have an established probability. Now let evidence e^* be obtained. Using &T and its probability, we infer that e^* gives $x_i = T_i(e^*) = x_i^*$, a degree of support also equal to two-thirds. *Surely that does not mean* *that this same evidence e*, and same theory* &T, *gives* $x_i = \bar{x}_i{}^*$ *a degree of support equal to one-third*. But that conclusion is required by the construal of degree of support as a probability, and is the source of one of the difficulties with rule (4) of the previous section.

What is required is a way of combining a "degree of support" for a claim with a prior probability distribution that has no effect on the degree of support that the evidence provides. Jeffrey's rule gives us no way to construct the combination effectively. It gives us at best a condition (3), which we might impose on the result of the combination, whatever it is. A theory proposed by A. Dempster and considerably elaborated in recent years by G. Shafer[1] has most of the desired properties.

To explain how the rudiments of Shafer's theory arise, consider a space of hypotheses having only two points, h^1 and h^2. Suppose there is a prior probability distribution Bel^1 over the space Ω. Given that $e_j = e_j{}^*$ and that no other theory of known probability (save extensions of &T) determines h^1 or h^2 as functions of the e_j, we take the degree of support, $Bel^2 (h^1)$, that the evidence provides for h^1 to be

Prob (&T)

if the $x_i{}^*$ satisfy h^1. Under these same assumptions, what is the degree of support that the evidence provides for h^2? *None at all*. The evidence provides no reason to believe h^2, but it does provide some reason to believe h^1. We set $Bel^2(h^2) = 0$. It seems a natural and harmless convention to suppose that any piece of evidence gives complete support to any sentence that is certain. Hence $Bel^2(\Omega) = 1$.

The function Bel^2, which represents the degrees of support that $e_j{}^*$ provides with respect to &T, does not satisfy the usual axioms of probability. Shafer calls such functions *Belief functions*. They include as special cases functions such as Bel^1, which satisfy the usual axioms of probability. In general, given Ω, a belief function Bel is a function $2^\Omega \rightarrow [0, 1]$ such that $Bel (\phi) = 0$, $Bel (\Omega) = 1$, and for any positive integer n and every collection $A_1, \ldots, A_n$ of subsets of Ω,
$$Bel (A_1 \cup \ldots \cup A_n) \geq \Sigma_I (-1)^{|I|+1} Bel (\bigcap_{i \in I} A_i). \quad \phi \neq I \subseteq \{1, 2, \ldots, n\}$$

Our problem is this: How should Bel^1 and Bel^2 be used to determine a new belief function in the light of the evidence, $e_j{}^*$? Shafer's answer is given most simply in terms of functions that are determined by the belief functions. Given Bel: $2^\Omega \rightarrow [0, 1]$, define

$$M(A) = \sum_{B \subseteq A} (-1)^{|A-B|} \text{ Bel } (B)$$

for all $A \subseteq \Omega$.

Shafer calls functions M *basic probability assignments*. A basic probability assignment reciprocally determines a belief function by

$$\text{Bel } (A) = \sum_{B \subseteq A} M(B)$$

for all $A \subseteq \Omega$.

Thus for the unit subsets of Ω—$\{h_1\}$ and $\{h_2\}$ in our case—Bel $(h_i) = M(h_i)$.

Shafer's version of Dempster's Rule for combining Belief Functions is

$$M(A) = M^1 \oplus M^2 = K^{-1} \sum_{i,j} \{ M^1 (B^i) M^2 (B^j) \mid B^i \cap B^j = A \}$$

where

$$K^{-1} = \sum_{i,j} \{M^1(B^i) M^2(B^j) \mid B^i \cap B^j \neq \phi\} \qquad B_i, B_j \subseteq \Omega.$$

The combination of Bel^1 and Bel^2 obtained by so combining M^1 and M^2 is denoted $\text{Bel} = \text{Bel}^1 \oplus \text{Bel}^2$ where $\text{Bel}(A) = \sum_i \{M(B^i) \mid B^i \subseteq A\}$.

Shafer terms a Belief Function such as Bel^2 a *simple support function;* the associated function M^2 is nonzero for a single set other than Ω.

Shafer's Belief Functions and Dempster's Rule of combination have the following properties:

(i) $\text{Bel}^1 \oplus \text{Bel}^2 = \text{Bel}^2 \oplus \text{Bel}^1$

(ii) If Bel^1 is a probability and Bel^2 a simple support function, then $\text{Bel}^1 \oplus \text{Bel}^2$ is a probability function.

(iii) If Bel^1 is a probability, $\text{Bel}^1 (A) \neq 0$, $\neq 1$, and Bel^2 is a simple support function with $M^2(A) \neq 0$, then $\text{Bel}^1 \oplus \text{Bel}^2(A) > \text{Bel}(A)$.

(iv) Under the same assumptions as in (ii), $\text{Bel}^1 \oplus \text{Bel}^2$ has the form of Jeffrey's rule of conditioning.

(v) Suppose $M^1(h) = \text{Prob}(T_1)$ and $M^2(h) = \text{Prob}(T_2)$, where M^1 and M^2 determine simple support functions and a joint probability exists for T_1 and T_2. Then $M^1 \oplus M^2(h) = \text{Prob}(T_1 \vee T_2)$ if and only if T_1 and T_2 are probabilistically independent.

In view of these features, the contexts in which it is most plausible to apply Shafer's theory seem to be those in which the following conditions obtain:

(i) The set of mutually incompatible theories $T_1, \ldots, T_n$ for which there are established probabilities do not jointly determine Prob(A/E) but do determine a lower bound for Prob(A/E), where A is a proper subset of the space of interest and E is the "direct" evidence.

(ii) Separate pieces of evidence E_2, E_3 ... E_n are statistically independent.

(iii) None of the theories $T_1, \ldots, T_n$ individually entails that the probability of any proper subset of Ω is unity.

Condition (iii) is, of course, a version of the conditions characteristic of bootstrap testing.

10. Application: Probabilistic Fault Detection

The conditions under which the combination of ideas just described actually obtain are not uncommon in engineering problems. A simple example can be found again in fault detection in computer logic circuits. Suppose that by testing circuit elements we know the probability of failure after n hours of use of a logic gate of any particular kind. Further suppose that the probabilities of failure for individual gates are independent. Finally, suppose that the probability that the gate failure results in a stuck-at-zero fault or a stuck-at-one fault is not known in each case, and the probability of a nonconstant fault is also not known.

With these assumptions, consider again the fragment of a logic circuit discussed in Section 5.

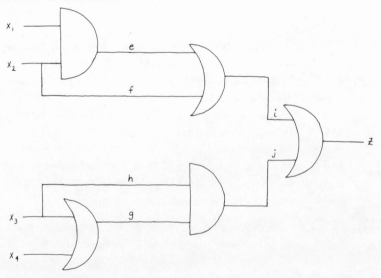

Suppose we are concerned as to whether the gate leading to line e is functioning normally, i.e., whether $e = x_1 \cdot x_2$. Using the bootstrap conception, we can, for given inputs, compute e as a function of output z. This function is determined by the graph G of all paths in the circuit leading from e to z.

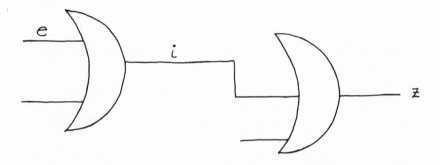

Likewise, from each of the input variables $x_1, \ldots, x_n$ whose values must be fixed at $(a_1, \ldots, a_n)$ in order for $e = f(z)$ to hold, there is a path leading from the variable to some node of G. Call the graph of all such paths S.

Assuming that the probabilities of failure of the circuit elements are independent, we can estimate the probability that $e = f(z)$ for inputs $a_1, \ldots a_n$ from the graphs S and G.

We have a prior probability that $e = x_1 \cdot x_2$, given by one minus the probability of failure of the gate leading to the e line. For a measured value z^* of z and inputs $(a_1 \ldots a_n)$ we also have a probability—that computed in the preceding paragraph—that $e = e^* = f(z^*)$. Assuming x_1 and x_2 known from measurement, for appropriate values of the measured variables we thus have a degree of support for $x_1 \cdot x_2 = e$, although we may have none for $x_1 \cdot x_2 \neq e$. If the measured value of z is not the value predicted by the normal circuit for the measured inputs (including x_1 and x_2) then the same procedure may support $x_1 \cdot x_2 \neq e$. In the more general case, the values of the lines leading to the e gate would not be known directly from measurement but would also have to be computed, and would for given inputs have probabilities less than one (determined, obviously, by the probabilities of the failure of gates in the graph of paths leading *from* input gates to the e-gate).

11. Appendix: On the Equivalence of Bootstrap and Boolean Derivative Tests of Circuit Faults

For any logic circuit without loops and with n input variables and m output variables, introducing a distinct variable for each line in the circuit generates, in a well-known way, a system of equations over a Boolean algebra. Thus, for any equation so generated, there is a well-defined class of values for input and output variables (i.e., a well-defined collection of data sets) that will provide bootstrap tests of that equation with respect to the system. And, given any equation expressing the line output of a logic gate as a function of the line inputs, there is a determinate class of input and output sets that provide bootstrap discriminations of that equation from the equation that puts the gate output equal to 1 (stuck-at-1 fault), and a determinate class of input and output sets that provide bootstrap discriminations of that equation from the equation which puts the gate output equal to 0 (stuck-at-0 fault).

Given a circuit with n input variables x_1 and m output variables z_j, let $e = e(x_k, \ldots, x_l)$ be the equation that specifies the value of line e as a function of the input variables $x_k, \ldots, x_l$, and let $z_j = z_j(e, x_1, \ldots, x_n)$ be the equation expressing the value of the jth output variable as a function of e and those input variables from which there is a path to z not passing through e. Then a set of inputs $(a_i, \ldots, a_n)$ provides a *Boolean derivative* discrimination of $e = e(x_k, \ldots, x_l)$ from the hypothesis that e is stuck at 0 if and only if:

$$(1) \quad e(x_k \ldots x_l) \cdot \left[\frac{dz_i(e, x_1, \ldots, x_n)}{de} + \ldots + \frac{dz_m(e, x_1, \ldots, x_n)}{de} \right]\Bigg|_{(a_1, \cdots, a_n)} = 1$$

And input values $(a_1 \ldots a_n)$ provide a Boolean derivative discrimination of $e = e(X_k \ldots X_l)$ from the hypothesis that e is stuck at 1 if and only if:

$$(2) \quad (1-e(x_k \ldots x_l)) \cdot \left[\frac{dz_i(e, x_1, \ldots, x_n)}{de} + \ldots + \frac{dz_m(e, x_1, \ldots, x_n)}{de} \right]\Bigg|_{(a_1 \cdots a_n)} = 1.$$

In these equations, $\dfrac{dz_i(e, x_1, \ldots, x_n)}{de}$, for example, signifies the "Boolean derivative" of the expression $z_i(e, x_1 \ldots x_n)$ with respect to e,[2] i.e., the function of $x_1 \ldots x_n$ that is equal to 1 if and only if a change in value of e changes the value of the expression.

The basic idea behind the Boolean derivative method is the natural and correct one that a fault is detectable by an input if, were the circuit normal, the value of some output variable would be different for that input than it would be if the circuit contained the fault in question (but no other faults).

The following equivalence holds: an input $(a_1 \ldots a_n)$ discriminates between the hypothesis that a circuit line e is "normal" (i.e., as in the circuit description) and stuck at 1 (0) according to the Boolean derivative method if and only if (i) there is a bootstrap test (in the sense of Schema I or III) with a data set including $(a_1 \ldots a_n)$ that tests, with respect to the system of equations determined by the circuit, an equation E, entailed by the system, giving e as a function of input values or of values of lines into the gate of which e is an output line and (ii) there is a bootstrap test with the same data sets that tests e $= 1$ (0) with respect to the system of equations determined by the circuit obtained from the original circuit by deleting all lines antecedent to (that is, on the input side of) e, and (iii) the data set confirms E with respect to the equations of the original circuit if and only if it disconfirms e $= 1$ (0) with respect to the equations of the modified circuit.

The equivalence follows from a few observations that may serve as the sketch of a proof. First, if the Boolean derivative condition is met, then some output variable z_j can be expressed as a function of e and input variables connected to z_j by lines not passing through e. If E_j is the equation expressing this relationship, then E_j will hold in both the original and the modified (i.e., stuck) circuits. Moreover, the Boolean derivative $\dfrac{dz_i(e, x_1, \ldots, x_n)}{de}$ equals 1 exactly for those values of the input variables which determine e as a function of z_j. It follows that every Boolean derivative discrimination is a bootstrap discrimination. Conversely, if conditions for bootstrap testing are met, there must be an equation

$$(3) \quad f(e, x_1 \ldots x_n, z_1 \ldots z_k) = 0$$

which constrains e, output variables $z_1 \ldots z_k$ and input variables $x_1 \ldots x_n$ connected to one or more of the z_1 by lines not passing through e, such that for input values $(a_1 \ldots a_n)$, equation (3) determines e as a function of $z_1 \ldots z_k$ and thus

$$(4) \quad e = g(z_1 \ldots z_k) \Big|_{(a_1 \cdots a_n)}$$

But each of the z_i is a function of the $x_1 \ldots x_n$ and e, and thus with $x_i = a_i$, of e alone:

(5) $e = g(z_1(e) \ldots z_k(e)) \Big|_{(a_1 \cdots a_n)}.$

Now taking the Boolean derivative of equation (5):

(6) $1 = \dfrac{de}{de} = \dfrac{d}{de} g(z_1(e), \ldots, z_k(e)) \Big|_{(a_1 \cdots a_n)}$

where $\dfrac{d}{de} g(z_1(e), \ldots, z_k(e)) \Big|_{(a_1 \cdots a_n)}$

is equal to the exclusive disjunction

(7) $g(z_1(e), \ldots, z_k(e)) \oplus g(z_1(e'), \ldots, z_k(e'))$

where $e' = 1-e$, and formula (7) has value 1 if and only if its two component subformulae have distinct values. But

$g(z_1(e), \ldots, z_k(e)) \neq g(z_1(e'), \ldots, z_k(e'))$

only if for some i, $z_i(e) \neq z_i(e')$, and hence only if for some i

(8) $\dfrac{dz_i(e)}{de} \Big|_{(a_1 \cdots a_n)} = 1.$

It follows that there is a Boolean derivative test with inputs $(a_i \ldots a_n)$.[3]

Notes

1. G. Shafer, *A Mathematical Theory of Evidence* (Princeton: Princeton University Press, 1980).

2. See for example, A. Thayse and M. Davio, "Boolean Calculus and Its Applications to Switching Theory," *IEEE Trans. Computers*, C-22, 1973, and S. Lee, *Modern Switching Theory and Digital Design* (Englewood Cliffs, N.J.: Prentice Hall, 1978).

3. Research for this essay was supported by the National Science Foundation, Grant No. SES 80 2556. I thank Kevin Kelly and Alison Kost for their kind help with illustrations.

Theory Comparison and
Relevant Evidence

What is the main epistemic problem concerning science? I take it that it is the explication of how we compare and evaluate theories, as a basis either for theory acceptance or for practical action. This comparison is clearly a comparison in the light of the available evidence—whatever *that* means. [1] My aim in this paper is very modest: to explore the logic of evidential support in the theory of Clark Glymour. [2] The question of how, and to what extent, overall theory comparison is to be related to the data from relevant tests, I broach in another paper. [3]

First I shall sketch a model of theory structure that covers what I take to be a large and significant class of scientific theories. With reference to that schema, I reconstruct notions of relevant evidence as introduced in Glymour's "bootstrap theory" of testing hypotheses. Then I shall explore the relations of support by the same evidence, between different hypotheses. The form chosen for the reconstruction is so as to make clear the application of Glymour's ideas to the case of a *developing* theory which is becoming more testable. Glymour himself and Michael Gardner have examined the development of atomic theory in the nineteenth and early twentieth centuries from this point of view. [4]

1. A Schema for (Simple, Quantitative) Theories

The semantic approach to theories, initiated by Beth and closely related to the approach introduced by Suppes, gives us an account that I shall here use as a framework for relations between tests, evidence, and hypotheses. [5] In doing this we broaden somewhat the initial basis for Glymour's own exposition, but we ignore his logistical reformulation prompted by compar-

This article is reprinted by courtesy of Adolf Grünbaum and Larry Laudan, general coeditors of the *Pittsburgh Series in the Philosophy and History of Sciences* published by the University of California Press. It will appear in the proceedings volume for the Workshop on the Limitations of Deductivism, held at the University of Pittsburgh Center for the Philosophy of Science in November 1980. The research for this paper was supported by National Science Foundation research grant SES-8005827, which is hereby gratefully acknowledged.

isons with Hempel.[6] Without attempting needless rigor, I shall describe a scientific theory of the sort I have in mind by describing what must be given to specify it exactly.

It has, to begin, a set W of possible *situations*. This is broadly conceived, so that the effect of its laws is to entail that many of these situations are not really possible.

Second, it has a set Q of *quantities*. For ease of present exposition I shall postulate that each quantity has a value (either a real value, or a real-valued function of time) in each possible situation. Quantity q has value $q(w)$ in element w of W.

Although it is crucial to Glymour's theory that distinct quantities may yet have the same value in all *really* possible situations, I shall take the set W to be so broadly conceived that a quantity q can be identified with (or via) the function that assigns to each possible situation in W, the values q has in that situation.

The set Q is closed under the formation of new quantities, by means of composition and transformation of functions taken as broadly as you like, but including at least composition and polynomials of the usual sort. Because of the identification in the previous paragraph, there is no division into "atomic" and "molecular" quantities. Thus q + q – q is identically the same quantity as q itself. A theory's laws may imply that q and q' always have the same value, and this will be stated as (q = q') as usual; that means only that in all the situations w allowed by the law, q(w) = q'(w). This is standard notation.

Formulas used to phrase hypotheses and laws describe relations between values of quantities. Thus Newton's second law says that at each instant of time, the (value of the quantity) impressed force equals the product of the (values of the quantities) mass and acceleration. Equality is one relationship among values; others may appear. I shall focus here on all the relations of equality and inequality ($=$, $\leq$, $<$, $\neq$, etc.) and use the letter R to stand for them. A *basic proposition* or *basic formula* is one of form (tRt'), where t and t' denote quantities. It is *true* in situation w exactly if t(w) bears R to t'(w).

The expression "solution of an inequality" is familiar and I shall take it for granted, but restrict its use as follows: a *solution* of basic formula A is a function s that assigns a value at least to each quantity that is denoted by a term, or well-formed part of a term, which appears in A, and there is some possible situation w such that

(a) $s(q) = q(w)$ for each q for which s is defined, and

(b) A holds in w.

For example, the function s that assigns values 3, 4, and 12/k to the pressure P, volume V, and temperature T of a gas, is a solution of the basic formula (PV = kT), provided however that space W of possible situations has some situation in which these are the correct values.

The domain of a solution of formula A includes the set of quantities denoted by well-formed parts of A; I shall call that set Q(A). Since the terms t, t′ in (tRt′) may be complex, that set will usually have more than two members. A formula A is called *logically valid* exactly if for each w in W, any map s that assigns $q(w)$ to q for each q in Q(A), is a solution of A. Similarly we may say that A *implies* B exactly if every solution of A that is also well defined on Q(B), is a solution of B.

Finally the theory has a set of *basic postulates*, each of which is a basic proposition. The *possibility space* Sp(T) of theory T is the set of possible situations in which all basic postulates are true. As a theory develops through the addition of postulates, this possibility space is narrowed; the set W of possible situations stays the same. Theorems are basic formulas implied by the postulates.

It will not cause confusion in this context to use the same letter T to denote the theory and its set of theorems, for I shall keep W and Q fixed. What the theory *says* is that any real situation is one in which all its theorems are true.

2. Characterizing Relevant Evidence

We must first describe the possible evidence that can be had. This is especially important here because for Glymour the relation between evidence E and hypothesis H depends in part on what the evidence might or could have been instead.

When a test is carried out, the values of certain directly measurable quantities are ascertained. What tests are possible, which quantities are directly measurable, we must here take as given. (In my own opinion, and I think also in Glymour's, this is a question of contingent, empirical fact.) The result of a test I shall call a *data set* or *data base*: it is a function E that assigns to those measured quantities the values ascertained. So any solution of a basic formula might, prima facie, be a data base; the two defining constraints on *possible data bases* E are, *first*, that each quantity in the domain of E is directly measurable, and *second* that there is a possible situation in which all the basic formulas q = E(q), for quantities q in the domain of E, are true. It will be convenient much of the time to identify E with that set of basic formulas.

An *alternative* E′ to data base E is a possible data base with the same domain as E. It will be noted that the theory itself, in delimiting the class W of possible situations, has implications concerning what evidence could in principle be had. A *hypothesis* is any basic formula. The first Glymourian mode of evidential support is the following.

> (2-1) E provides *weakly relevant evidence* for H *relative to* theory T exactly if E has some alternative E′ and T some subset T_o such that:
>
> (1) T∪E∪ {H} has a solution.
> (2) T_o∪E′ has a solution.
> (3) All solutions of T_o ∪ E are solutions of H.
> (4) No solutions of T_o ∪ E′ are solutions of H.

As a simple example, let A, B, C be directly measurable quantities, and let T have postulates

$$P_1 \quad A = x + u$$
$$P_2 \quad B = u - z$$
$$P_3 \quad C = x + z$$

Let hypothesis H be postulate P_3, and let the evidence E be {A = 3, B = 4, C = –1}. Using subtheory T_o = {P_1, P_2} and alternative evidence base E′ = {A = 3, B = 4, C = 1}, we can verify at once that E provides weakly relevant evidence for P_3 relative to theory T. The different values given to C in E and E′ show at once that clause (4) holds. To verify clauses (1) and (2) we note that

$$s: A = 3, \ B = 4, \ C = -1, \ x = 10, \ u = -7, \ z = -11$$
$$s': A = 3, \ B = 4, \ C = 1, \ x = 10, \ u = -7, \ z = -11$$

are solutions of T∪E and of T_o ∪E′ respectively. The crucial clause (3) also holds because T_o∪E implies at the same time that A – B = C (this comes from E) and A – B = x + z (that comes from T_o), hence the two right-hand sides must have the same values, as P_3 says. I shall return to this example at several points below.

The above relation of weak evidential support already accomplishes one of Glymour's aims: it is hypothesis-specific, allows for selective support, and gets us out of methodological holism. But Glymour takes equally seriously the idea of the *empirical determination* of specific quantities. A test of a hypothesis about temperatures, relative to a theory in which temperature equals mean kinetic energy of the constituent molecules, can yield weakly relevant evidence for a hypothesis about molecular kinetic energies. But since in measuring temperature we deal only with the mean, this is a test in which no specific molecular energy is determined, except perhaps in some absolutely minimal sense. More full-blooded, more and more precise empirical determination of exact values of theoretical quantities plays an extremely important role in the building up of evidential support for the theory, according to Glymour.

As my phrasing already suggests, this aspect of tests admits of degrees, or at least of comparisons of more and less. I shall list three possible amendations for (2-1); the first of these was suggested by Glymour in correspondence.

(2-2) If q belongs to $Q(A)$ then the following are possible:
   (1) The set of values assigned to q by solutions of $T_0 \cup E$ is properly contained in the set of values assigned to q by solutions of T_0.
   (2) All solutions of $T_0 \cup E$ assign a value to q in the same interval I.
   (3) All solutions of $T_0 \cup E$ assign the same value to q.

One of these conditions should be added as fifth clause (2-1), to arrive at a stronger criterion of relevant evidence. Next we can strengthen the criterion by adding as sixth clause a corresponding condition for the alternative data base E'. To mark the extreme, in which every quantity in the formula is calculable from the data base via the theory, I shall give it a name too.

(2-3) E provides *strongly relevant evidence* for H relative to theory T exactly if E has some alternative E' and T has a subtheory T_0 such that
   (a) clauses (1)-(4) of (2-1) hold
   (b) clause (3) of (2-2) holds, and holds also when E is replaced by E' for all q in $Q(H)$.

In section 4 I shall give examples of strongly relevant evidence.

Thus the evidence is relevant to A if it allows selective testing of A (relative to the theory), but the degree of relevance is determined by how closely the evidence allows us to calculate (relative to the theory again) the values of the quantities that appear in this formula. The reader is asked to think up names for intermediate degrees of relevance.

3. Significance of Glymour's account

How good is my reconstruction of Glymour's account? I take it to be good; good but not perfect. So I shall explain some points of divergence and also say why I believe the present account to share the considerable virtues that can be claimed for Glymour's own.

Most obviously omitted from my version is the notion of computation. I have been unable to find a precise reconstruction to which the use of that notion makes a difference; and I believe that Glymour is now also of the opinion that this had a didactic use only.[7]

Second, Glymour gives his account twice over, once for theories stated in equational form, and once for those formulated in a first-order language. I do not find the latter idealization very interesting, and believe that it actually has definite harmful effects on philosophical discussions. But it must be presumed that, although Glymour does not say so explicitly, his account for equational theories needs some emendation corresponding to clause (2) in the statement of the *Bootstrap Condition* for the first-order case.[8] I shall not discuss such minor variations in further detail; as Glymour himself says, we should just be flexible in their admission as problems arise.

My reasons for stating the account in several definitional stages are two. First, I want to allow for approximations. Even if the evidence (data base) is totally precise (and I take allowances for imprecise data to introduce no difficulties of principle), it may not determine the *exact* value of other quantities in the hypothesis. Second, as a theory develops, as in the example of the atomic theory in the nineteenth century, more and more quantities may become empirically determinable; conversely, a good deal of testing is already being done, while some of the crucial quantities still remain indeterminate. For example, Gardner states that Herapath derived the equation $PV = Nm v \frac{2}{3}$, and used this to make "the first calculation ever published of molecular speeds." I have no reason to doubt that his contemporaries saw it that way, and considered him to have answered the objection that molecular speed is an empirically indeterminable quantity. Yet it is equally clear that the most the formula will help us deduce is a mean value, or some characteristics of the distribution, of that parameter in the total body of gas characterized by P and V. Between 1803, when Dalton proposed the theory, and a century later, when Einstein worked on Brownian motion, the developing theory became ever *more testable*, and this must be reflected in our account.

As long as some hypothesis of a theory is not supported by strongly relevant evidence, relative to that theory, that can be raised as an objection to the theory. Glymour raises this sort of point to explain the felt deficiency of "de-occamized" theories; and theories that admit of distinctions said not to have any physical meaning (e.g., phase of the wave function). As long as the theory is still developing, however, we are better advised not to turn our backs on it if we cannot immediately get strongly relevant evidence, but to look to more tenuous sources of evidential support as well. And finally, in the theory's days of triumph, when the development is thought to be complete, mathematicians may be turned loose to remedy such apparent deficiency (e.g., replacement of wave functions by statistical operators to represent states) because it is no longer felt that any and all potential resources should be husbanded for the rainy days ahead.

As I said, the virtues that may be claimed for Glymour's theory are considerable. By explicating selective testing of hypotheses in bootstrap fashion, he has illuminated an important and pervasive aspect of scientific methodology. By focusing our attention on the empirical determination of values, for most directly measurable quantities, in such tests, he has shown how a developing theory may become more and more testable, and how more and more strongly relevant evidence may become available.

But while lauding these achievements, I want to add two criticisms. These concern virtues that cannot be claimed, and should instead be marked down as open problems.

The theory of evidence has a number of standard problems. One is the apparent additional value evidence takes on if it is collected after the theory has predicted it. I say "apparent" because perhaps there is only the value any evidence has when it is clear that the theory has not been "cooked up" to account for it. Glymour provides a formidable problem to Bayesian approaches that is more or less a variant of that familiar one: how can a Bayesian see *any* confirmation for a theory in evidence that was assimilated before the theory was proposed? Attempts to meet this objection by reference to probabilities formed prior to the gathering of that evidence fail for a number of reasons. However this may be, we must not ignore the fact that Glymour's theory has itself nothing to say on this point.

A second such standard problem is that a *variety* of evidence appears to provide better support for a theory than a narrow spectrum of evidence. Again I say "appears," for it is not clear just how objective the classification of phenomena needed to make this distinction is. (Do the sudden shocks felt by lovers kissing on a rug, Benjamin Franklin's kite, balloons sticking to the wall, St. Elmo's fire, and the Leyden jar form a narrow or a broad spectrum of phenomena?) Glymour takes it seriously, objects to various accounts that have been given of it, and offers his own.[9] But his proffered justification of the methodological requirement of variety of evidence, consists in the observation that without that requirement, his account would lead us to say that a hypothesis is well confirmed when in fact it is not. Specifically, he uses Kepler's ubiquitous laws once more to show a shortcoming of the bootstrap conditions that for its remedy require a demand for variety of evidence. I don't quite know how to distinguish this sort of merit from an objection.

In this connection a point may be made concerning tactics. Just by explicitly *building in* the requirement that all relevant quantities be determinable by the computation used in confirmation (i.e., opting for what I call strongly relevant evidence in his definition of confirmation), Glymour automatically precludes his acount from providing an explanation of why scientists should consider it a shortcoming if not all relevant quantities are empirically determinable.

We all have convictions concerning the importance of these notable features of evidence, and use them to gauge evidential support. But why? What are the basic aims of the activity of testing, experimenting, evidence gathering that must explain, if anything does, our predilection to gauge evidential support in these ways? Bayesians pride themselves on being able to offer some rationale, and I have sketched an approach to this problem in my other paper on Glymour's theories.[10] Glymour's own account is much closer to the mire and blood of research practice; but it does not, it seems to me, justify these general features of scientific methodology.

It is time to return to the exploration of relevant testing and evidence. I shall do so by means of a series of logical thought experiments.[11]

4. Gauging Evidential Support

In the debate over scientific realism, one epistemological question is always lurking in the background: can the evidence give greater support to one theory than another even if the two are empirically equivalent? Glymour has argued the affirmative by attempting to give us reason to think that an extension of a theory, which makes no new empirical predictions, can be better supported than the original.[12] This is very audacious; but it accords with the quite familiar phenomenon that a lengthy, circumstantial story is often more plausible than a bare statement of some of its implications. On the other hand, it is in apparent conflict with the opinion, certainly widespread among philosophers, that a longer story has more ways of being false, and is therefore less likely to be true. This is a puzzling situation and needs to be clarified. I shall return to it in the next section, after a look at the logic of Glymourian testing.

While recognizing that his theory of evidence does not immediately lead to an account of theory comparison, Glymour offers some criteria for comparing entire theories that, he says, emerge from the bootstrap theory.[13] The notions of being *testable* and *tested* that he uses in this discussion can be defined as follows.

(4-1) Hypothesis A is [*strongly*] *testable* [respectively, *tested*] *with respect to* theory T exactly if there exists some data-set [respectively, we have in our possession some data-set] that provides [strongly] relevant evidence for A relative to T.

(4-2) Theory T is *optimally* [*strongly*] *testable* [respectively, *tested*] exactly if it is axiomatized by a set of hypotheses each of which is [strongly] testable [respectively, tested] with respect to T itself.

Glymour shows (and I shall give further examples below) that a subtheory may be less well tested than the whole theory of which it is part. Indeed, in some cases the subtheory is not testable at all. Conversely, any subtheory or hypothesis will be at least as well tested with respect to a larger, logically stronger theory than with respect to a smaller one.

(4-3) If T implies T′ and E is consistent with T, and E provides. [weakly, strongly] relevant evidence for consequence A of T′ relative to T′, then E also provides this for A relative to T.[14]

Thus for any given hypothesis, we may increase the relevant evidence (*tout court*, relative to the theory we accept) without doing new experiments, by introducing stronger postulates. Of course, it may be objected that in such a case our *overall* theory will then have less relevant evidence, or at least be less tested, than before we accepted these stronger postulates. That this objection is mistaken follows exactly from the fact that, on the contrary, a whole theory may be better tested than a given subtheory.

I do not mean this as a reductio. It is a credo of Wilfrid Sellars, and also of Gilbert Harman, that all inference consists in placing (or attempting to place) ourselves in a position where we have a better and more explanatory account of the world, and that when we have *that* our epistemic position is more secure.

Glymour provides us with an instructive example of the situation I have just described in the abstract, pointing to celestial mechanics: "Seventeenth century astronomers were able to confirm Kepler's first law only by using his second, and they were able to confirm his second only by using his first."[15] An artificial example using three postulates is easily constructed; in fact the example of section 2 above will do. Recall that A, B, C are the directly measurable quantities:

$$P_1 \quad A = x + u$$
$$P_2 \quad B = u - z$$
$$P_3 \quad C = x + z$$

For each of the three postulates we can find a data base that provides weakly relevant evidence for it relative to the whole theory. In section 2 we looked at relevant evidence for P_3 relative to this theory. Yet neither P_3 nor P_2 is testable relative to the subtheory formed by dropping the first postulate.

So if someone originally wished to advocate the theory whose postulates are just P_2 and P_3, there is no doubt that his advocacy would be substantially improved if he decided to accept P_1 as well. Before he does so, his theory simply cannot pass, or fail, any test at all. Nor can it be of more interest than the hypothesis that gravity is a form of love, not even if he tells us that his terms B, C, x, u, z denote such eminently interesting and quantifiable magnitudes as performance on an I.Q. test, number of cars owned, aptitude for sailing, femininity quotient, and so forth. Depending on what these terms are, or how they are interpreted, the theory may be meaningful, possibly true or false, audacious, or even outrageous; but it is not testable. The evidence *cannot* give us reason to accept it, in this sense.

In commentary on this paper, Glymour has suggested a plausible way to state this fact in general: the evidence can give us less reason to accept a theory T taken as a whole, than to accept a larger theory T', of which T is a part, taken as a whole. And in one straightforward sense, that is just what the preceding examples show, and cannot be denied by anyone. But it seems to me that the situation is more puzzling yet than these examples bring out. For even relative to the large theory, one of its larger parts may receive more suport than one of the smaller parts thereof. Relevant evidential suport is not inherited by consequences.[16]

Recall that Q(B) is the set of quantities denoted by expressions found in B. Let us examine the following question:

(4-4) If A implies B, and E provides [strongly] relevant evidence for A relative to T, must E provide [strongly] relevant evidence for B?

The answer is no, simply because B may be irrefutable by any data base alternative to E; for example if B is logically valid. The form of this counter example shows at once that adding the proviso-clause $Q(B) \subseteq Q(A)$ to (4-4) will not get us a positive answer either.

Would it help to add "provided B is not logically valid"? No, for it may be that any subtheory T_o needed in the calculation leading from an instance of E (or its alternative) to values for quantities in A, will imply B and hence make refutation impossible. Here is an example. Let f_X be the a characteristic function of set X. That means that for each quantity q we have another quantity $f_X q$ that takes value 1 exactly when q takes a value in X, and that takes value 0 otherwise. We can furthermore use multiplication; thus $q(f_X q)$ will take value q or *zero* depending on whether q takes a value inside X or outside it. Let us now attempt to test two hypotheses

H_1 $q = 1$
H_2 $0 \le q \le 1$

and suppose that there is only one usable subtheory, consisting of a single postulate of the theory (plus the theorems implied by that postulate alone), namely

P_o $A(f_{[0, \ 1]}A) \le q \le A$

where A is directly measurable and q is not.

If evidence E gives A a value x; then with the help of P_o we calculate $x \le q \le x$ if x lies in [0, 1], but we can conclude only that $0 \le q \le x$ if x lies outside that interval (thus P_o rules out negative values for A, as well as for q, and has therefore some immediate empirical import). In the first case the alternative data base $E' = \{A = 1/2\}$ can be cited to show that E provides strongly relevant evidence for hypothesis H_1. But there is no data base at all that can provide even weakly relevant evidence for H_2. The reason is that we can find no E' which, together with T_o implies that $q \nleq 1$; hence no alternative to E of the sort needed for relevant evidence.

The astonishing consequence is that in H_2 we have a hypothesis that is certainly not logically valid, that can be established on the basis of theory plus evidence (namely as a consequence of H_1), and that cannot even possibly receive the support of even weakly relevant evidence relative to that theory.

It may be thought that in any such case, either the subtheory used in the calculations will itself be untestable, or else the hypothesis will be capable of support from data bases used to test that subtheory. Combining the strategies of the two sorts of examples we have had, I shall now show that this surmise too is incorrect. Let the directly measurable quantities be A and B and the "theoretical" quantity q, and let the two hypotheses to be considered again be H_1 and H_2 above. But now let the usable subtheory consist of two postulates, namely P_o above plus:

P_{oo} $Bf_{[0, 1]}B \le q \le B.$

As data bases, consider:

E = {A = 1, B = 1}
E' = {A = 1, B = 1/2}
E'' = {A = 1/2, B = 1}

Relative to any theory including P_o and P_{oo}, we find that E provides strongly relevant evidence for each; hence the little theory whose sole postulates are these is optimally testable. (For P_o plus (A = 1) implies that $q = 1$, which together with B = 1 implies P_{oo}; while the formulas $q = 1$, B = 1/2 have no solution that is also a solution of P_{oo}. This establishes the claim of strong evidence for P_{oo}; *mutatis mutandis* for P_o using E and E''.)

Exactly similarly, E provides strongly relevant evidence (relative to the theory whose postulates are these two) for hypothesis H_1. But the logical consequence H_2 thereof again fails to be even weakly testable relative to this theory. No possible evidence can provide relevant evidence for it.[17]

It would be, for me, a welcome surmise that we are dealing here only with a minor technical point, and that the defined notions of relevant evidence are merely meant to "generate" the correct relation of derelativized evidential support. The suggestion that what is needed in addition is some sort of "ancestral," or "inherited support," relationship, is surely one that must ocur at once to any reader at this point in the story. But I do not believe that this would be welcome to Glymour, given the uses he makes of these features of his account.

5. Comparison of Theories

As Glymour, and a number of others, have pointed out in discussion, a longer story is often more plausible or credible than a short one.[18] At the same time, even the most rudimentary (comparative, nonquantitative) discussions of probability entail that no matter how much A helps to support B, the conjunction (A & B) can be no more likely to be true than B alone (in the light of whatever evidence we have). We seem to have a difficult dilemma: either the way we reckon up the weight of evidence may favor hypotheses or theories less likely to be true, or else even the most rudimentary notions of probability are radically mistaken.

The dilemma is real, and we should embrace the first horn, not feel impaled by it. As I have argued elsewhere,[19] theory comparison and acceptance are a matter of decision making by, and balance striking between, conflicting criteria. The conflict between the desire for more informative theories (predictive power, explanation, empirical strength) with that for likelihood of truth is the main example. A well-designed test does not speak merely to the second desideratum: only powerful theories are in a position to pass powerful tests. Tallying up the support of relevant tests is a way of gathering reasons for the acceptance of theory on several counts at once. (It follows as a corollary that acceptance is not simply belief, for nothing counts as a reason for belief, as such, that detracts from the likelihood of truth.) To reconcile ourselves further to this choice of one horn of a dilemma, here is a more circumstantial account of just how longer, more circumstantial, stories may be more credible than short ones.

If tests are a way of gathering support of several sorts, then it is not surprising that a theory with features that detract from likelihood of truth may do better. In a certain laboratory, small water samples may be more likely to be hot than large ones at a certain time. Yet more of the large ones may yield a reading above 90°C than the small ones on massive, cold thermometers. In this case the thermometer reading reflects several features of the sample, and although the thermometer is the official method of registering hotness, a certain feature that in this case makes hotness less likely, also makes an accurate temperature test more feasible.

This takes care of one way in which theories that are more credible may yet be less likely to be true (for "more credible," in this case, read "better tested.") The second way is brought out by an example due to Nancy Cartwright.

A rumor that undergraduate women will be permitted to live off-campus as of age 21 (circulated at Pittsburgh in 1963) may not find much credence. But if this rumor is enlarged to include a supposed reason for this change in relations, such as that the university finds itself in the difficult situation of having overadmitted undergraduates, it will be much more believable.

The intuitive force of the example is clear. I am not sure that it supports Glymour's contention, because although the conclusion is the same, the apparent cause of the increased plausibility is quite different. In Glymour's case, the enlarged theory is more testable. In Cartwright's example, the enlarged theory is more plausible because a reason or cause for the main prediction is included. Although that reason is itself independently confirmable (and presumably still only an unconfirmed rumor), it shows how the asserted change could have come about.

We can imagine therefore that the standard reaction to a hypothesis has several stages. The first stage is the question, "How could that be?" Until we have satisfied ourselves that there is indeed some plausible way the event in question could come about, we refuse to go to the testing and evaluation stage at all. This transition from one stage to the next, however, does not require enlarging the theory: it requires only that some way of enlarging the account so as to embed the theory in a plausible story is possible. We can imagine that the undergraduates' first reaction is incredulity, because they can see no way the university, not known for its liberal innovations, could have come to such a decision. This lack on their part is merely one of imagination, however (in more technical cases it may be our well-known lack of logical omniscience), and the theory can be taken seriously when that obstacle is removed.

Theory comparison, theory choice, theory acceptance are not merely a matter of seeing which is most likely to be true. Glymour's penetrating analysis of testing has shown, in my view at any rate, that neither is evidential support merely a matter of gauging likelihood of truth in the light of the evidence. In this way, Glymour's analysis destroys the whole basis for what used to be known as confirmation theory. The correct reading is not that testing can provide us with more reason to believe an audacious theory than it provides for belief in the empirical adequacy thereof, but rather that it can give us other sorts of reasons for acceptance.

Notes

1. Evidence itself is an abstraction that should be more controversial than it is; the confrontation between orthodox and Bayesian views may yet make it so. From several points of view, all we have to play the role of evidence is some distribution of comparative certainties

and uncertainties over a set of propositions. Use of an idealization that takes such propositions as representing the evidence needs justification, especially if that distribution is a function of feedback from theories considered in its light.

2. *Theory and Evidence* (Princeton: Princeton University Press, 1980). See also his *Explanations, Tests, Unity, and Necessity. Nous* 14 (1980): 31-50, and Bootstraps and Probabilities. *Journal of Philosophy*, 77 (1980): 691-649 (with *erratum* 78 (1981), page 58).

3. Glymour on Evidence and Explanation, this volume.

4. *Theory and Evidence*, pp. 226-263, and M. Gardner, Realism and Instrumentalism in the Nineteenth Century. *Philosophy of Science* 46 (1979): 1-34. As far as the main thrust of Gardner's paper is concerned, I regard it of course as based on naive identification of debates about whether to accept the atomic theory, with ones about whether or not to believe it to be true. I call this naive not because of identification itself is, but because Gardner is ostensibly addressing the issue of realism versus (what he calls) instrumentalism, which is trivialized by that identification.

5. Semantic analysis of physical theory harbors several approaches, some more extensionalist and some less (Suppes's work being an example of the former). In *The Scientific Image* (Oxford: Oxford University Press, 1980) I have given a number of references in notes 22 and 23 to Chapter 3 (page 221), and note 29 to Chapter 6 (page 228).

6. See *Theory and Evidence*, pp. 111-123.

7. Specifically, the following definition is equivalent to (2-3):

> D*. E provides *strongly relevant evidence* for A relative to T exactly if there is a set of quantities Q^*, an alternative E' to E, and a subset T_q of T for each q in Q^* (whose union is a subset T_o of T) such that:
>
> (1) $T \cup E$ has a solution
> (2) All solutions of $T_q \cup E$ assign the same value, r_q, to q
> (3) All solutions of $\{q = r_q : q \, \varepsilon Q^*\}$ are solutions of A
> (4) $T_o \cup E'$ has a solution
> (5) All solutions of $T_q \cup E'$ assign the same value r'_q, to q
> (6) No solution of $\{q = r'_q : q \varepsilon Q^*\}$ is a solution of A
> (7) $Q(A) \subseteq Q^*$

The idea of this definition is that for each q, we compute a precise value from E via a set T_q of theoretical hypotheses; and this same computation leads from E' to an alternative value for q; the former imply an instance of A and the latter an instance of some contrary of A. But different subsets of T_o cannot lead from E (or from E') to different precise values of q, on pain of inconsistency. Second, placing a constraint on a set of solutions is really equivalent to saying that they are all extendable to some quantity so as to yield the same value for *that*. (For example, the constraint $q \geq 0$ is equivalent to implication that $f_{[0,\infty]}q$ has a value 1 (where f_X is the characteristic function for X).)

8. See *Theory and Evidence*, p. 131; this clause serves to evade the difficulty discussed on p. 132. Suppose for example that T has postulates (where A, B, C are measurable):

> T1. $E - 1 = 0$
> T2. $E = A \cdot B$
> T3. $C = 1$

The data base $I = \{A = 1, B = 1\}$ provides strongly relevant evidence for T1 (using T2). But also, using only T2 and T3, I can provide relevant evidence for the hypothesis H: $E - C = 0$. This is analogous to the problem example discussed on page 132; a problem since we don't suppose that introducing theoretical parameter E in this fashion allows us to confirm a hypothesis about the color of swans by looking at black shoes. If we restrict ourselves to equational theories, all basic propositions can be put in the form $f(—) = 1$ using characteristic functions, and conjunctions can be captured: $X = 1$ and $Y = 1$ exactly if $X \cdot Y = 1$. Thus we can provide a parallel to clause (v). If we allow inequalities, we can use that manoeuvre, via characteristic functions applied to quantities as in the preceding note.

9. *Theory and Evidence*, pp. 139-142.

10. Glymour on Evidence and Explanation, this volume.

11. It is clearly important to see how Glymour's account accords with the history of science, and more specifically whether it throws light on puzzling developments. Glymour supports his account of testing this way through a large and varied class of historical examples. To investigate logical interconnections, thought experiments and fictional histories seem to me to be more appropriate. Especially in the case of the relations between theory comparison and evidence, the historical example method leaves much to be desired. To be shown several examples in which a later theory (a) is in retrospect superior, and (b) represents a gain in evidential support by a certain criterion, established little or nothing about that criterion. Should anyone argue for the curative powers of Vitamin C in that way, Glymour could be the first to show his error.

I do not mean to suggest that Glymour's own support for his theory takes that naive form. On the contrary, he gives us logical analyses of the structure of supporting arguments that cite evidence, in the case of figures as diverse as Copernicus and Freud, so perceptive and captivating that we can immediately perceive the evident value of his methodological insights. But even the best historical support must remain an illustration: we are no more able to deduce the principles of methodology from observed phenomena of scientific activity than Newton could deduce the laws of motion from observed physical phenomena.

12. *Theory and Evidence*, p. 161-167; see also his paper in *Nous* cited in note 2 above, and my response in the paper cited in note 3.

13. *Theory and Evidence*, pp. 152-155.

14. This follows at once from the definition. Note that consistency of T and E means simply that T∪E has a solution.

15. *Theory and Evidence*, p. 141.

16. This appears to contradict the opinion stated in *Theory and Evidence*, sentence straddling pp. 153-154, whence I derive the notion of optimally tested theory.

17. I must add here that I have not exploited a possibility that Glymour mentions: that a hypothesis A itself be used in the calculation of quantities, from a data base, to provide relevant evidence for A. Any defects or surprising features of the account following therefrom, I would tend to discount, because Glymour would lose nothing if he decided after all to forbid it. The example he gives to support it seems to me to involve an equivocation. Suppose we test $PV = rT$ by:

(a) first measuring P, V, T at time t, and calculating a value r_o for r by means of that hypothesis

(b) secondly measuring P, V, T at a later time, and confirming that $PV = r_oT$.

We certainly have a test here of something, but I submit that it is a test of the entailed hypothesis that the value of PV/T is constant in time. If it be insisted that r itself is a physical quantity, then we are testing a little theory consisting of two postulates:

1. (t) (t') (r(t) = r(t'))
2. (t) (P(t)V(t) = r(t)T(t))

and we can use evidence plus either hypothesis to confirm the other. In neither case do we confirm one hypothesis via a calculation which uses that hypothesis. The same point is made by Aron Edidin, this volume.

18. Wesley Salmon, Susan Hollander, Nancy Cartwright, and Ian Hacking (in the order in which they made these comments).

19. Glymour on Evidence and Explanation, this volume.

——————————— Aron Edidin ———————————

Bootstrapping without Bootstraps

Clark Glymour has persuasively outlined the advantages of an account of the confirmation of scientific hypotheses in terms of what he calls a bootstrap strategy. The stragey relates sentences in an observational vocabulary to hypotheses in a theoretical vocabulary as follows: Auxiliary hypotheses are found that, when conjoined with the observation-sentences in question, entail in a certain way theoretical sentences that, in turn, confirm the hypotheses in some standard way. When the strategy succeeds, the hypotheses are confirmed by the observation-sentences relative to the auxiliary hypotheses (or any theory that contains them).[1]

Despite its many virtues, though, the account of the bootstrap strategy presented in *Theory and Evidence* needs to be modified. The two modifications I propose leave intact the basic structure of the strategy, although one would render the name "bootstrap strategy" somewhat inappropriate. On the other hand, the modifications dramatically alter the relation of Glymour's strategy to Bayesian confirmation theories.

The first modification addresses a problem that arises from the fact that Glymour's strategy goes only part way toward connecting observational evidence with hypotheses. The strategy itself gets us from observation-sentences to intermediate theoretical sentences, which are connected to the hypotheses in question in some other way:

$$\text{Observation sentence} \xrightarrow[\substack{\text{Auxiliary}\\ \text{hypotheses}}]{\substack{\text{Bootstrap}\\ \text{strategy}}} \text{Intermediate theoretical} \rightarrow \text{Hypothesis} \atop \text{sentence}$$

In what Glymour calls the Hempelian version of the strategy, the intermediate theoretical sentences are instances of the hypotheses.[2] In other versions, other sorts of sentences might play the role of intermediate. But whatever sentences are used, an observation sentence can confirm a hypothesis by way of such an intermediate only if the intermediate itself

Research for this paper was supported in part by the National Science Foundation.

counts in favor of the hypothesis, relative to the auxiliary hypotheses used by the strategy. Moreover, this support must not be undermined by the observation sentence itself.

The problems that can arise in such a context are especially striking for the Hempelian version of the strategy. It is well known that under certain conditions an instance of a hypothesis can actually count against the hypothesis.

For example, let $A = (a_1, a_2, \ldots)$ and $B = (b_1, b_2, \ldots)$ be two sequences of objects. One or both may be infinite, or both may be finite. Define them in such a way that no object is repeated in either sequence, i.e.,

$$a_n = a_m \text{ iff } n = m$$
$$b_n = b_m \text{ iff } n = m$$

Let R be a relation.

Consider the hypothesis that every member of A is related by R to at least one member of B. That is,

(1) $(x) (x\varepsilon A \supset (\exists y) (y\varepsilon B \ \& \ xRy))$.

Then the sentence

(2) $a_2\varepsilon A \ \& \ b_1\varepsilon B \ \& \ a_2Rb_1$

is an instance of the hypothesis. It is an instance on Hempel's account, and it seems clear that it will be an instance of the hypothesis on any satisfactory account. And it seems that this instance does confirm the hypothesis.

But suppose that we also have good reason to believe the following: First, that no two elements of A are related to the same element of B, and, second, that no element of A is related to an element of B with a higher subscript than its own, i.e.,

(3) $(x) (y) (\mathbf{z}) (x\varepsilon A \& y\varepsilon A \& \mathbf{z}\varepsilon B \& xR\mathbf{z} \& yR\mathbf{z} \supset x = y)$;

(4) $(x) (y) (a_xRb_y \supset x \geq y)$.

If we have good reason to believe (3) and (4), then (2), although it is an instance of the hypothesis (1), actually gives us good reason to *deny* the hypothesis. Indeed, the conjunction of (2), (3), and (4) is *inconsistent* with the hypothesis. (1) and (4) together entail that a_1 is related to b_1, for b_1 is the only element of B whose subscript is less than or equal to that of a_1. But (2) and (3) together entail that a_1 is *not* related to b_1. For by (2), a_2 is related to b_1. By the original description of A and B, $a_1 \neq a_2$. And (3) asserts that no two distinct members of A are related to the same member of B; a_2 has preempted b_1, so b_1 is not available for a_1.

Actually the problem raised by (1) - (4) is taken care of by a clause of Glymour's Booptstrap Condition. Recall that confirmation by Glymour's strategy is relative to a theory. His formulation requires that the theory in question be consistent with the conjunction of the hypothesis being tested and the evidence statement.[3] Now what we don't want is to be able to confirm (1) by way of (2) relative to any theory that includes (3) and (4). But suppose that evidence statement E is such that (2) can be derived from E and auxiliary hypotheses from the theory. Then the theory and E together entail (2). But if the theory contains (3) and (4), it follows that the theory and E together entail the negation of (1), so Glymour's condition is not satisfied.

It is, however, possible to weaken (3) or (4) in such a way that the original problem is revived. Suppose we replace (3) with the claim that it's highly improbable that two distinct members of A are related to any given member of B.

(3′) $(z)(z \varepsilon B \supset \text{Prob}((\exists x)(\exists y)(x \varepsilon A \& y \varepsilon A \& x \neq y \& x R z \& y R z) < .00001))$

Now the conjunction of (2), (3′), and (4) is *not* inconsistent with (1), so it is possible to satisfy Glymour's condition even if the theory contains (3′) and (4) and the theory and the evidence jointly entail (2). But (4) entails that (1) and (2) can both be true only if both a_1 and a_2 are related to b_1, and (3′) entails that the probability that that's the case is less than 1/100,000. So if we have good reason to believe (3′) and (4), then (2) counts heavily against (1) even though (2) is an instance of (1).

The core of Glymour's strategy doesn't concern the relation between the intermediate theoretical sentences derived by the strategy and the hypothesis to be confirmed. It seems appropriate, then, simply to require that the intermediate confirm the hypothesis relative to the auxiliary hypotheses used in the derivation, in a way that isn't undermined by the observational evidence.

The second modification is the one that, as it were, cuts the bootstraps off the strategy. A noteworthy feature of Glymour's strategy is that there is no restriction on the set of auxiliary hypotheses used in computations save that they be consistent with the evidence and the hypothesis being tested. In particular, the set of auxiliary hypotheses may entail or even include the hypothesis being tested. This doesn't trivialize the whole procedure because the strategy is arranged to ensure that the computations do not guarantee a positive outcome.[4]

But the use of such sets of auxiliaries causes other problems for the strategy, and these problems are best solved by prohibiting the use of sets of auxiliaries that entail the hypothesis being tested.

To see that this is so, we may begin by considering the following case. The hypothesis to be tested is

(H) A = B.

We have no way of measuring A or B, but we do have means of measuring quantities C and D. We also have these supplementary hypotheses:

$(h_1)(= H$ itself) A = B
(h_2) B = C   and
(h_3) B = D.

We can compute a value for B by measuring D and using h_3. We can compute a value for A by measuring C and using h_2 and h_1. There is no guarantee that the resulting values will be equal, since there are possible values for C and D such that C $\neq$ D. But of course we haven't confirmed that A = B. All we've done has been to compute values for B in two different ways and then obtain a value for A directly from one of the values for B via the hypothesis that A and B are equal.

Glymour suggests that the problem with this computation is that although there are values of the measurable quantities that would disconfirm the hypothesis, there are no such values that are consistent with the supplementary hypotheses other than H itself. He writes, "the possibility of producing a counterexample...in this case is spurious, for any such conterexample would have to be obtained from observational data inconsistent with the auxiliaries used to obtain it."[5]

The obvious solution is to require that there be possible values of the measurable quantities that would disconfirm the hypothesis *and* that are consistent with the auxiliary hypotheses other than the one being tested. This is Glymour's approach, incorporated in his Bootstrap Condition.[6]

It is important to understand why, when we require that there be possible disconfirming values that are consistent with the auxiliary hypotheses, we must make an exception of the hypothesis being tested. Suppose that there is a set of possible values for the measurable quantities involved in a set of computations, but that this set of possible values is inconsistent with the hypothesis being tested. Far from yielding only a spurious possibility of disconfirmation, the existence of such a set ensures in the strongest way that a positive outcome is not guaranteed.[7]

We must, then, permit computations for which the only possible set of disconfirming values is inconsistent with the hypothesis being tested. But we can't stop there. Any set of values that is inconsistent with the hypothesis being tested is also inconsistent with any auxiliary hypothesis that entails the hypothesis being tested. We must, then, also permit computations for which the only possible set of disconfirming values is inconsistent with any auxiliary hypothesis which entails the hypothesis being tested. If we do so, however, counterexamples are easy to construct.

In what follows, it will be convenient to deal with sentences rather than equations. If we replace H and h_1, h_2, and h_3 with analogous first-order formulae, the computations look like this (where the positions above and below the arrows indicate stages in the computation, and the sentences to the right of the arrows are auxiliaries):

$$H':(x)(Ax \equiv Bx)$$

$$
\begin{array}{ccc}
Bx & & Ax \\
\uparrow \ h_2': & & \uparrow \ h_1'(=H'): \\
\ (x)(Bx \equiv Cx) & & \ (x)(Ax \equiv Bx) \\
Cx & & Bx \\
& & \uparrow \ h_3': \\
& & \ (x)(Bx \equiv Dx) \\
& & Dx
\end{array}
$$

Glymour suggests that the problem with these computations is that the only values for Dx and Cx that would yield a counterinstance of H' would also be inconsistent with the conjunction of h_2' and h_3', the two auxiliary hypotheses that do not entail H'.

But a counterexample to the modified Bootstrap Condition may be obtained if we now alter these computations by conjoining H' to each of the auxiliary hypotheses. In the new set of computations, there is no auxiliary hypothesis that does not entail H'. The modified consistency condition is satisfied vacuously: since every auxiliary hypothesis entails H', the condition applies to none of them. But the new computation clearly inherits the inadequacy of the old. The original computation doesn't confirm H', and merely conjoining H' itself to each auxiliary hypothesis doesn't change matters. The new computation satisfies the Bootstrap condition with the expanded exception to the consistency condition, but it does not confirm H'.[8]

A second sort of counterexample, which will be useful later, may be constructed as follows: find a single hypothesis that entails all the original auxiliary hypotheses, and thus entails H'. Replace each auxiliary hypothesis with this one. Now there is only one auxiliary hypothesis, and it entails H'. Duplicate the computations in the original case in this way: a computation from the original set will be replaced by one that uses only the single auxiliary hypothesis but proceeds by first deriving from the single hypothesis those hypotheses that appeared in the original computation and then proceeding as before. This computation, too, satisfies the modified Bootstrap Condition, and it too fails to confirm H'.

The counterexamples can be avoided by prohibiting the use in computations of hypotheses that entail the one being tested. And this is a natural response to the problems they pose. In the original case, the hypotheses other than H' provided two different ways to compute values for Bx, but no way to compute a value for Ax. This has the effect of ensuring that values for Cx and Dx that yield counterinstances of Ax will be inconsistent with these hypotheses, for such values will have to yield two distinct, incompatible values for Bx. But given this structure for the computations, H' or hypotheses that entail it will be required in order to get a value for Ax out of one of the values for Bx. So it is only the presence of H' or auxiliary hypotheses that entail H' that enable the objectionable computations to yield an instance of H'. By prohibiting the use as auxiliaries of hypotheses that entail the one being tested, the counterexamples are avoided.

I have said that Glymour's strategy should be modified by prohibiting the use of sets of auxiliary hypotheses that entail the hypothesis being tested. It may seem that the arguments to this point support only a weaker conclusion: that the use of *individual* hypotheses that entail the one being tested should be prohibited. Actually they support the stronger conclusion as well. Note first that the problem in the original example doesn't go away if we change the computation so that no single auxiliary hypothesis entails H'. If we replace h_1^i by two separate hypotheses:

$$(x)\ (Bx \equiv A'x)\quad \text{and}\quad (x)\ (A'x \equiv Ax)$$

the computation on the right looks like this:

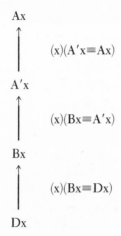

Clearly nothing essential is changed.

But there's a more general argument. Recall the second sort of counterexample, in which the auxiliary hypotheses were replaced by a single, stronger hypothesis that entails each of them. Now consider any set of computations that uses a set of auxiliary hypotheses that jointly entail the hypothesis being tested. To any such set of computations, there corresponds (in the way indicated in the counterexample) an equivalent set of computations using a single auxiliary hypothesis that is itself equivalent to the conjunction of the first set of hypotheses. If the latter set of computations is ruled out, the former should be as well. But if individual hypotheses that entail the hypothesis being tested may not be used, the latter set of computations *is* ruled out, because its single auxiliary hypothesis does entail the one being tested. It follows that sets of computations using auxiliary hypotheses that jointly entail the one being tested should be excluded.

It may be that there is a way of altering the Bootstrap condition that avoids the counterexamples but falls short of prohibiting the use of sets of auxiliaries that jointly entail the hypothesis being tested. But I think that the cases I've decribed create a strong prima facie case in favor of the restriction I propose.

Philosophers who believe that testing a hypothesis essentially involves putting the hypothesis in jeopardy have a further reason to prohibit the use of sets of auxiliaries that entail the hypothesis being tested. For unless this restriction is adopted, it will be possible to confirm hypotheses via sets of computations such that no values of the evidential quantities would *disconfirm* the hypothesis relative to any theory. Indeed, this will be the case whenever the set of auxiliaries used in the computation entails the hypothesis being tested. The Bootstrap condition requires that there be values of the evidential quantities that yield a counterinstance of the hypothesis by way of the computations. But those values only disconfirm the hypothesis if they confirm its negation. As was noted in the discussion of disconfirmatory instances above, they can confirm the negation of the hypothesis relative to a theory only if the theory is consistent with the negation of the hypothesis and also includes all of the auxiliary hypotheses used in the computations. But if those auxiliaries entail the hypothesis, no theory that includes them is consistent with the negation of the hypothesis. It follows that if the hypotheses used in a set of computations entail the hypothesis being tested, those computations cannot disconfirm the hypothesis relative to any theory at all. If the computations satisfy the Bootstrap Condition they won't guarantee a positive outcome—they might fail to confirm the hypothesis—but they do guarantee that there won't be a negative outcome—they cannot disconfirm the hypothesis. And it might well be thought that computations that shield a hypothesis from disconfirmation in this way cannot confirm it.

The case against the use of sets of auxiliaries that entail the hypothesis being tested is strengthened by the apparent absence of any good reason to include them. Since Glymour on several occasions notes that his strategy permits the use of such computations, and indeed uses such a computation in the very example with which he introduces his strategy, it might be thought that the restriction cuts to the very heart of the strategy. That is not so.

Although Glymour claims that cases violating the restriction abound in the scientific literature, it is striking that no such case appears in any of the chapters concerning applications and historical examples of the use of his strategy.[9] The cases he has in mind seem to be cases in which the hypothesis being tested is used to determine the value of a *constant* occurring within it. In this they resemble the example Glymour uses to introduce his strategy. Let us then consider this example.

To be tested is the hypothesis that for any sample of gas, so long as no gas is added to or removed from the sample, the product of the pressure and volume of the gas is proportional to its temperature. This amounts to the claim that, under the given conditions,

$$PV = kT,$$

where k is an undetermined constant. If we can measure P, V, and T, but cannot measure k, "the hypothesis may be tested by obtaining two sets of values for P, V, and T, using the first set of values together with the hypothesis [itself]...to determine a value for k, i.e.,

$$k = \frac{PV}{T},$$

and using the value k thus obtained together with the second set of values for P, V, and T either to instantiate or to contradict the hypothesis."[10]

It is clear, though, that in this case the hypothesis can be tested in Glymour's manner without itself being used in the computation. The hypothesis that Glymour formulates as the claim that under certain conditions PV = kT may equivalently be formulated as the claim that under those conditions,

$$\frac{PV}{T} = \frac{P'V'}{T'}$$

for any two sets of values for pressure, volume, and temperature. From the same two sets of values that were used in Glymour's computation, we may now instantiate or contradict the hypothesis directly, without the need to compute values for any additional quantities.

Moreover, the same procedure can be used whenever it is necessary to use the hypothesis being tested to compute the value of a constant occurring within it. In such a case, the hypothesis must be equivalent to a statement of the form

$$k = f(Q_1, \ldots, Q_n),$$

where values for $Q_1, \ldots, Q_n$ are either measurable or computable from observations by way of other auxiliaries, and k is an undetermined constant. But then, the hypothesis is also equivalent to a statement of the form

$$f(Q_1, \ldots, Q_n) = f(Q'_1, \ldots, Q'_n)$$

for any two sets of values of the quantities, so the hypothesis can be instantiated or contradicted directly.

It seems that we have yet to see an example of an ineliminable use of the hypothesis being tested in a legitimate computation. Moreover, one of Glymour's own arguments concerning variety of evidence suggests that computations using the hypothesis being tested don't yield very good tests of the hypothesis anyway. One reason variety is desirable is that you don't have much support for hypotheses A and B if you've only tested A by using hypotheses that entail B and you've only tested B by using hypotheses that entail A. Glymour develops this point in some detail in his discussion of Kepler's laws[11] and reiterates it in his discussion of criteria of theory choice,[12] where he suggests that a theory containing such a pair of hypotheses is ceteris paribus less well tested than one that does not. But if the only available test of A uses hypotheses that entail A, the result is a degenerate case of the problem (for A = B). (Note that this does *not* seem to be a problem for Glymour's example concerning the hypothesis PV = kT. I take this to confirm my claim that the use of the hypothesis in that computation is inessential.)

It would seem, then, that the use of sets of auxiliary hypotheses that entail the hypothesis being tested is dispensable. Since dispensing with the use of such sets also enables us to avoid various counterexample to the unaltered strategy, I conclude that the use of such sets should be prohibited.

Before leaving the ideal gas law too far behind, it will be worthwhile to discuss one feature of Glymour's strategy that should *not* be changed. Glymour notes that "the consequence condition, which requires that every logical consequence of any collection of confirmed hypotheses be confirmed as well, is not met." But he claims that "there is no difficulty... in simply imposing the consequence condition independently, so that all logical consequences of whatever consistent class of sentences is confirmed by the bootstrap condition are taken to be confirmed as well."[13]

But the example we've just been considering, concerning the hypothesis that PV = kT, suggests that imposing the consequence condition would itself have unfortunate consequences. It seems clear that at least two sets of values of pressure, volume, and temperature are needed to test this hypothesis. Both Glymour's computation and my modified version use two sets of values. No single set of values can test the hypothesis by way of the Bootstrap Condition alone, because no single set of values can yield a counterinstance. But if the Consequence Condition is added to the Bootstrap Condition, then *any* single set of values will not only test but confirm the hypothesis. Let (a, b, c) be such a set of values. Then the set (a, b, c) confirms the hypothesis that

$$PV = \frac{ab}{c}T.$$

But that hypothesis entails that for some k, $PV = kT$, which is just to say that it entails $PV = kT$ where k is an undetermined constant. If every logical consequence of a confirmed hypothesis is itself confirmed, then, the single set of values a, b, and c confirm that $PV = kT$.

I have argued that Glymour's strategy of confirmation should be modified in two ways. The satisfaction condition should be altered to require that the conjunction of the instance obtained and the evidence used count in favor of the hypothesis relative to the auxiliary hypotheses used in the computation, and sets of auxiliary hypotheses that entail the hypothesis being tested should be excluded. I have argued further that neither of these modifications seriously violates the intent of the strategy. But if these two modifications are adopted, the resulting strategy is easily derivable from a Bayesian confirmation theory.

Consider a set of computations that satisfies the modified Bootstrap Condition. If E is the evidence statement and $h_1, \ldots, h_2$ the auxiliary hypotheses, then the conjunction of E and h_1 through h_2 entails an instance of the hypothesis being tested. Moreover, the instance in question counts in favor of the hypothesis relative to the evidence and auxiliary hypotheses. So we have

(1) $E \& h_1 \& \ldots \& h_n \to S,$

and, if the prior probability of E is less than 1,

(2) $p(H/S\&E\&h_1\& \ldots \&h_n) > p(H/h_1, \ldots, h_n),$

where H is the hypothesis being tested, S is an instance of H, and E is the evidence-statement used in the computation. Since the auxiliaries can't entail H, this is no empty identity.

Now (1) entails that (if coherence is preserved)

(3) $p(H/S\&E\&h_1\& \ldots \&h_n) = p(H/E\&h_1\& \ldots \&h_n)$

(2) and (3) together entail that

(4) $p(H/E\&h_1\& \ldots \&h_n) > p(H/h_1\& \ldots \&h_n).$

But (4) is just the formal way of saying that E is positively relevant to H, relative to the auxiliaries $h_1, \ldots, h_n$. Thus if E confirms H relative to theories that entail $h_1, \ldots, h_n$ by the modified version of Glymour's strategy, E is positively relevant to H relative to $h_1, \ldots, h_n$ on a Bayesian theory. What was advanced as an alternative to the Bayesian theory turns out to be derivable from the theory.

The principal advantage Glymour claims for his strategy is that it explains how evidence selectively confirms some hypotheses but not others. He argues that Bayesian theorists cannot do the same without recourse to ad hoc restrictions on prior probabilities. But if I'm right about the need to modify the strategy as I suggest, and about its subsequent derivability from the Bayesian theory, then it may be that the principal benefit of the strategy is that it shows how Bayesians can explain the selective relevance of evidence without ad hoc restrictions.

Notes

1. The strategy is described in greater detail in Glymour's contribution to this volume, and in his *Theory and Evidence* (Princeton; Princeton University Press, 1980), Chapter 5, esp. pp. 130-131. I discuss the various virtues of Glymour's account, as well as certain problems which are examined in greater detail below, in "Glymour on confirmation," *Philosophy of Science* 48 (1981).

2. This is the version presented in Chapter V of *Theory and Evidence*. Carl Hempel's account of confirmation by instances appears in Studies in the Logic of Confirmation. In *Aspects of Scientific Explanation* (New York; Free Press, 1965).

3. *Theory and Evidence*, p. 130 (clause i. of the Bootstrap Condition).

4. Thus, in the Hempelian version of the strategy, there must be possible values of the evidential quantities that yield a counterinstance of the hypothesis via the computations actually used to derive an instance. See *Theory and Evidence*, pp. 130-131 (clause iv. of the Bootstrap Condition).

5. *Theory and Evidence*, p. 117n.

6. Ibid., clause iv., pp. 130-131.

7. This is in fact the case in the example which Glymour uses to illustrate his strategy. See *Theory and Evidence*, p. 111.

8. Paul Horwich has noted that this counterexample can be avoided by allowing the hypothesis being tested to be used as an auxiliary, but prohibiting the use of any other auxiliaries that entail it. But this added condition remains vulnerable to other counterexamples. The following computation satisfies the proposed restriction but inherits the inadequacy of the original example:

$$
\begin{array}{ll}
\text{Ax} & \\
\uparrow & \\
\quad (x)(Ax \equiv Bx) & \\
& \text{Bx} \\
& \uparrow \\
(x)((Ax \equiv Bx) \supset (Bx \equiv Cx)) \& Cx & \\
\uparrow & \quad (x)(Bx \equiv Dx) \\
\quad (x)((Ax \equiv Bx) \supset (Bx \equiv Cx)) & \\
\text{Cx} & \text{Dx}
\end{array}
$$

9. *Theory and Evidence*, Chapters VI-IX.

10. Ibid., p. 111.

11. Ibid., pp. 140-141.

12. Ibid., p. 153.

13. Ibid., p. 133.

Explanations of Irrelevance

On the Relative Merits of Bayesianism, the Hypothetico-Deductive Method, and Glymour's "Bootstrap" Theory of Confirmation

The singular feature of Clark Glymour's "bootstrap" account of evidence, that which primarily distinguishes it from both Bayesianism and the hypothetico-deductive (henceforth, H-D) method, is that a whole theory is not necessarily confirmed just because one of its parts is: neither evidence for p, nor even the discovery of p, is sufficient to confirm the conjunction $p \wedge q$. This is the intended result of his requirement that confirmation of a hypothesis involve derivations of instances of *every* quantity in it. For example, a set of observations might support

$$y = x^2 + z$$

but not the conjunction

$$y = x^2 + z \text{ and } x = s{\cdot}t$$

and, therefore, not

$$y = (s{\cdot}t)^2 + z$$

because values of s and of t may not be calculable from the data.

Therefore our assessment of Glymour's proposal boils down to whether we think this special feature is an advantage. Is it in fact intuitively plausible to maintain that the discovery of p provides no evidence for $p \wedge q$? Would this permit the accommodation of intuitive relevance judgments that are inexplicable by means of hypothetico-deductive principles? Are accounts of various other aspects of scientific methodology facilitated by this characteristic of the bootstrap approach?

Here I should mention parenthetically a point on which I disagree with Aron Edidin.[1] He maintains—and I sympathize with this—that p clearly does confirm $p \wedge q$. He does not, however, infer that Glymour's strategy is misguided; instead he suggests that Glymour does not purport to have

55

characterized the *only* legitimate method of confirmation. Consequently we should conclude that there are two methods: first Glymour's, and second, the confirmation of a conjunction by evidence for one of its conjuncts.

But this strikes me as a vain attempt to have one's cake and eat it. If we supplement Glymour's theory with the conjunction criterion, we are led straight back to the hypothetico-deductive method: the extra thesis with which Edidin hopes to salvage Glymour's account is the very principle that Glymour has struggled so ingeniously to avoid. Moreover, it seems to me that Glymour, whether or not he says so explicitly, does take himself to be providing the only method of confirmation (at least for nonstatistical hypotheses). Otherwise he could hardly claim to provide explanations for our intuitions of irrelevance. For if he were offering merely a sufficient condition of confirmation—if he left open the possibility of alternative methods—then the fact that his condition is violated in certain cases could not justify the conclusion that these are not cases of confirmation. Explanations of irrelevance require a necessary condition of confirmation.

Returning to our questions about Glymour's proposal, perhaps the strongest thing to be said in favor of the opposite view, that p does confirm p∧q, is that it fits the following principle: a theory is confirmed when the data ought to enhance our confidence in it. No one is prepared to deny, I suppose, that in general the discovery of p should diminish some of the uncertainty attaching to p∧q and ought therefore, to augment its credibility. Consequently, if we adopt Glymour's theory, we are compelled to recognize that data may boost the credibility of a theory without confirming it. This result constitutes a two-pronged objection to the bootstrap approach. First, the view we have been compelled to abandon—the correlation between confirmation and enhanced credibility—has great virtues of simplicity and at least rough correspondence with scientific practice. Second, its rejection creates an urgent need for some alternative account of the relationship between confirmation and rational belief. Unless this is provided we are left with an emasculated account of confirmation, incapable of fulfilling the very role in epistemology for which a theory of evidence was desired.

But perhaps these difficulties are just the price we have to pay for a decent account of relevance. One of Glymour's main complaints about the H-D method, to be rectified by means of the bootstrap condition, is that it fosters a misleading and unnecessary holism. When a theory containing many hypotheses accurately accommodates a set of data, then according to the H-D method the *whole* theory is confirmed. But this is unacceptable, says Glymour, for the theory may well contain superfluous parts that deserve no credit at all.

As it stands this objection is inconclusive. Nevertheless, I think Glymour is moved by a genuine and major flaw in the H-D method, a flaw, however, that is inherited by his own bootstrap theory. Let me try to substantiate these claims.

The objection, as just stated, is that there are cases in which a theory T is confirmed by evidence E according to the H-D method, even though that evidence is in fact irrelevant to some part T' of T. We should note, in the first place, that this possibility does not reveal any internal incoherence in the H-D method. In its standard formulation, which Glymour employs, the method does not entail that what confirms a theory should also confirm its parts, and we are by no means required by the evidence for T to have any more confidence in the truth of T'.

But this response does not do justice to Glymour's concern. For he is right to complain that the H-D method inadequately specifies how credit for the accurate prediction of E should be distributed amongst the component hypotheses (consequences) of T. The H-D method is not totally silent on this question; it does tell us that any part that entails E is confirmed. But this is not enough. Clearly there will be components of T that do not entail E but are nevertheless confirmed by its discovery. The real substance of Glymour's criticism is that this problem cannot be solved by supplementing the H-D principle,

(H-D) E (directly) confirms T iff T entails E

with the so-called *consequence condition*

(C) If E confirms T and T entails T', then E confirms T'.
For consider an arbitrary statement p. Given (H-D), E confirms E∧p; and given (C), E confirms p. Thus (H-D) supplemented with (C) yields absurd consequences; but in the absence of some such additional principle the H-D condition for confirmation is much too restrictive, dictating incorrectly that a statement is confirmed only by data that it entails. We then could not explain, for example, why the observation of many black ravens should confirm the belief that future ravens will be black.

So let us examine a potentially helpful liberalization of the (H-D) principle. In the first place, one might not insist that the evidence be entailed by what is confirmed, but only that the data E be divisible into two parts $(E_1 \wedge E_2)$ in such a way that one part, E_1 (the "initial conditions"), entail, with the aid of the hypotheses in question, T, the other part, E_2. Second, we should appreciate that confirmation often takes place relative to the assumption that certain previously established, "background" theories, B, are true. These ideas may be incorporated into the H-D strategy, as follows:

(H-D*) E (directly) confirms T relative to B iff

(1) $E \equiv (E_1 \wedge E_2)$
(2) $(B \wedge T \wedge E_1) \rightarrow E_2$
(3) $(B \wedge E_1) \nrightarrow E_2$

For example, the hypothesis $y = x^2$ would be confirmed relative to the theory $y = z + 1 \wedge x = w^3$ by the data $w = 1 \wedge z = 0$. Simply let E_1 be $w = 1$ and E_2 be $z = 0$; then it is easy to see that

$$(y = z + 1 \wedge x = w^3 \wedge y = x^2 \wedge E_1) \rightarrow E_2,$$

whereas

$$(y = z + 1 \wedge x = w^3 \wedge E_1) \nrightarrow E_2.$$

We must bear in mind that confirmation relative to B should significantly affect the credibility of T only to the extent that B is a theory that has already been well supported. Otherwise such relative confirmation of T will provide no reason to increase our confidence in it. This natural constraint on the application of the method removes one of Glymour's objections to (H-D*). "Given a true sentence *e*, almost every sentence S is confirmed by *e* with respect to some *true* theory. More precisely, let *e* be a true not logically valid sentence and S any sentence at all that is both consistent and not a logical consequence of *e*; then S is confirmed by *e* with respect to (S⊃*e*), which is true." (*Theory and Evidence*, p. 36) This is correct, but quite innocuous. For if S is an arbitrary truth, then the so-called background theory, (S⊃*e*), cannot have been independently established. Therefore this relative confirmation of S should not enhance its credibility.

Nevertheless, (H-D*) is still seriously inadequate: for although our modifications of (H-D) are good ones, this new version of the hypothetico-deductive method is subject to the same objections as the old one. Taken to provide necessary and sufficient conditions for confirmation, it remains much too restrictive: we still can't account for the confirmation of future predictions by past data. And adding the consequence condition (C) will produce the same absurd results as before.

Turning now to Glymour's bootstrap theory to see if it provides relief from this difficulty, let us first consider one of his examples. We all agree that observation of at least *two* planets is needed in order to confirm Kepler's 3rd Law, which states that the square of a planet's period is proportional to the cube of its mean distance from the sun. Glymour maintains that a special merit of his approach is its capacity to account for this intuition and others like it; and he insists that the hypothetico-deductive method cannot do so. However, it seems, on the contrary, that the (H-D*) story is quite straightforward: viz., given the periods $t = t_1$ and $t = t_2$, and the average orbit radii $d = d_1$ and $d = d_2$, for two planets, then

$$(t = t_1 \wedge d = d_1 \wedge t = t_2) \nrightarrow d = d_2$$

and Keplers 3rd Law would indeed be confirmed iff

$$(t = t_1 \wedge d = d_1 \wedge t = t_2 \wedge K3) \rightarrow d = d_2$$

moreover there is no previously established background theory B and no set of possible observations O, of just a single planet, such that

(1) $O \equiv O_1 \wedge O_2$
(2) $(B \wedge K3 \wedge O_1) \rightarrow O_2$
(3) $(B \wedge O_1) \nrightarrow O_2$.

However, although it is clear how (H-D*) could be employed to derive this case of evidential irrelevance, this does not really undermine Glymour's complaint. For we have already conceded that the *unsupplemented* (H-D*) account is too restrictive: it does not provide a satisfactory *necessary* condition for confirmation. And no theory of confirmation may be thought to explain our intuitions of evidential *irrelevance* unless it does contain an acceptable necessary condition for confirmation. In other words, although the unsupplemented H-D* method would imply that observations of a single planet could not confirm Kepler's 3rd Law, we cannot regard this as a good explanation of our intuition, since we have already seen that the unsupplemented H-D* method is inadequate in precisely the respect that it yields too many irrelevance judgments and fails to recognize many legitimate cases of confirmation.

How does the bootstrap method fare with this problem? Glymour explains the intuition by referring to the fact that observations of a single planet will not permit us to compute values of all the quantities in Kepler's 3rd Law. It remains to be seen, however, whether his account is any better than the explanation provided by the unsupplemented H-D* method. Glymour's strategy would be similarily undermined if the bootstrap principles are similarly inadequate, dictating conditions for confirmation that are too stringent.

Indeed, it might seem obvious that this will be so, since the bootstrap conditions for confirmation are even more restrictive than those of the H-D* method. But this is too glib. We have seen that the pure H-D* method cannot be rescued by supplementing it with the consequence condition, for this would result in too much confirmation: if E confirms T, then for almost any p, E also confirms T∧p, and would, given the consequence condition, also confirm p. But it is striking that this argument relies overtly upon the thesis that E confirms anything that entails it. So since that is the very feature that Glymour's strategy is designed to avoid, we might reasonably hope that his approach can be supplemented coherently with the consequence condition and thereby avoid the incorrect attributions of irrelevance to which it would otherwise be committed.

Perhaps this is what Glymour has in mind when he says:

> The consistency condition, which requires that all of the hypotheses confirmed by a given piece of evidence (with respect to a fixed theory) be consistent with one another, just fails to be realized; it does not hold in certain odd cases, although it could be made to hold at the cost of some additional complexity in the formulation of the bootstrap condition. The consequence condition, which requires that every logical consequence of any collection of confirmed hypotheses be confirmed as well, is not met. There is no difficulty however, in simply imposing the consequence condition independently, so that all logical consequences of whatever consistent class of sentences is confirmed by the bootstrap condition are taken to be confirmed as well. (*Theory and Evidence*, pp. 132-133)

My reason for quoting the first part of this passage is for its bearing upon how we should construe the last sentence. Does it mean that the consequence condition may be applied (1) only to consistent subsets of what is confirmed, or (2) only when the consistency condition is satisfied? Something is needed to deal with the allegedly rare cases in which incompatible statements are confirmed. (1) is required to avoid the conclusion, in those cases, that any arbitrary p is confirmed. (2) is needed to prevent the indirect confirmation of contradictory statements.

I think the second, stronger, qualification is desirable (it entails the first), and also closer to the text. But I don't believe that much depends upon the choice. For what I shall argue is that the consistency condition is almost never satisfied. Therefore, given the strong qualification (2), the consequence condition will almost never be applicable, and so the problems that called for it will remain unresolved. Moreover it is no use retreating to the weaker construal (1) of the qualification. For, given that construal, we shall be swamped with cases in which contradictory claims are confirmed by the same evidence—and that is as bad as no evidence at all.

To see how easily the consistency condition is violated, consider the following standard examples.

First, the case of incompatible hypotheses concerning the functional relationship between the values of quantities x and y: let H_1 be $y = x$, and H_2 be $y = x^2$, and E be the data set $\{y = 0, x = 0; y = 1, x = 1\}$.

Second, the grue problem: let H_1 be "All emeralds are green," and H_2 be "All emeralds are either sampled and green, or unsampled and not green," and let E be the data "$a_1, a_2, \ldots, a_n$ are sampled green emeralds."

As far as I can see, in each of these examples E confirms both H_1 and H_2, according to Glymour's conditions. Nor should this be surprising, since Glymour concedes in Chapter VIII ("The Fittest Curve") of *Theory and Evidence* that his bootstrap method makes no progress towards a solution to the problem of simplicity. What I am suggesting is that this failure has unrecognized ramifications. It means, first of all, that we cannot obtain much benefit from the consequence condition as long as we are only permitted to use it when the consistency condition is satisfied; and second, if that restriction is lifted, we shall be saddled with the absurd confirmation of contradictory theses: e.g., "Unexamined emeralds are green" and "Unexamined emeralds are not green" will both be supported by our evidence.

It seems, in conclusion, that both Glymour's strategy and the H-D method are seriously defective. Both of them imply that data are irrelevant in cases when we are inclined to say that the hypothesis is confirmed. Neither can accommodate the most valuable and philosophically challenging cases—those in which the data are not entailed by what is confirmed. Neither may be rescued by tacking on the consequence condition. And the fundamental reason for these difficulties is that neither theory provides a solution to the problem of simplicity, including the so-called new riddle of induction. Thus the glaring deficiences of the H-D method are not mitigated by bootstrapping, and so there appears to be no good reason to insist upon it: to thereby abandon the principle "p confirms p∧q" and to swallow the difficulties this would entail.

Finally, a few remarks about Bayesianism. On the question of relevance judgments, Bayesianism is in more or less the same boat as Glymour and the H-D method. The core of the view is that a system of rational beliefs must conform to the probability calculus. If this so-called coherence is also taken to be *sufficient* for rationality, then examples such as the Kepler case can be dealt with easily. Clearly, mere coherence will not require that one's degrees of belief concerning observational properties of a *single* planet be affected by the supposition of Kepler's 3rd Law, although they may well have to be modified by the supposition of his 1st or 2nd Laws. It seems unlikely, however, that coherence on its own is sufficient for rationality. Coherence does not dictate, for example, that given the evidence, one ought to be confident that unexamined emeralds will be green and not grue. Further constraints will be needed to completely characterize our inductive practice, and until these are supplied the Bayesian approach cannot do much better than its competition in systematically accounting for our intuitions of relevance.

Nevertheless, there is some reason to prefer Bayesianism. From that perspective we can explain the widespread accuracy and therefore the initial plausibility of the *sufficient* conditions of confirmation implicit in the alternative approaches; and, in so doing, we can reveal circumstances in which those criteria will lead us astray. Let me first indicate how the bootstrap sufficient condition is validated by (H-D*). Then I shall give a Bayesian rationale for the sufficient condition implied by (H-D*) (and therefore indirectly justify the bootstrap criterion). This rationale depends upon certain assumptions; and in circumstances where they are violated both in (H-D*) and the bootstrap criteria will become misleading.

Suppose E confirms T relative to B, according to the bootstrap criterion, and suppose B does not entail T. Then (E ∧ B) entails an instance of T, and there is a possible data set E^1 such that $(E^1 \wedge B)$ entails $-T$. Let F be the disjunction of all the possible data sets that would *not* yield, via B, a negative instance of T. And let M be the statement that some data were obtained. Then:

$$E \rightarrow (M \wedge F)$$
$$E \equiv M \wedge F \wedge (M \wedge F \rightarrow E).$$

Also, $M \wedge B \wedge T$ entails that some data were obtained that do not yield a negative instance of T; whereas $M \wedge B$ (without T) does not entail this; i.e.,

$$[M \wedge B \wedge T \wedge (M \wedge F \to E)] \to F$$

and

$$[M \wedge B \wedge (M \wedge F \to E)] \nleftrightarrow F.$$

Then, splitting the evidence E into

$$E_1 = M \wedge (M \wedge F \to E)$$
$$E_2 = F.$$

We can see that the H-D criterion of confirmation is satisfied. Thus whenever E bootstrap-confirms T relative to B, it will also be the case that E H-D*-confirms T relative to B.

A Bayesian explanation for the frequent success of these criteria may proceed as follows. From the probability calculus, we have

$$P(E \wedge T) = P(E \wedge T \wedge B) + P(E \wedge T \wedge -B)$$
$$P(T/E)P(E) = P(E/T \wedge B)P(T/B)P(B) + P(E/T \wedge -B)P(T/-B)P(-B).$$

For the sake of simplicity let us assume that T and B are independent of one another; i.e.,

$$P(T/B) = P(T/-B) = P(T)$$
$$\therefore P(T/E)P(E) = P(T)[P(E/T \wedge B)P(B) + [P(E/T \wedge -B)P(-B)].$$

If the H-D* criteria of confirmation are satisfied, then

$$E \equiv E_1 \wedge E_2$$

and

$$B \wedge T \wedge E_1 \to E_2$$
$$\therefore P(E/T \wedge B) = P(E_1/T \wedge B).$$

Therefore the degree of confirmation $P(T/E)/P(T)$—the factor by which the discovery of E should increase the credibility of T—becomes

$$(1) \quad \frac{P(E_1/T \wedge B) \, P(B)}{P(E)} + \frac{P(E/T \wedge -B) \, P(-B)}{P(E)}.$$

In many cases, the initial conditions, E_1, are more or less independent of $(T \wedge B)$, and the evidence E is roughly independent of $(T \wedge -B)$. i.e.,

$$P(E_1/T \wedge B) \simeq P(E_1)$$

and

$$P(E/T \wedge -B) \simeq P(E).$$

Then the degree of confirmation $P(T/E)/P(T)$ becomes approximately

(2) $\dfrac{P(E_1)P(B)}{P(E_1 \wedge E_2)} + P(-B).$

This is always greater than or equal to one, and increases as our confidence increases in the background theory B. It reaches a maximum of $P(E_1)/P(E_1 \wedge E_2)$ when we are certain that B is true, and a minimum of one when we are sure that B is false.

On the other hand, there are circumstances in which the assumptions leading to (2) are mistaken. In particular, it may happen that the mere occurrence of E_1—the initial condition of our experiment—itself tends to disconfirm T. Consequently $P(E_1/T \wedge B)$ could be substantially lower than $P(E)$, so that $P(T/E)/P(T)$—given by formula (1)—is less than one and T is disconfirmed. For example, let our evidence E consist in the discovery that some randomly selected person is a great philosopher from Ohio. And let T be the hypothesis: "All great philosophers come from Ohio." In this case the H-D*, and the bootstrap, conditions for confirmation are met. One might quite correctly take E to count *against* T, however, for roughly the following reason. The fact that we happened to come across a great philosopher indicates that there are probably quite a few of them in the world, and so it would be a remarkable and unlikely coincidence if they all came from Ohio. Or, in other words, the supposition that T is true—that all great philosphers do come from Ohio—suggests that there are only a handful of great philosophers, and so substantially reduces the subjective probability that one will be encountered; consequently, the actual discovery of such an individual would tend to cast doubt on the original supposition.

A central component of Glymour's confirmation theory is the principle that a hypothesis may be confirmed only by data that permit the computation of values of every quantity in it. At the beginning of this paper I noted a counterintuitive consequence of that principle. It entails that p does not confirm p∧q; thus data may increase the (epistemic) probability of a hypothesis without confirming it. The Bayesian analysis I have just given complements and reinforces this criticism, for it shows that Glymour's confirmation criterion may be satisfied without any increase in the probability of the hypothesis (indeed, even when that probability is diminished). Thus confirmation turns out to be neither necessary nor sufficient for the enhancement of credibility. This anomaly need not automatically disqualify the bootstrap theory, but it does reveal an important deficiency; for the main point of confirmation theory is to improve our understanding of scientific belief. Perhaps some alternative account of the relationship between confirmation and rational belief will one day be designed; but until then, the epistemological significance of bootstrap confirmation cannot be established.

Note

1. A. Edidin, Glymour on Confirmation, *Philosophy of Science* 48 (1981); see also Edidin's chapter in this volume.

II. THE BAYESIAN PERSPECTIVE AND THE PROBLEM OF OLD EVIDENCE

Why Glymour Is a Bayesian

In the third chapter of his book *Theory and Evidence*, Clark Glymour explains why he is not a Bayesian. I shall attempt to show, on the contrary, that he is a Bayesian, more so than many who march under that banner.

1. Bootstrapping and Bayesian Inference

The central problem his book addresses is to explain how findings in one (observational) vocabulary can evidence propositions stated in a different (theoretical) vocabulary. The solution offered is that a hypothesis is confirmed with respect to a theory by deducing instances of that hypothesis from the evidence and assumptions of the theory, where these assumptions may include the very hypothesis under consideration. (It is the latter feature that leads Glymour to label the procedure "bootstrapping.") Confirmation is thus a ternary relation linking a bit of evidence e to a hypothesis h by way of a background theory T. In addition, Glymour requires that the observation or experiment that issues in e be such that it could have issued in a disconfirming, rather than a confirming, instance. In short, the experiment must place the hypothesis in jeopardy.

Both features are nicely illustrated by Glymour's discussion of the hypothesis h, common to Ptolemaic and Copernican astronomy, that a planet's period increases with its distance from the center of motion. His point is that h is testable (hence confirmable) relative to the Copernican theory but is not relative to the Ptolemaic. For in Copernican astronomy, observed planetary positions can be used to determine the relative distances of the planets from the sun. And using the earth's known period of 365.2425 days, the directly observable synodic periods (the times between successive superior conjunctions when earth, sun, and planet all lie on a line) determine the sidereal periods (or time for a complete circuit of the sun), and so the latter may be inferred from the observations as well. We find, for example, that the maximal elongation from the sun is larger for Venus than Mercury, whence Venus's orbit of the sun must contain

Mercury's. Then *h* predicts that Venus will have the longer (sidereal) period, and hence the longer synodic period. This prediction is borne out by observation. If, however, the observed synodic periods satisfied the reverse inequality, we would have instead a counterinstance of *h*. Relative to Copernican theory, then, the observed positions and synodic periods do place *h* in jeopardy. But as there is no like determination of the relative sizes of the planetary orbits in Ptolemaic astronomy, *h* cannot be tested or confirmed relative to that theory. Instead, Ptolemaic astronomers simply *assumed h* in order to fix the order of the planets.

That the hypothesis *h* stands in this relation to the two theories is clearly a result of the fact that relative distances from the center of motion are deducible from observations in the Copernican theory but not in the Ptolemaic. That is to say, it results from the greater *simplicity* or *overdetermination* of the Copernican theory. As we will see, greater overdetermination renders a theory more highly confirmable on Bayesian grounds. This already suggests a relation between Glymour's account of evidence and a Bayesian account very different from opposition, but let us look more closely.

For *e* to confirm *h* relative to T, Glymour first requires that *e* be an instance of *h* in Hempel's sense. Hempel's satisfaction criterion effectively equates con*firm*ing observations with con*form*ing observations, and is of course strongly at odds with a Bayesian account of confirmation based on positive relevance. From a Bayesian point of view, Hempel's "positive instances" are confirmatory only when they happen to be consequences or verified predictions of the hypothesis. This suggests opposition, and yet it is surely very striking that Glymour's examples are all most naturally interpreted as Hempelian instances of this more restricted kind! This is perfectly clear in the example from astronomy, in which we can imagine first ascertaining that the maximal elongation is greater for Venus than Mercury. Then the hypothesis *h* relating period to orbital radius predicts that Venus will be found to have a longer synodic period than Mercury (to overtake Earth less frequently). Therefore, using Copernican theory, the synodic periods are restricted to satisfy a simple inequality. Similarly, in the examples of theories formulated as equations (pp. 112 ff.), overdetermination of the theory expresses itself in the fact that different subsets of the equations can be solved to yield different determinations of a theoretical quantity, and the predictions are then of the form that these different determinations will agree. In all of these cases, talk of deducing an instance of a hypothesis from theoretical assumptions can be translated without loss into talk of verifying a prediction based on the hypothesis and the subsidiary assumptions.

Consider next the (Popperian) requirement that the observation or experiment place the hypothesis in jeopardy. As Glymour phrases it (p. 127), "the deduction is such that it does not guarantee that we would have gotten an instance of the hypothesis regardless of what the evidence might have been." That is, the relevant observation might have issued in other outcomes from which a counterinstance or disconfirming instance of the considered h would have been deducible, using T, as in the example from astronomy. We may think of the evidence e, therefore, as representing a particular subset of the allowed values of the observable quantities. Glymour's first condition is that the actually observed values do indeed fall in the allowed subset. His second (Popperian) condition is that the complement of the allowed subset be nonempty. If we equate possible outcomes with those of positive probability, his account of "e confirms h relative to T" comes to this:

(1.1) $P(e/h,T) = 1$ and $P(e/T) < 1$.

This looks very much like hypothetico-deduction (see Glymour's own formulation on p. 168).

More to the point, the two conditions of (1.1) are sufficient that e confirm h relative to T on a Bayesian account of confirmation. The second condition of (1.1) is definitely needed. Indeed, if we wish, more generally, to incorporate cases in which e is not a consequence of h and T, it is natural to replace (1.1) by the weaker condition:

(1.2) $P(e/T) < P(e/h,T)$,

which merely expresses the positive relevance of h to e against the theoretical background T. (Notice that (1.1) entails (1.2), but not conversely, although the second part of (1.1) is entailed by (1.2).)

Glymour hankers after conditions that further constrain the confirmation relation. As we shall see, his chief objection to hypothetico-deductive and Bayesian approaches is that they are too liberal, admitting as confirmatory items of evidence that we should not countenance as such. From this viewpoint, it is ironic that the Bayesian reconstruction of the bootstrapping argument just offered is far more restrictive than the one based on Hempelian confirmation, for Hempel's criterion, we have seen, is far less austere than the positive relevance criterion (1.2). And inasmuch as Glymour's own examples seem to depend only on Hempelian instances that happen to be verified predictions or consequences of the hypothesis, one would think that Glymour himself would prefer the Bayesian analysis of bootstrapping to the Hempelian.

In fact, he does express misgivings about Hempel's satisfaction criterion (in his closing chapter), pointing out that it does not permit confirmation of sentences having only infinite models or confirmation by "partial instances" (e.g., of "everything bears the relation R to everything" by "*a* bears R to *b*"). Yet these criticisms suggest that Hempel's criterion is too narrow, whereas one would have thought that it is too broad, as shown, for example, by the paradoxes of confirmation. At any rate, in a paper that has since appeared (Glymour 1981), Glymour expands some of the replies he offered to the version of this paper presented at the conference. He shows how to connect bootstrapping to a Bayesian account (essentially as above) but continues to insist that Bayesian methods are too permissive. The main thrust of the paper is to deny what I had argued in my original presentation: that bootstrapping reduces to a nuts-and-bolts form of Bayesian confirmation theory.

2. Is Bayesian Methodology Too Weak?

Glymour is hardly the first to press this line of criticism or urge that an adequate methodology should impose additional strictures of a non-Bayesian kind. Before I take up his specific objections in detail, it may be well to look briefly at some earlier criticisms of a similar kind. This will not only set Glymour's reservations in better perspective, but it will allow us to highlight additional parallels between his account of evidential support and the present author's.

(a) High content versus high probability

Perhaps the most noteworthy previous attempt to show that Bayesian methodology is too liberal comes from Sir Karl Popper. His chief criticism seems to be that Bayesians cannot account for the demand for content. For if high probability is the *ens realisimum* of inquiry, it is best attained by putting forth theories of low content that run a minimal risk of exposure to contrary or falsifying evidence. That confirmation is easily attained if sought, is a recurring theme in Popper. This accounts, he thinks, for the otherwise surprising "success" of Freudians and Marxists. And let us admit that this criticism has real bite when applied to Hempelian confirmation. For if consistency with our hypotheses is all we demand, then confirmation is indeed easy to come by. The moral Popper draws is that genuine confirmation or support can be obtained only by running risks. And we run risks; first by putting forward "bold conjectures" or theories of high content, and second by subjecting our theoretical conjectures to stringent tests and searching criticism. In fact, Popper carries this line of thought right to its logical conclusion, insisting that confirmation can result only from a *sincere* attempt to overthrow or refute a conjecture. (We have already seen that this "nothing ventured nothing gained" philosophy is incorporated in Glymour's account of confirmation.) In resting content with nothing short of a sincere attempt at refutation, Popper enters the shadowy realm of the psycho-logistic. Although this may seem somewhat out of character, it is important to recognize this strain in his thinking, for we shall encounter it below in the writings of Popper's follower Imre Lakatos. From a Bayesian standpoint, it would be most natural to equate a stringent or sensitive test with one that has a low probability of issuing in a conforming outcome if in fact the conjecture is false. But Popper has been at best ambivalent about attempts to capture what he is saying in probabilistic terms.

Let me now sketch a Bayesian response to Popper's criticism, one that I have developed elsewhere in greater detail (especially in chapters 5-7 of Rosenkrantz 1977), although the present treatment contains important additions and qualifications.

To begin with, Popper's notion of content seems unduly narrow. Roughly, he equates a statement's content with the set of "basic state-ments" it logically excludes. In practice, though, a theory or model does not *logically* exclude any outcome of a relevant experiment. This is patently true of a probability model. As determined by a suitable statistical criterion of fit, the outcomes will be in only more or less good agreement with such a model. This will also be true of a deterministic model, for empirical study of such a model is always coupled with a probabilistic theory of errors of observation. Moreover, any theory, probabilistic or deterministic, will typically have adjustable parameters that must be estimated from the data used to test the theory. And the number of parameters that must be estimated is surely relevant to any assessment of a theory's content.

A natural way of extending Popper's notion to accommodate degrees of fit and numbers of parameters is to measure a theory's content (or simplicity, or overdetermination) *relative to a contemplated experiment* by the proportion of possible outcomes of the experiment that the theory "fits" by the lights of a chosen probabilistic criterion. I term this proportion the theory's *sample coverage* for the given experiment. And for theories with adjustable parameters, sample coverage is just the union of the sample coverages of the special cases of the theory obtained by assigning definite values to all free parameters. The smaller its sample coverage (i.e., the narrower the range of experimental findings it accommodates in a probabilistic sense), the greater a theory's content. And, I hasten to add, the contemplated experiment relative to which sample coverage is computed may be a composite experiment comprising several applications of the theory, or even its entire intended domain of applications.

The concept of sample coverage captures a good deal of what is packed into our ordinary understanding of content or simplicity. Thus quantitative theories are simpler (have more content) than their qualitative counterparts, and unifications of theory (e.g., of electricity and magnetism, Mendelian genetics and cytology, or quantum theory and relativity) represent (usually major) simplifications of theory, for experiments formerly regarded as independent then appear as highly dependent. Above all, we complicate a theory when we enlarge its stock of adjustable parameters, for each parameter we add extends the range of possible findings that the theory can accommodate. (It doesn't follow, however, that we can compare the content of two theories merely by counting parameters.) The explication of content in terms of sample coverage and the relativization to an experiment help us to avert some familiar difficulties, such as irrelevant conjunction (which I discuss below in connection with Glymour's critique of hypothetico-deductivism). But the really essential point is that by using a Bayesian index of support, we can show that simpler theories are more confirmable by conforming data—they have, so to speak, higher cognitive growth potential. And this already provides a partial answer to Popper's charge that Bayesians cannot explain our preference for content or simplicity.

To illustrate the connection, consider a composite hypothesis H with special cases $h_1, \ldots, h_n$. (H is the disjunction of the mutually exclusive h_i's.) Applying Bayes's formula,

$$P(H/e) = \Sigma_i P(h_i/e)$$
$$= \Sigma_i P(e/h_i) P(h_i) / P(e)$$

$$= [P(H)/P(e)][\Sigma_i P(e/h_i)\underline{P(h_i)}].$$
$$P(H)$$

I call this the *generalized Bayes formula* and the bracketed quantity, which mirrors the evidence, the *average likelihood* of H. Thus

$$(2.1) \quad P(H/e) = \underline{P(H)}[\Sigma_i P(e/h_i) \underline{P(h_i)}]$$
$$P(e) \qquad P(H)$$

expresses the posterior probability of H as the product of its prior probability by the average likelihood divided by P(e), which I term the *expectedness* of e. (Note: P(e) must always be computed relative to the considered partition of hypotheses.) In practice, of course, one has a continuum of special cases corresponding to different settings of a real-valued parameter (or vector of parameters), and then the summation of (2.1) gives way to an integral. Where the parameter is *freely* adjustable (i.e., where the theory itself gives no clue as to its value), an "uninformative" parameter distribution should be employed. In this way we impose the maximum penalty for lack of content. But in any case it is clear that this penalty will be higher when there are more special cases over which to average the likelihood. A simple example will make this clear.

Ptolemaic astronomy tells us that the center C of Venus's epicycle lies (at all times) on the line ES joining Earth and Sun, but it imposes no further constraint. (Even the constraint that C lies always on ES is rather ad hoc; it does not grow organically out of a geocentric conception but is inferred from observation.) Applied to Venus, the Copernican theory may be considered as the special case of the Ptolemaic that locates C at the point S, the center of the sun. Reflect for a moment on the contrast: one theory confines C to a line, the other to a single point of that line! To see the connection with support, let us look first at the situation in qualitative terms. Qualitatively, there are just three possibilities: (a) C lies close to S with S inside the epicycle, (b) S lies between E and C on line ES with S outside the epicycle, or (c) C lies between E and S with S outside the epicycle. As telescopic observation of the phases of Venus first disclosed, possibility (a) is realized. Hence the Copernican special case has an average likelihood of 1, and the Ptolemaic theory has an average likelihood of 1/3. This gives a "Bayes factor" (or ratio of average likelihoods) of 3:1 in favor of Copernicus. This is not very impressive, but if, in quantitative terms, we could show that C = S (within the limits of observational accuracy), the

Bayes factor in favor of Copernicus would be effectively infinite. For the average likelihood of the Ptolemaic theory would be close to zero (we would be integrating over the entire line ES and only special cases corresponding to settings of C close to S would have appreciable likelihoods). Historically, of course, the phases of Venus did not (and could not) show that C = S. I am drawing this comparison only to illustrate the incomparably greater cognitive growth potential of a simpler theory.

Notice that I have been using the average likelihood to compare a theory with a special case of itself. I see nothing wrong with that. Of course, if we wanted to compare the two in terms of probability, we should have to take logical differences, equating (in our example) the Copernican special case with the hypothesis C = S and the Ptolemaic alternative with C $\neq$ S. As removal of a single point does not affect an integral, the relevant average likelihoods would be the same. Failure to see this possibility seems to be most of what lies behind Popper's oft-repeated equation of simpler hypotheses with *less* probable hypotheses, and the consequent denial that one can account for the importance of simplicity by connecting it to probability.

To resume the main thread of argument, we have given a direct and compelling Bayesian reason for valuing high content and simplicity. Some Popperians will scoff, nevertheless, saying that we are just mimicking Popper's methodology in Bayesian terms, trying, as it were, to recreate the flavor of the gospel in the vulgar tongue. For Bayesians still seek high probability first and foremost, even if, coincidentally, the way to obtain it is to find the simplest theory that can be squared with the "hard" data. But the charge is unfounded. Granted that probability is the *yardstick* by which Bayesians compare rival conjectures, it doesn't follow that high probability is the *goal* of any scientific inquiry. The yardstick is simply the means by which we measure our progress towards the goal, whatever the goal may be. And for my own part, I am quite comfortable with Popper's identification of that goal as the attainment of ever more truthlike theories, i.e., of theories that are closer and closer to the truth. Moreover, highly truthlike theories are just those that combine a high degree of content with a high degree of accuracy—in I.J. Good's happy phrase, they are "improbably accurate"—and a precise explication can be given along Bayesian lines by equating a theory's truthlikeness with its expected support, i.e., its support averaged over the outcomes of a relevant experiment. Then a theory is close to the truth when it is strongly supported by those outcomes of the experiment that are highly probable, conditional on the truth. Insofar as Bayesian support is a (determinate) blend of accuracy and content, the same will be true of our concept of truthlikeness. Again, the probabilistic explication appears to escape notorious difficulties associated with its more narrowly deductive Popperian counterpart (see Rosenkrantz 1980 for a fuller account), but these·matters are somewhat peripheral to our present concerns.

Up to this point in our story, it may well appear that I am just offering a sort of Bayesification of Popper's notion of content. Significant differences emerge, however, when our accounts of the role simplicity plays in theorizing are compared.

Popper connects simplicity with falsifiability and quotes with approval William Kneale's remark that "the policy of assuming always the simplest hypothesis which accords with the known facts is that which will enable us to get rid of false hypotheses most quickly." (Popper 1959, p. 140) There is, to be sure, a pervasive equivocation in Popper on "falsifiability," which is used in both a semantical sense (namely, the number of basic statements a theory excludes) and a pragmatic sense (namely, the ease with which a false conjecture can be exposed as such). And it is not generally true that conjectures that are more falsifiable in the semantic sense are more readily disposed of. But perhaps this is a quibble. The more serious criticism levelled at Popper is that mere elimination of false pretenders does not necessarily leave one closer to the truth. For in theorizing, one seldom has an exhaustive list of theoretical possibilities at hand. Indeed, there is a certain temptation to stand on its head Popper's taunt that confirmation is easily obtained if sought, and maintain that it is rather falsification that is easily obtained if sought. One can easily imagine all sorts of Goodmanesque (gruelike) alternatives to well-established hypotheses that would be easy to falsify. At the very least, Popper unduly neglects considerations of plausibility in theory construction; and more than that, there is something seriously askew in his view that interesting truth is most efficiently attained via elimination of false conjectures. Perhaps we can best appreciate my misgivings by turning forthwith to a Bayesian account of these matters.

We must recognize, to begin with, that *Bayesian* confirmation or support is *not* easily obtained. For it requires both accuracy and simplicity. In fact, the ideal case is that in which the theory fits all and *only* those experimental outcomes that actually occur (e.g., just the actually observed frequencies with which the planets retrogress). From this perspective, it is not at all surprising to find that "particle physicists are in the habit of thinking that anything not expressly forbidden by nature is compulsory." (Calder 1979, p. 186) And in the same vein, C. Lanczos writes:

> In 1929 he [Einstein] talked of the "Promethean age of physics," in which one is no longer satisfied with the discovery of the laws of nature, but one wants to know why nature is the way it is *and cannot be anything else.* . . . The impressive feature of Einstein's gravitational theory was that if one wanted to characterize a Riemannian geometry by the simplest set of field equations, one automatically arrived at Einstein's gravitational equations, which gave a complete explanation of Newtonian gravity, without the necessity of a special force of gravitation. (1967, pp. 185-186)

There is much more to efficient theorizing, however, than fitting all and only what occurs. For one thing, the "hard facts" vary in hardness, and it will often be impossible to accommodate all the mass of partially conflicting data. And, in any case, it seems advisable to begin with special cases of the complex system or process of study and "put in our ingredients one at a time." (Bartlett 1975)

What are some of the things to be said for starting with a deliberately oversimplified model? First, there is mathematical *tractability*. We can construct and explore the properties of simple models rather easily; highly complicated models may require techniques that lie beyond the present reach of mathematics. Second, there is *economy of effort*. About his search for an adequate model of DNA, James Watson writes:

> We could thus see no reason why we should not solve DNA in the same way. All we had to do was to construct a set of molecular models and begin to play—with luck, the structure would be a helix. Any other type of configuration would be much more complicated. Worrying about complications before ruling out the possibility that the answer was simple would have been damned foolishness. Pauling never got anywhere by seeking out messes. . . (1968, pp. 47-48)

And later he adds:

> Finally over coffee I admitted that my reluctance to place the bases inside partially arose from the suspicion that it would be possible to build an almost infinite number of models of this type. (p. 139)

A third advantage that springs to mind is *feedback*. A workable model of even a highly schematic version of the system studied provides information about how the full system works in special circumstances or when certain variables are controlled or confined to subsets of their allowable ranges, and this allows the model builder to see precisely how his simplified model breaks down when complicating factors are introduced. This provides insight into what sorts of complications will most dramatically improve goodness-of-fit.

In sharp contrast to Popper's account, then, far from aiming at rejection of false theoretical alternatives, theoreticians seek a model that works tolerably well in circumscribed contexts (a sort of "first approximation") and then ("putting in their ingredients one at a time") seek ways of complicating or refining the picture to capture "second-order effects" or finer details. In short, the development of a theory occurs less by eliminative induction than by successive approximation or "structured focusing." And Popper's account is weakest in describing what might be called the "developmental phase." Popper and Lakatos demand that a "progressive" modification of theory increase testability and content and that some of the excess content be corroborated. But, almost by definition, a complication of theory will increase sample coverage and thereby reduce content, so that, in effect, Popperian methodology condemns any complication of theory out of hand, no matter how much it improves accuracy! The more liberal Bayesian approach, on the other hand, qualifies a complication as "progressive" just in case the accuracy gained is enough to offset the loss of content, as determined by the precise yardstick of average likelihood. Bayesians may speak, accordingly, of *support-increasing* or *support-reducing* complications of theory. Persistent failure to find a support-increasing complication to account for discrepant data certainly looms as a difficulty for any theory (and its proponents), but no automatic rejection is implied.

To illustrate these rather abstract remarks, consider again the problem of the planets. The heliocentric scheme has all the planets orbiting the sun in simple closed curves. To capture the salient features of planetary motion, we might begin with an oversimplified heliocentric model based on uniform coplanar sun-centered circles. This model is not very accurate, but its major simplifications already stand out quite clearly. For the relative sizes of the orbits can be determined from observation of just a few positions per planet, and all the major irregularities of planetary motion—the number and frequency of retrogressions, variations in apparent brightness and diameter, and so forth—are accounted for at one stroke as effects of the earth's motion (as Copernicus emphasized in Book I of *De Revolutionibus*). Moreover, the theory fits *only* the behaviors actually

observed. The contention that the complexity of the full system Copernicus proposed obscured these simplifications strikes me as highly questionable. Astronomers like Brahe and Kepler distinguished quite clearly
between the simplifications inherent in the heliocentric picture and the
complexities of Copernicus's own filling out of the details. (Kepler even
accuses Copernicus of not being Copernican enough in needlessly complicating his system by placing the center of planetary motion at a point near,
but distinct from, the center of the sun.) And, in point of fact, a rather
minor complication of the oversimplified model based on eccentric circles
with motion uniform about an equant point and orbital planes slightly
inclined to the ecliptic but all passing through the sun, would have
produced unprecedented accuracy.

Kepler's refinement of the picture clearly embodies the methodological
principles stated here. Thus, in complicating a model to improve fit, one
should complicate it minimally. Kepler's ellipses are minimal complications of circles, and, in addition, his second law retains the feature of
uniform circular motion that the radius vector sweeps out equal areas in
equal intervals of time. Finally, his third law represents a quantitative
sharpening (or simplification) of the empirical rule-of-thumb (discussed
earlier) that a planet's period increases with its distance from the center of
motion. Because this law relates the motions of different planets, Kepler's
laws as a whole provide a model of the planets that is, I would surmise,
comparable in simplicity to the model based on uniform circles. (Newton's
gravitation law represents an additional simplification, imposing dynamical
constraints that exclude various kinematically possible systems of Kepler-
ian orbits as unstable.) In any case, the vastly improved accuracy of
Kepler's model renders it support-increasing. And, in addition, Kepler's
model lends itself to a natural causal or physical interpretation in a way that
Ptolemaic and Tychonic models did not. Planets speed up as they approach
the sun and planets closer to the sun go round faster, pointing clearly to the
sun as a causal agent.

Let us look now at Glymour's position on these matters, for again, we
find much substantive agreement in the face of proclaimed disagreement.
First, Glymour is one of the very few codefenders of the view, espoused in
chapter 7 of my 1977 book, that the Copernican theory really is simpler
than the Ptolemaic and that its greater simplicity has evidential value. "On
several grounds," he writes, (1980, p. 198), "Copernican theory is superior
to Ptolemaic astronomy: there are properties of the bodies of the solar

system that are presupposed by both theories, but that are indeterminable in Ptolemaic theory whereas they can be determined within Copernican theory." (1980, p. 198) And he goes on to urge that this virtue rendered the Copernican alternative the better confirmed of the two.

Similar agreement with my view that simplicity has evidential force is found in his discussion of the classical tests of general relativity, a discussion ostensibly designed to show that Bayesians cannot account for the judged relative importance of the different tests. (pp. 277 ff.) After pointing out that the anomalous advance of the perihelion of Mercury could be accommodated by a number of theories, he writes:

> Perhaps the most common and influential objection to these contend-
> ers against general relativity was that, unlike Einstein's theory, they
> saved the phenomena only by employing an array of arbitrary
> parameters that had to be fitted to the observations. Eddington,
> barely concealing his contempt, objected against ether theories of the
> perihelion advance and gravitational deflection that they were not "on
> the same footing" with a theory that generated these phenomena
> without any arbitrary parameters. It was pointed out that Poincaré's
> extension of Lorentz's theory could, by proper adjustments of a
> parameter, be made consistent with an infinity of perihelion advances
> other than the actual one. Conversely, Einstein's derivations of the
> phenomena were praised exactly because they involved no arbitrary
> parameters—and, the exception that proves the rule, also criticized
> because they did. (p. 284)

No one who has digested even the very sketchy discussion of average likelihood in this paper will have the slightest difficulty accounting for such judgments in Bayesian terms. Glymour's own essential agreement with the deliverances of Bayesian analysis comes out most clearly in his chapter VIII on curve-fitting. The application of the average likelihood index of support to polynomial regression is taken up in chapter 11 of Rosenkrantz (1977), and its performance is compared with that of various non-Bayesian (or "orthodox") tests. Glymour does not discuss either Bayesian or orthodox approaches to curve-fitting, but he does offer a way of assessing the severity of a test that uses a notion quite reminiscent of sample coverage:

> How is severity to be compared? Suppose that we have the hypotheses
> H and G and that, given prior data, the range of outcomes on a new
> experiment that would be in contradiction with H is properly
> contained in the range of possible outcomes that would be in
> contradiction with G... In this sort of case, I think it is natural and
> proper to regard G as more severely tested than H. (pp. 333-334)

And later he adds, "one can try to develop, not just the crude comparison of severity of tests I have used, but a *measure* of severity of tests. . . ." (p. 339) If "consistency with the data" is understood in a probabilistic sense admitting of degrees (really the common usage in the sciences), sample coverage provides just such a measure. Then Glymour's suggestion that polynomials of lower degree "are preferred to more complex families that also fit the data because . . . the data provide more and severer tests of the simpler hypothesis than of the more complex one" (p. 335) will follow readily from a Bayesian analysis in terms of average likelihoods, if by "fit the data" we understand "fit the data equally well."

It is unfortunate that a misreading of my (possibly obscure) 1976 paper on simplicity prevented Glymour from appreciating that Bayesian analysis delivers precisely what his own intuitions demand. He says that I fail to show "that in curve-fitting the average likelihood of a linear hypothesis is greater than the average likelihood of a quadratic or higher degree hypothesis." But of course I don't want to show *that*, for it isn't true! What can be shown is that the average likelihood of the quadratic family will be higher than that of the linear family when the data fit the quadratic hypothesis *sufficiently better* than the linear one, whereas the latter will enjoy higher average likelihood when the two families fit equally well.

Obviously it has not been my intention to attack Glymour's intuitions about simplicity. By and large, I see in him a kindred spirit, one who recognizes both the central role simplicity plays in the deliberations of theoreticians of all stripes and its objective evidential force. His tendency to think that Bayesians cannot account for its role and force is perhaps understandable in light of the extent to which the subjectivist form of the Bayesian approach has dominated the scene, until quite recently. (Indeed, many writers still use "Bayesian" and "subjectivist" interchangeably.) Unlike objectivists, such as Sir Harold Jeffreys, subjectivists have laid very little stress on average likelihood or on models with adjustable parameters, quite possibly because the need to adopt a parameter distribution or weighting function when employing the average likelihood index somewhat vitiates the subjectivists' claim to be able to separate cleanly the "subjective element" (the prior) from the "public element" (the import of the data). At any rate, no theory of evidence that fails to handle models with free parameters or account for the felt diminution of support that results from adding parameters or accommodating more outcomes that might have been but were not observed, can be taken very seriously. Simplicity looms as the central problem in the whole field of theory and evidence (and a glance at Glymour's index would tend to vindicate this judgment). To Popper must go much of the credit for keeping the issue of simplicity alive at a time when methodologists of a positivist persuasion were more inclined to dismiss it as a will-of-the-wisp or consign it to the limbo of the "purely pragmatic" or the "merely aesthetic."

(b) Novel predictions and old evidence

Another old chestnut, closely related to the Popperian demand for placing our conjectures in jeopardy, is the maxim that hypotheses are more strongly confirmed by their ability to predict facts not already known. Some would go even further and say that theories are not confirmed at all by already known facts or previously available data. Yet, to all appearances, Bayesian methodology is at odds with this principle. For if we think of support or likelihood as a timeless relation between propositions (akin to logical implication in this respect), then $P(E/H)$ does not depend on whether or not E was known prior to proposing H.

Scientists have, though, a curious ambivalence about this time-honored precept. Specifically, they never fail to pay it lip-service and never fail to disregard it in practice whenever it tends to weaken the evidence for their own theories. Almost all the empirical support for Dalton's atomic theory, including the laws of constant and multiple proportions, was already known, yet it was cited as evidence for the theory. And in his popular account of relativity (Einstein 1916), Einstein quite expressly states that all the facts of experience that support the Maxwell-Lorentz theory of electromagnetic phenomena also support the special theory of relativity, since the latter "has crystallized out of" the former. Einstein cites in particular the experiment of Fizeau as having "most elegantly confirmed by experiment" the relativistic version of the law of addition for velocities, even though that experiment had been performed more than fifty years earlier (before relativity was even a twinkle in Einstein's eye) and had moreover been explained by the Maxwell-Lorentz theory. In point of fact, there was no evidence of a "novel" sort to which Einstein could point, since it was not then technically feasible to accelerate small particles to speeds approaching that of light or to perform mass-energy transformations in the laboratory. What Einstein did point out instead was the ability of relativity theory to account for "two classes of experimental facts hitherto obtained which can be represented in the Maxwell-Lorentz theory only by the introduction of an auxiliary hypothesis"—in other words, he pointed to the greater simplicity of relativity. About the negative result of the Michaelson-Morely experiment he writes:

> Lorentz and Fitzgerald rescued the theory from this difficulty by assuming that the motion of the body relative to the aether produces a contraction of the body in the direction of motion, the amount of contraction being just sufficient to compensate for the difference in time mentioned above. Comparison with the discussion of Section XII shows that also from the standpoint of relativity this solution of the difficulty was the right one. But on the basis of the theory of relativity the method of interpretation is incomparably more satisfactory. According to this theory there is no such thing as a "specially favored" (unique) coordinate system to occasion the introduction of the aether-idea, and hence there can be no aether-drift, nor any experiment with which to demonstrate it. Here the contraction of moving bodies follows from the two fundamental principles of the theory, without the introduction of particular hypotheses; and as the prime factor involved in this contraction we find, not the motion in itself, to which we cannot attach any meaning, but the motion with respect to the body of reference chosen in the particular case in point. (1916, p. 53)

The situation was not so very different in the case of general relativity. Einstein laid great stress on the equality of inertial and gravitational mass, a brute fact in the old physics, but a necessary consequence of the general principle of relativity in the new physics. Here too the difference is one of overdetermination and has nothing to do with novelty per se. And of course the advance of the perihelion of Mercury, predicted by general relativity, had long been established and measured with precision (the predicted advances in the perihelia of other planets were too small to be detectable).

Faced with these and other obvious exceptions to the precept, Lakatos and Zahar (1975) fall back on a modified form of it that accounts a prediction "novel" when it was not consciously used to arrive at the theory (so that the theory explains it in passing). If taken quite literally, this proposal would require us to read a theorist's mind before being able to assess the evidence for his theory (see Michael Gardner's contribution to this volume). In any event, they are able to argue on this basis that the stations and retrogressions of the planets, the brightness of a planet at perigee, and the bounded elongation of an inner planet from the sun, etc., all count as "novel" predictions of the Copernican theory, though not of the Ptolemaic. They observe that "although these facts were previously known, they lend much more support to Copernicus than to Ptolemy, within whose system they

were dealt with only in an ad hoc manner, by parameter adjustment"
(Lakatos and Zahar, 1975, p. 376). That is, the Ptolemaic theory could
account for these effects of the earth's motion only by fitting additional
parameters or making additional assumptions. Could it be more clear that
the real appeal here is, not to "novelty," but to overdetermination?

There is, to be sure, another sense of "novelty" that plays a more
important role: namely, a prediction is novel when it is unexpected on rival
theories (or on rival theories of comparable simplicity). And, of course,
Bayesians have no difficulty accounting for the force of predictions that are
"novel" in this sense.

The solution of the problem Glymour poses about old evidence (1980,
pp. 86-92) should also be clear. The puzzle is this: if an item e of evidence is
already known, then it must have probability one, and consequently, even
if a hypothesis h entails it, $P(h/e) = P(e/h)P(h)/P(e) = P(h)$, using Bayes's
formula, and no confirmation is registered. (This is a sort of obverse of the
charge that Bayesians are unable to account for the peculiar force of novel
predictions.) On *objectivist* Bayesian grounds, however, the likelihoods
$P(e/h_i)$ of the alternative hypotheses are timeless relations, and of course
$P(e)$ must be computed relative to a considered partition of hypotheses,
$h_1, \ldots, h_n$ by the partitioning formula, $P(e) = P(e/h_1)P(h_1)$
$+ \ldots + P(e/h_n)P(h_n)$. And this quantity will be less than one, unless e is a
necessary truth. For purposes of comparing hypotheses, then, the probability of old evidence is *not* one, and may even be quite small. This only
shows, of course, that old evidence poses no difficulty for an objectivist
Bayesian position—a point that Glymour readily conceded at the conference. (For a subjectivist's way of handling the problem, see Daniel
Garber's contribution to this volume.)

What does cry out for explanation is our conviction that the ability of
general relativity to fit the already measured advance of the perihelion of
Mercury can afford just as striking a confirmation (and seem quite as
"miraculous") as the ability of that theory to predict the precise magnitude
of the deflection of starlight passing close to the sun. While I was listening
to Glymour describe Einstein's vicissitudes in finding a covariant theory
that would account for the advance of Mercury's perihelion, the solution of
this puzzle suddenly became quite clear. The point is that Einstein's
success was not assured. What is generally overlooked is that one is not
interested in finding any old theory to explain an anomaly; one seeks, in
practice, a (reasonably simple) theory *of specified form*. Thus Einstein

sought a theory that satisfies the general principle of relativity. And we can think of such a quest in the following way. One is interested, at bottom, in the hypothesis that there *exists* a (not unduly complicated) theory of such-and-such form capable of accommodating the data from a certain class of experiments, only some of which have already been performed. That there does exist a theory of the required form that fits the output of an already performed experiment of the class in question affords, on straightforward Bayesian grounds, a more or less striking confirmation of the existential hypothesis in question. And the longer or more tortuous the derivation, and the more different (and tenuous) the theoretical assumptions involved, the more striking the confirmation (of all the implicated principles) will be (as in Bohr's derivation of the Balmer series for hydrogen).

(c) Projectibility

Hypothetico-deductive accounts face the difficulty that an observation may be a consequence of more than one hypothesis, and, in particular, of a "counterinductive" or "unprojectible" hypothesis. Examining an emerald before time t and finding it green is a consequence of "All emeralds are grue," as well as of "All emeralds are green." And this seems disturbing, inasmuch as the grue hypothesis licenses the prediction that emeralds not examined before time t are blue, hence emeralds of a different color. Since consequences of a hypothesis are confirmatory on a Bayesian account, some restriction of the Bayesian confirmation relation seems called for. And, quite apart from this concern, we have been witnessing, since the early 1950s, a search for a basis for excluding such "counterinductive inferences."

To be sure, Bayesian inference blocks this alleged paradox at many points. For one thing, there is no Bayesian consequence condition that would allow one to confirm the prediction of blue emeralds after time t. And, more generally, there are ways of handling irrelevant conjunction. Yet these considerations do not seem to go to the heart of the matter. For the more serious issue here, in my view, is *whether* (not how) we should drive a wedge between "projectible" (or "lawlike") and "unprojectible" hypotheses.

The grue hypothesis belongs to a class we might label *bent* or *crooked*. Such hypotheses posit a breakdown of a *straight* counterpart in some nonlocal region of space or time. The grue hypothesis is, admittedly, a rather extreme case in that it posits a sharp discontinuity, but presumably those who view such hypotheses as absolutely unconfirmable would regard their continuous or gradualistic modifications as equally unsavory.

And yet science is riddled with bent or crooked hypotheses, and this should certainly make us wary of any proposal to banish them wholesale. Nelson Goodman's own attempt to do so, the entrenchment theory, would list, among others, the hypotheses of relativity theory among the unprojectible! For example, the Einsteinian hypothesis "All particles subject to constant force have linearly increasing relativistic momentum" is "overridden," in Goodman's sense, by its Newtonian counterpart, "All particles...have linearly increasing momentum." For the latter had, circa 1905, much the better entrenched consequent predicate and was, up to that time, unviolated, supported, and unexhausted. In effect, the hypotheses of special relativity posit departures from their Newtonian counterparts that become experimentally detectable only at speeds close to the speed of light. They are, in this respect, perfectly representative bent hypotheses. It is no defect of Bayesian methodology that it gives such hypotheses a hearing.

From a Bayesian point of view, lawlikeness admits of degrees and is chiefly a function of simplicity and theoretical assimilability (as reflected in a prior distribution). I am quite content to let it go at that, for I am convinced that there is no fundamental distinction to be drawn between hypotheses that are projectible or confirmable and those that are *absolutely* unconfirmable. (I argue this point at greater length in Rosenkrantz 1982, pp. 86-91)

3. Informal Assessments and Epistemic Utilities

A good theory can explain the salient facts without recourse to special assumptions of an arbitrary kind. I have been urging that the Bayesian theory of evidence is a theory of precisely this sort. It dispenses with ad hoc prescriptions and so-called epistemic utilities. *Genuine* epistemic utilities, like content, are automatically reflected in support (and, in effect, this provides a criterion for distinguishing the genuine from the spurious among them). From a strict Bayesian point of view, support is all in all. It is not surprising to find, therefore, at least one sympathetic reviewer of my 1977 book (Jaynes 1979) wondering why I even bother with the adhockeries that disfigure so much of the literature of scientific method. Why, indeed, do I attempt a precise explication of simplicity when, however defined, simplicity matters only insofar as it is reflected in support? At the other extreme, some critics of the Bayesian approach question its applicability to actual scientific evidence. Glymour raises such doubts in his recent paper when he writes:

> I am inclined to doubt that, in many situations, we have either objective probabilities or subjective degrees of belief of a sufficiently global kind upon which we can rely to relate evidence to theory. When theories are proposed for novel subject matters (as in some contemporary social science) or when new theories are seriously considered which deny previously accepted fundamental relationships..., we may be at a loss for probabilities connecting evidence to theory. (1981, p. 696)

These two questions may seem unrelated, not to say oppositely directed, but, in essence, they elicit the same reply.

Although most criticism of the second kind focuses on the alleged arbitrariness of prior probabilities of theoretical hypotheses (the passage from Glymour tends that way), the real difficulty, more frequently, is to compute the relevant likelihoods—a point that Patrick Suppes has emphasized on numerous occasions. It often happens that we can calculate conditional outcome probabilities for a "null hypothesis" of chance or randomness, but we cannot calculate them for the hypotheses (of association, or causal connection) of real interest to us. For a very simple example, consider R.A. Fisher's celebrated case of the tea-tasting lady, who claims an ability to discriminate whether the tea or milk was infused first in a mixture of milk and tea. Fisher's design calls for the lady to classify eight cups, of which four are milk-first and four are tea-first (and the lady knows this). It is then easy to find the probability that she classified r of the eight cups correctly, given that she is merely guessing; but there is no way to calculate these probabilities on the supposition that she has some skill. The prevalence of cases like this one explains the widespread use of tests of statistical significance. Such tests are used to make rather informal assessments of evidence even in cases in which no well-defined alternative hypotheses are in view.

Now my answer to both points can be given at once. First, epistemic utilities are important in precisely those contexts in which Bayesian methods cannot be applied for inability to compute the relevant likelihoods. (My earlier, often outspoken, criticism of epistemic utilities is here softened to this extent.) At the same time, however, our informal assessments in these cases are (and ought to be) guided by the methodological insights that formal Bayesian analysis affords in the contexts in which it does apply.

To begin with, we might seek a qualitative analogue of the average likelihood. The latter, you recall, is a determinate blend of accuracy and content; it measures, roughly speaking, the improbability of a theory's accuracy. The ideal case is that in which the theory fits all and *only* those possible outcomes that actually occur. Demands of accuracy and simplicity alike narrow the range of outcomes that a theory can accommodate. Now in cases in which likelihoods cannot be computed, we may still have an intuitive rank ordering of experimental outcomes as agreeing more or less well with the theoretical conjecture of interest. Then we can mimic average likelihood in a qualitative way by the proportion of possible outcomes that (by the intuitive yardstick) fit the hypothesis *at least as well* as the outcome observed. The size of this proportion will again reflect accuracy and simplicity in a determinate way, and, moreover, in a way that tends to yield assessments qualitatively similar to those yielded by average likelihood where both methods apply (see the last section of Rosenkrantz 1976 on this). I call this proportion the *observed sample coverage*. In principle, any two hypotheses, whether mutually exclusive or not, can be compared by this informal measure. More generally, using a suitable null hypothesis, we can compute the *chance probability* of agreement with the hypothesis of real interest as good as (or better than) that observed.

To illustrate, if someone claims an ability to detect water with a hazel prong and boasts of a ninety percent rate of success, we should not be impressed unless his success rate is materially higher than that achieved by digging at random in the same area (i.e., the chance rate). If that cannot be shown, his accuracy is not improbable and his claim is unsubstantiated.

Informal assessments of evidence are often aimed at establishing improbable accuracy. I recently came across a beautiful example in Thor Heyerdahl's interesting book, *Early Man and the Ocean* (1979, chapter 3). The hypothesis of interest is that the cultural flowering that occurred in ancient Meso-America had sources (Sumerian, Egyptian, Hittite, or Phoenician) in the Near East. Heyerdahl protests the tendency of "isolationists" to dismiss the parallels between these cultures singly, rather than confronting them collectively, for there is a compounding of improbabilities. That one or two such parallels should arise by mere coincidence does not strain credulity, but the probability of finding well over a hundred by chance seems infinitesimal.

Heyerdahl's point is well taken, and even the partial list of over fifty parallels he compiles is nothing if not impressive (1979, pp. 84-92). Yet, the evidence from cultural parallels could be marshalled more convincingly by introducing missing ingredients of the informal Bayesian paradigm I have sketched. What we lack is a sense of how much similarity typifies cultures between which there has been no contact. We also need some assurance that dissimilarities are being systematically taken into account.

To this end, we need a well-defined sample space, in effect, an ethnographic survey of many cultures based on a single workable typology, and then we need a measure of similarity between cultures based on this typology. A computer could then be programmed to calculate the proportion of pairs of surveyed cultures manifesting a degree of similarity at least as great as that of the pair for which contact is hypothesized. That proportion (the observed sample coverage) estimates the probability that two cultures *chosen at random* would manifest at least as much similarity (i.e., the chance probability).

Such comparisons are, of course, no better than the typology and similarity measure on which they are based. Imagine that given items of the typology are treated as branching classification trees. As a first step toward measuring similarity with respect to that item, proceed down the tree to the last branch point at which the two cultures A and B of a comparison agree, then compute the proportion of surveyed cultures (*including* the pair A,B) which proceed at least that far down the same branch of the tree. Then the square of this proportion (necessarily positive) estimates the probability that two cultures chosen at random would agree on the given item to at least that level of specificity. In this way, our measure of similarity reflects both the specificity and statistical rarity of a shared custom or artifact, and dissimilarities are systematically taken into account. This desideratum stands out very clearly in Heyerdahl's discussion, which will suggest other desiderata and ways of refining our measure. My purpose here is no more than to indicate the general lines along which one might proceed.

As for prior probabilities, admittedly they are of little importance in preliminary investigations where we lack a sharply delimited set of theoretical alternatives. Observed sample coverage can still be applied to assess support in such contexts, without regard to alternative hypotheses. But where the theoretical possibilities have been effectively narrowed, we can expect the informal, qualitative counterparts of prior probabilities, which I will call "initial plausibilities," to play a major role. Indeed, Heyerdahl's famous voyages were mounted to explode the supposed implausibility of certain migration routes. His point is that routes that seem implausibly long in miles may actually be short when powerful ocean currents and trade winds are taken into account. His voyages demonstrated the feasibility of a journey across the Atlantic or across the Pacific from Peru to Polynesia in the highly seaworthy wash-through reed vessels or balsa rafts of the Egyptians and Incas (highly specialized constructions whose occurrence in all three places is itself one of the important bits of evidence pointing to contact between these cultures). Finally, by using the procedure of the last paragraph, one could hope to rule out alternative migration routes.

My suspicion is that informal counterparts of the three main elements of a formal Bayesian analysis—prior probabilities, likelihoods, and alternative hypotheses—figure importantly in nearly all informal assessments of evidence, and that more explicit use of the informal Bayesian index of support (the observed sample coverage) would often render assessments of this sort more systematic and more objective.

4. Glymour's Misgivings

With this much background, we can turn at last to Clark Glymour's reservations about Bayesian methods (some of which have already been touched on in passing) and the additional constraints he wishes to impose.

I think of Bayes' theorem as a refinement of the hypothetico-deductive approach. We seek hypotheses conditional on which actually occurring outcomes have high probability while nonoccurring outcomes have low probability. More precisely, Bayes's formula implies that a hypothesis h_i of a partition $h_1, \ldots, h_n$ is confirmed by an outcome e just in case e has a higher probability on h_i than it has on the average, relative to the members of the partition (i.e., iff $P(e/h_i) > P(e/h_1)P(h_1) + \ldots + P(e/h_n)P(h_n)$). And, by the same token, a member h_i of a partition of hypotheses is not disconfirmed by outcomes that are highly improbable on h_i unless those outcomes are substantially more probable on alternative hypotheses of the partition. It is widely conceded that this scheme characterizes, in a general way, both the precepts and practice of working scientists and model-builders. Glymour too concedes it, yet he denies that hypothetico-deduction is a sound scheme in all respects (1980, pp. 29 ff.).

His main fear is that it cannot handle irrelevant conjunction. If e is held to confirm h just by virtue of h's entailing it, then, equally, e must confirm h&k as well, where k is any hypothesis you like. Again, if degree of confirmation is measured by the ratio P(h/e):P(h) = P(e/h):P(e) of posterior to prior probability, then if e is a consequence of h, P(e/h) = P(e/h&k) = 1, and e will accord h&k precisely the same degree of confirmation it accords h alone. That seems objectionable when k is extraneous, and even more objectionable when k is probabilistically incompatible with h in the sense that P(k/h) is low. Personally, I have always considered this reason enough to reject the ratio measure in favor of the difference measure:

(4.1)   dc(e, h) = P(h/e) − P(h)

writing dc(e, h) for the *degree of confirmation e* accords h. This measure is easily seen to satisfy the following condition:

(4.2)   dc(e, h&k) = P(k/h)dc(e, h) when e is a consequence of h.

And this little theorem seems to deliver precisely what intuition demands. For, on the one hand, we certainly don't want to say that a consequence of h should *dis*confirm h&k. But neither should it confirm h&k as strongly as h. Indeed, the degree of compatibility of k with h should control the rate of depreciation, and this is what (4.2) says.

The difference measure can be applied to conclude, for example, that examining a sample of emeralds for color before time t and finding them green *(e)* accords "All emeralds are green" a higher degree of confirmation than "All emeralds are grue." For the former hypothesis is the conjunction of h: "All emeralds examined before time t are green," with k: "All emeralds not examined before time t are green," whereas the latter is the conjunction of h with k': "All emeralds not examined before time t are blue." Given our background knowledge that emeralds do not change color all at once, either individually or as a class, P(k/h) >> P(k'/h). And the asymmetry in question is language independent. By contrast, the Hempelian account of confirmation registers confirmation for both h&k and h&k', and leaves us unable to discriminate between them. Worse still, because that account satisfies the consequence condition, e will also confirm k'—the dreaded counterinductive inference. Irrelevant conjunction is, therefore, very much a two-edged sword.

Notice too how our explication of content handles irrelevant conjunction. On Popper's account, conjoining an extraneous hypothesis represents a simplification, since more states of the world are then logically excluded. But, in our probabilistic version, this is not so, for content is relativized to a contemplated experiment. Thus conjoining, say, a hypothesis about the velocity of neon light in beer to a Mendelian model of a given mating experiment will have no effect on the latter's sample coverage for that experiment. No simplification results, but prior probability is necessarily reduced.

I come next to "deoccamization" (see Glymour 1980, pp. 30-31). At first blush, one is tempted to say that a deoccamized theory (one in which a parameter is replaced throughout by a function of several other parameters) differs only notationally from the theory it deoccamizes. To the extent that two theories fit the same outcomes of the same experiments to the same degree, I regard them as equivalent. And so it troubles me not at all that a theory and a deoccamization of it may have the same sample coverage or the same support. The only considerations that would lead anyone to prefer one such notational variant to another one are, I should think, considerations of elegance or of a heuristic nature. And I see no reason to issue prescriptions on this matter.

There is nevertheless something about Glymour's position that troubles me. He leaves it as an exercise for the reader to show that deoccamization will reduce a theory's testability. (pp. 143-144) But let the theoretical term t of theory T be replaced throughout by the sum $t' + t''$ of two new parameters, t' and t'', yielding the deoccamization T' of T. (To borrow one of his examples, "force" in classical mechanics might be uniformly replaced by the sum of "gorce" and "morce.") Now it seems to me that any instance of a hypothesis h of T deducible from observations and T is ipso facto an instance of the corresponding hypothesis h' of T'. For any determination of t is likewise a determination of $t' + t''$. So T and T' have, on Glymour's own showing, the very same evidence. I think he escapes this conclusion only by imposing a further requirement, namely, that for a hypothesis to be tested by given data, *every* theoretical parameter of that hypothesis must be determined. It is not enough that $t' + t''$ be determined from the observations; each of t' and t'' must be determined (a reading suggested by Glymour 1980, p. 357).

It will come as no surprise that I consider this requirement overly stringent. In fact, I think it goes against the grain of Glymour's own approach. For if observations determine a sum of two theoretical quantities, why shouldn't we be willing to count that as a test of any hypothesis in which they occur—albeit a weaker test? After all, two different determinations of such a sum must yield the same value, and this constrains the data. That all quantities be individually determinable is the ideal case, but our admiration for the ideal should not lead us to disparage the good.

Glymour himself concedes that "pure deoccamization perhaps never occurs in science, but what does sometimes occur is deoccamization together with additional, untested claims about the new quantities." (p. 364) The clear implication is that such quasi-deoccamization is as much to be shunned as the real thing. I wonder about that too. Where there are additional claims, there is additional content, even if it lies beyond the reach of present experimental techniques. A theory like Einstein's, which says that mass (or energy) is really a sum of two terms, rest energy and energy of motion, one of which becomes appreciable only at speeds close to that of light, seems to be a theory of exactly this sort. When it was proposed, there was no way to test it. And similarly, particle physics is riddled with hypotheses stating that an elementary particle is really made up of a pair of such particles, but where the new predictions that follow are presently inaccessible to experimentation. Consider the following illustrative passage about "charm" from Nigel Calder's popular book *The Key to the Universe:* "Nor could the gipsy itself help in settling the issue in favor of charm. Supposing that the new particle did indeed consist of the charm/anti-charm combination, the charm was thoroughly hidden because it was self-cancelling. With zero net charm the gipsy could not be expected to show direct signs of charmed behavior." (p. 111) This looks very much like another case of quasi deoccamization, one that should give us pause.

To be sure, Glymour's comments on my presentation (repeated in Glymour, 1981) make it plain that he does not object to deoccamization when there are positive reasons for thinking that the new quantities have distinguishable denotata. He avers that "the demand for bootstrap confirmation [wherein every quantity is individually determined] is, I am sure, at best *prima facie* and indefeasible." But then it is left for the rest of us to wonder what all the hoopla is about if, as he admits, *pure* deoccamization never occurs. The only substantive issue that divides us is whether or not to insist that every theoretical quantity be individually determined in any test of a hypothesis. And this requirement strikes me as highly representative of those that cry out for justification, either in terms of a more comprehensive methodology or theory of rationality or as facilitating the achievement of cognitive objectives.

Many actual cases of bootstrapping seem to violate this additional stricture. Newton was able to test his gravitation law by comparing "the force requisite to keep the moon in her orb with the force of gravity at the surface of the earth" and finding them "to answer pretty nearly." By equating the centripetal force acting on the moon with gravitational force (and neglecting the sun), one obtains:

$$m_M v^2/R = Gm_E m_M/R^2;$$

and equating the moon's velocity v with the circumference of its orbit, $2\pi R$, divided by its period T, one has the following expression for T:

$$T^2 = 4\pi^2 R^3/Gm_E$$

where m_E is the mass of the earth and G is the gravitational constant. In this test of the law, Newton was not able to determine G and m_E separately, but he could determine their product as $Gm_E = gr^2$, where r is the earth's radius and g the acceleration of free fall, using the obvious relation $mg = Gm_e m/r^2$. This gives a theoretical determination of the moon's period which could be checked against observation. Would Glymour deny the force of the very test that apparently clinched the matter for Newton?

Here is another example. When Venus is at maximal elongation from the sun, earth, Venus, and sun lie on a right triangle and measurement of the angle SEV at E yields the ratio VS:ES of the orbital radii. On the other hand, at inferior conjunction, when E, V, S lie on a line in that order, we have ES = EV + VS. Assuming that ES is known, we have a determination of EV, the distance from the earth to Venus at inferior conjunction. Now Venus is fairly close at inferior conjunction, and we might hope to determine this distance directly by triangulation. One slight hitch is that we can't really observe Venus at inferior conjunction, since its orbit is nearly coplanar with the earth's orbit. But we can measure its apparent diameter at points very close to inferior conjunction, so let us ignore this difficulty for the sake of argument. The more serious problem is that we lack an independent determination of the actual diameter of Venus. Still undaunted, we make the natural but wholly untested assumption that the diameter of Venus does not differ appreciably from that of the earth. Now, for the punch line, imagine that our two independent determinations of EV agree within experimental error. Would this confirm the (heliocentric) hypothesis that the center of Venus's epicycle is at S? Here the apparent diameter determines only the product of EV by the actual diameter. Still, I am inclined to think that some confirmation is registered, if only because the apparent diameter is *determined* by an assumption about the actual diameter within the heliocentric theory but not within the geocentric theory. In fact, any other epicycle of Venus (compatible with its observed angular variations) containing the sun must intersect the sun-centered epicycle, and at the points of intersection we would have conflicting predictions of apparent diameter. Still, all I want to claim is that some slight confirmation of all the implicated hypotheses would be registered by agreement of our two determinations of EV. Does Glymour disagree? In defense of bootstrapping he writes:

> One claims that if certain principles of the theory are true, then certain empirical data in fact determine an instance of some theoretical relation, . . . This is some reason to believe the hypothesis, but a reason with assumptions. Of course it is possible that the assumptions—the hypotheses used to determine values of theoretical quantities—are false and a positive instance of the hypothesis tested is therefore spurious, or a negative instance equally spurious. But this does not mean that the test is circular or of no account. (1980, p. 352)

And that is why I said earlier that the requirement that all quantities be separately or independently determinable goes against the grain of Glymour's own conception of bootstrapping. In the case before us, we achieve this only by making a wholly untested assumption. But that does not make our test "of no account."

Glymour's remaining objection to the Bayesian account of confirmation is that it does not satisfy the *consequence condition:* that whatever confirms a hypothesis H confirms any consequence K of H. His intuitions tell him that this holds in at least some cases. But presumably his intuitions also allow that hypotheses are confirmed by their consequences or verified predictions in at least some cases. And he knows that this principle cannot be combined with the consequence condition to yield a non-trivial confirmation theory, unless one of the conditions is suitably restricted. The Bayesian theory restricts the consequence condition, satisfying it only for those consequences K of H such that $P(K/H) >> P(K/notH)$. (Such K might be called "explained consequences," inasmuch as alternative explanatory hypotheses are effectively excluded.) True, this opens the door to irrelevant conjunction, but the alternative to admitting that a consequence E of H confirms the conjunction of H with any H′ is, we saw, far less palatable.

And failure of the consequence condition removes much of the sting anyway, for even though E confirms H&H', it may disconfirm H'. Moreover, on the difference measure, $dc(E, H) = P(H/E) - P(H)$, the degree to which E confirms H&H' drops to zero when H is inconsistent with H'. No other confirmation theory, I submit, can steer a safer passage between the implausibilities of the various corner positions.

Wherever one looks for substantive disagreement between the deliverances of the bootstrapping and the Bayesian accounts of confirmation, one fails to turn them up, *unless* additional strictures that fly in the face of much scientific practice and cry out for justification are introduced.

References

Bartlett, M.S. 1975. Epidemics, In *Probability, Statistics and Time*, New York: Wiley, pp. 111-123.

Calder, N. 1979. *The Key to the Universe*. New York: Viking.

Calder, N. 1980. *Einstein's Universe*. Harmondsworth: Penguin.

Einstein, A. 1916. *Relativity*. New York: Crown. (Reprinted 1961)

Glymour, C. 1980. *Theory and Evidence*. Princeton: Princeton University Press.

Glymour, C. 1981. Bootstraps and Probabilities. *Journal of Philosophy* LXXVII: 691-699.

Heyerdahl, T. 1979. *Early Man and the Ocean*. New York: Doubleday.

Jaynes, E.T. 1979. Review of Rosenkrantz (1977). *Journal of the American Statistical Association* 74: 740-741.

Lanczos, C. 1967. Rationalism and the Physical World. *Boston Studies in the Philosophy of Science*, V. III, Dordrecht: Reidel.

Lakatos, I., and Zahar, E. 1975. Why Did Copernicus's Research Programme Supercede Ptolemy's? In *The Copernican Achievement*, Berkeley and Los Angeles: University of California Press, pp. 354-383.

Morrison, D.F., and Henkel, R.E., eds. 1970. *The Significance Test Controversy*, Chicago: Aldine.

Popper, K. 1959. *Logic of Scientific Discovery*. London: Hutchinson.

Rosenkrantz, R.D. 1976. Simplicity as Strength. In *Foundations of Probability and Statistics and Statistical Theories of Science*, v. 1, ed. W.L. Harper and C.A. Hooker, Dordrecht: Reidel, pp. 167-196.

Rosenkrantz, R.D. 1977. *Inference, Method and Decision*, Dordrecht: Reidel.

Rosenkrantz, R.D. 1980. Measuring Truthlikeness. *Synthèse* 45: 463-487.

Rosenkrantz, R.D. 1982. Does the Philosophy of Induction Rest on a Mistake? *Journal of Philosophy* LXXIX, 78-97.

Watson, J. 1968. *The Double Helix*. New York: Athenaeum.

— Daniel Garber —

Old Evidence and Logical Omniscience in Bayesian Confirmation Theory

The Bayesian framework is intended, at least in part, as a formalization and systematization of the sorts of reasoning that we all carry on at an intuitive level. One of the most attractive features of the Bayesian approach is the apparent ease and elegance with which it can deal with typical strategies for the confirmation of hypotheses in science. Using the apparatus of the mathematical theory of probability, the Bayesian can show how the acquisition of evidence can result in increased confidence in hypotheses, in accord with our best intuitions. Despite the obvious attractiveness of the Bayesian account of confirmation, though, some philosophers of science have resisted its manifest charms and raised serious objections to the Bayesian framework. Most of the objections have centered on the unrealistic nature of the assumptions required to establish the appropriateness of modeling an individual's beliefs by way of a point-valued, additive function.[1] But one recent attack is of a different sort. In a recent book on confirmation theory, Clark Glymour has presented an argument intended to show that the Bayesian account of confirmation fails at what it was thought to do best.[2] Glymour claims that there is an important class of scientific arguments, cases in which we are dealing with the apparent confirmation of new hypotheses by old evidence, for which the Bayesian account of confirmation seems hopelessly inadequate. In this essay I shall examine this difficulty, what I call the problem of old evidence. I shall argue that the problem of old evidence is generated by the

Earlier versions of this paper were read to the Committee on the Conceptual Foundations of Science at the University of Chicago and to the conference on confirmation theory sponsored by the Minnesota Center for Philosophy of Science in June 1980. I would like to thank the audiences at both of those presentations, as well as the following individuals for helpful conversations and/or correspondence concerning the issues taken up in this paper: Peter Achinstein, Jon Adler, John Earman, Clark Glymour, James Hawthorne, Richard Jeffrey, Isaac Levi, Teddy Seidenfeld, Brian Skyrms, William Tait, and Sandy Zabell. Finally, I would like to dedicate this essay to the memory of David Huckaba, student and friend, with whom I discussed much of the material in this paper, who was killed in the crash of his Navy training flight in February of 1980 while this paper was in progress.

requirement that the Bayesian agent be logically omniscient, a requirement usually thought to follow from coherence. I shall show how the
requirement of logical omniscience can be relaxed in a way consistent with
coherence, and show how this can lead us to a solution of the problem of old
evidence.

Since, as I. J. Good has conclusively shown, there are more kinds of
Bayesianism than there are Bayesians[3], it will be helpful to give a quick
sketch of what I take the Bayesian framework to be before entering into the
problem of old evidence. By the Bayesian framework I shall understand a
certain way of thinking about (rational) belief and the (rational) evolution of
belief. The basic concept for the Bayesian is that of a degree of belief. The
degree of belief that a person S has in a sentence p is a numerical measure of
S's confidence in the truth of p, and is manifested in the choices S makes
among bets, actions, etc. Formally S's degrees of belief at some time t_0 are
represented by a function P_0 defined over at least some of the sentences of
S's language L.[4] What differentiates the Bayesian account of belief from
idealized psychology is the imposition of rationality conditions on S's
beliefs. These rationality conditions are of two sorts, synchronic and
diachronic. The most widely agreed upon synchronic condition is
coherence:

> (D1) A P-function is *coherent* iff there is no series of bets in accordance
> with P such that anyone taking those bets would lose in every possible
> state of the world.

Although there are those who would argue that coherence is both
necessary and sufficient for S's beliefs to be rational at some given time, I
shall assume only that coherence is necessary. One of the central results of
Bayesian probability theory is the coherence theorem, which establishes
that if P is coherent, then it is a (finitely additive) probability function on
the appropriate group of objects (i.e., the sentences of S's language L).[5] In
the discussions below, I shall assume that an individual's degrees of belief
have at least that much structure. Although there is little agreement about
rational belief change, one way of changing one's beliefs is generally
accepted as rational by most Bayesians, conditionalization. One changes
one's beliefs in accordance with conditionalization when, upon learning
that q, one changes one's beliefs from P_0 to P_1 as follows:

$$P_1(p) = P_0(p/q)$$

where conditional probability is defined as usual. There are some who take conditionalization as the sine qua non of the Bayesian account of belief, but I shall regard it as one among a number of possible ways of changing rational belief, a sufficient but not necessary condition of diachronic rationality.[6] Despite this proviso, though, conditionalization will have a major role to play in the discussion of confirmation that follows.

There are two competing ways of thinking about what the Bayesian is supposed to be doing, what I call the thought police model and the learning machine model. On the thought police model, the Bayesian is thought of as looking over our shoulders and clubbing us into line when we violate certain principles of right reasoning. On this view, the axioms of the theory of probability (i.e., coherence) and, perhaps, the dynamical assumption that we should change our beliefs in accordance with conditionalization are the clubs that the Bayesian has available. On the learning machine model, on the other hand, the Bayesian is thought of as constructing an ideal learning machine, or at least describing the features that we might want to build into an ideal learning machine.[7] Unlike others, I do not see a great deal of difference between these two ways of thinking about the enterprise. The Bayesian thought policeman might be thought of as clubbing us into behaving like ideal learning machines, if we like. Or, alternatively, we can think of the ideal learning machine as an imaginary person who behaves in such a way that he never needs correction by the Bayesian thought police. The two models thus seem intertranslatable. Nevertheless, I prefer to think of the Bayesian enterprise on the learning machine model. Although this has no theoretical consequences, I think that it is a better heuristic model when one is thinking about the confirmation of hypotheses from a Bayesian point of view.

1. The Problem of Old Evidence

In the course of presenting his own ingenious account of the confirmation of scientific hypotheses by empirical evidence, Clark Glymour offers a number of reasons why he chooses not to follow the Bayesian path. Many of Glymour's arguments are worth serious consideration; but one of the problems Glymour raises seems particularly serious, and seems to go to the very foundations of the Bayesian framework. Glymour writes:

> Scientists commonly argue for their theories from evidence known long before the theories were introduced. Copernicus argued for his theory using observations made over the course of millenia . . . . Newton argued for universal gravitation using Kepler's second and third laws, established before the *Principia* was published. The

argument that Einstein gave in 1915 for his gravitational field equations was that they explained the anomalous advance of the perihelion of Mercury, established more that half a century earlier . . . . Old evidence can in fact confirm new theory, but according to Bayesian kinematics it cannot. For let us suppose that evidence e is known before theory T is introduced at time t. Because e is known at t, $\text{Prob}_t(e) = 1$. Further, because $\text{Prob}_t(e) = 1$, the likelihood of e given T, $\text{Prob}_t(e, T)$, is also 1. We then have:

$$\text{Prob}_t(T, e) = \frac{\text{Prob}_t(T) \times \text{Prob}_t(e, T)}{\text{Prob}_t(e)} = \text{Prob}_t(T)$$

The conditional probability of T on e is therefore the same as the prior probability of T: e cannot constitute evidence for T . . . . None of the Bayesian mechanisms apply, and if we are strictly limited to them, we have the absurdity that old evidence cannot confirm a new theory.[8]

Before trying to understand what is going wrong for the Bayesian and seeing what can be said in response, it will be worth our while to look more closely at the problem itself. There are at least two subtly different problems that Glymour might have in mind here. One of these problems concerns the scientist in the midst of his investigations who appears to be using a piece of old evidence to increase his confidence in a given theory. If we adopt a Bayesian model of scientific inquiry, then how could this happen? How could an appeal to old evidence ever raise the scientist's degree of belief in his theory? This is what I shall call, for the moment, the *historical* problem of old evidence.[9] But there is a second possible problem lurking in Glymour's complaints, what might be called the *ahistorical* problem of old evidence. When we are first learning a scientific theory, we are often in roughly the same epistemic position that the scientist was in when he first put the theory to test; the evidence that served to increase *his* degrees of belief will increase *ours* as well. But having absorbed the theory, our epistemic position changes. The present appeal to Kepler's laws does not any more actually *increase* our confidence in Newton's theory of universal gravitation, nor does the appeal to the perihelion of Mercury actually *increase* our confidence in general relativity any more. Once we have learned the theories, the evidence has done its work on our beliefs, so to speak. But nevertheless, even though the old evidence no longer serves to increase our degrees of belief in the theories in question, there is still a sense in which the evidence in question remains good evidence, and there is still a sense in which it is proper to say that the old evidence confirms the theories in question. But if we are to adopt a Bayesian account of confirmation in accordance with which e confirms h iff P (h/e) > P (h), then how can we ever say that a piece of evidence, already known, confirms h?[10]

Now that we have a grasp on the problems, we can begin to look for some possible ways of responding. One obvious response might begin with the observation that if one had *not* known the evidence in question, then its discovery *would have* increased one's degrees of belief in the hypothesis in question. That is, in the circumstances in which e really does confirm h, if it *had been* the case that P (e) ≤ 1, then it *would also have been* the case that P (h/e) > P (h). There are, to be sure, some details to be worked out here.[11] If P (e) were less than one, what precisely would it have been? What, for that matter, would all of the *rest* of the P-values have been? If such details could be worked out in a satisfactory way, this counterfactual gambit would offer us a reasonably natural solution to the *ahistorical* problem of old evidence. This solution amounts to replacing the identification of confirmation with positive statistical relevance with a more subtle notion of confirmation, in accordance with which e (ahistorically) confirms h iff, if e had been previously unknown, its discovery would have increased our degree of belief in h. That is, e (ahistorically) confirms h iff, if P(e) (and, of course, P(h)) were less than one, then P(h/e) would be greater than P(h). In what follows I shall assume that the ahistorical problem of old evidence can be settled by some variant or other of this counterfactual strategy.[12]

It should be evident, though, that however well the counterfactual strategy might work for the *ahistorical* problem of old evidence, it leaves the *historical* problem untouched. When dealing with Einstein and the perihelion of Mercury, we are not dealing with a *counterfactual* increase in Einstein's confidence in his theory: we are dealing with an *actual* increase in his degree of belief. Somehow or other, Einstein's consideration of a piece of old evidence served to increase his confidence in his field equations, not counterfactually, but actually. This is something that the counterfactual solution cannot deal with.

How, then, are we to deal with the historical problem of old evidence, the cases in which considerations involving old evidence seem actually to *raise* an investigator's confidence in one of his hypotheses? We can put our finger on exactly what is going wrong in the Bayesian account if we go back and examine exactly when a piece of old evidence does seem to confirm a new hypothesis. It is appropriate to begin with the observation that Glymour's conclusion is not *always* implausible. There are, indeed, some circumstances in which an old e *cannot raise* the investigator's degree of belief in a new h. For example, suppose that S constructed h *specifically* to account for e, and knew, from the start, that it would. It should not *add* anything to the credibility of h that it accounts for the evidence that S knew all along it would account for. In this situation, there is not confirmation, at least not in the relevance sense of that term.[13] The evidential significance of the old evidence is, as it were, *built into* the initial probability that S assigns to the new hypothesis. Where the result is paradoxical is in the case in which h was concocted *without* having e in mind, and only later was it *discovered* that h bears the appropriate relations to e, i.e., that h (and perhaps some suitable auxiliaries) entails e, that e is a positive instance of h, or the like. Just *what* the relationship in question is a matter of some debate. But it seems clear that in the cases at hand, what *increases* S's confidence in h is not e itself, but the *discovery* of some generally logical or mathematical relationship between h and e. In what follows I shall often assume for simplicity that the relation in question is some kind of logical entailment. But although the details may be shaped by this assumption, the general lines of the discussion should remain unaffected.

With this in mind, it is now possible to identify just which part of the Bayesian framework is generating the problem. In the Bayesian framework, coherence is almost always taken to imply that the rational subject S, the constraints on whose degrees of belief the Bayesian is trying to describe, is *logically omniscient*. Since logical (and mathematical) truths are true in all possible states of the world, if P is to be coherent, then coherence must, it seems, preclude the possibility of S's accepting a bet against a logical truth. Consequently, coherence seems to require that S be certain of (in the sense of having degree of belief one in) all logical truths and logical entailments. Now for logically omniscient S it is absolutely correct to say that old evidence e does not increase his confidence in a new hypothesis h. Because of S's logical omniscience, S will see immediately, for every new hypothesis, whether or not it entails his previously known evidence (or, perhaps, bears the appropriate logical relations to it). No hypothesis ever enters S's serious consideration without his knowing explicity just which of his past observations it entails. So every new hypothesis S takes into consideration is, in a clear sense, *based* on the previously known observations it entails: the initial probability assigned to every new hypothesis already takes into account the old evidence it entails. For no hypothesis h and evidence e can the logically omniscient S ever *discover*, after the fact, that h entails e. And, as I have suggested above, in such a circumstance, it is perfectly intuitive to suppose that the previously known evidence does not confirm the new hypothesis in the sense of raising its probability. The historical problem of old evidence, then, seems to be a consequence of the fact that the Bayesian framework is a theory of reasoning for a logically omniscient being.

It has generally been recognized that the Bayesian framework does not seem to allow the Bayesian agent to be ignorant of logical truths, and thus does not allow a Bayesian account of logical or mathematical reasoning. Although this has been considered a weakness of the framework, it has usually been accepted as an idealization that we *must* make in order to build an adequate account of the acquisition of *empirical* knowledge. What the problem of old evidence shows is that this idealization will not do: without an account of how the Bayesian can come to learn *logical* truths, we cannot have a fully adequate theory of *empirical* learning either. So if we are to account for how old evidence can raise the investigator's degree of belief in new hypotheses, we must be able to account for how he can come to know certain logical relations between hypothesis and evidence that he did not know when he first formulated the new hypothesis.

The problem of old evidence is not of course the only reason for seeking an account of logical learning consistent with Bayesian principles. There is an even deeper concern here. With the assumption of logical omniscience, there is a philosophically disturbing asymmetry between logical and empirical knowledge in the Bayesian framework. Although it may be unfortunate that we lack omniscience with respect to empirical truths, the Bayesian account makes it *irrational* to be anything but logically omniscient. The Bayesian agent who is *not* logically omniscient is incoherent, and seems to violate the only necessary condition for synchronic rationality that Bayesians can agree on. This is an asymmetry that smacks of the dreaded analytic-synthetic distinction. But scruples about the metaphysical or epistemic status of that distinction aside, the asymmetry in the treatment of logical and empirical knowledge is, on the face of it, absurd. It should be no more *irrational* to fail to know the least prime number greater than one million than it is to fail to know the number of volumes in the Library of Congress.[14]

The project, then, is clear: if the Bayesian learning model is to be saved, then we must find a way to deal with the learning of logical truths within the Bayesian framework. If we do this correctly, it should give us both a way of eliminating the asymmetry between logical and empirical knowledge, and a way of dealing with the problem of old evidence. This is the problem taken up in the following sections.

2. Two Models of Logical Learning

A solution to the problem of old evidence requires that the Bayesian be able to give an account of how the agent S can come to know logical truths that he did not previously know. In this section I shall present and discuss two possible Bayesian models of logical learning. Because of the immediate problem at hand, the models will be formulated in terms of a particular kind of logical truth, those of the form "p logically entails q," symbolized by "$p \vdash q$," although much of what I say can be extended naturally to the more general case. In this section I shall not discuss the precise nature of the logical implications dealt with here (i.e., truth-functional entailment vs. first order quantificational entailment vs. higher order quantificational entailment, etc.), nor shall I discuss the nature of the underlying language. These clarifications and refinements will be introduced as needed in the succeeding sections. But even without these refinements, we can say some interesting things about the broad paths we might follow in providing a Bayesian account of logical learning.

The two models of logical learning that I would like to discuss are the *conditionalization model* and the *evolving probability model*. On the conditionalization model, when S learns that $p \vdash q$, he should change his beliefs from P_0 to P_1 as follows:

$$P_1(-) = P_0 (-/p \vdash q)$$

On the evolving probability model, on the other hand, when S learns that $p \vdash q$, he is required to change his beliefs in such a way that $P(q/p) = 1$, and to alter the rest of his beliefs in such a way that coherence is maintained, or at least in such a way that his beliefs are as coherent as they can be, given his imperfect knowledge of logical truth.[15]

Which, if either, of these models should the Bayesian adopt? The conditionalization model has obvious attractions, since it fits neatly into the most popular Bayesian account of belief change in general. But however attractive it might be on its face, the conditionalization model has one obvious difficulty. I pointed out earlier that coherence seems to require that all logical truths get probability one. Consequently we are left with an unattractive choice of alternatives. It *seems* as if we must either say that the conditionalization model fails to allow for any logical learning, since in the case at hand P_1 must always equal P_o; or we must radically alter the notion of coherence, so that logical truths can get probability of less than one. Let us then set the conditionalization model aside for the moment and see if we can make do with evolving probability.

The evolving probability model does not have the obvious difficulties that the conditionalization model has. It does, however, require a major change in the way we think about coherence. If we adopt the evolving probability model, then we are implicity removing coherence as a synchronic constraint on rational belief. The best that we can say is that an individual ought to regard coherence as an ultimate goal. That is, the evolving probabilist seems forced to the position according to which the synchronic condition for rationality is not coherence itself, but only that the rational individual try to become as coherent as he can. Although this is intuitively not unattractive, it does have at least one unattractive consequence. If it is not required that an individual be coherent at any given time, then it would seem that nothing very strong could be said about the general characteristics that a rational individual's beliefs would have to satisfy at any given time. All of the wonderful theorems of the mathematical theory of probability would not apply to the rational investigator, but would only apply at the limit, at the end of inquiry, when his beliefs became fully coherent. But although this is somewhat unattractive, we could probably learn to live with this consequence if the evolving probability model turned out to be otherwise adequate to the task.

Unfortunately, though, it does not. Even if we could accept the required weakening of the constraint of coherence, there are three other problems that should give us serious pause. For one, the evolving probability model as stated gives us very little guidance as to how we ought to change our beliefs upon discovering that $h \vdash e$. If the *only required* changes in our beliefs upon learning that $h \vdash e$ are to alter $P(e/h)$ to one and restore coherence, we can always find a way of changing our beliefs consistent with the evolving probability model that will raise, lower, or leave $P(h)$ unchanged. Suppose at t_o, $P_o(E/h) < 1$. That is, $P_o(e \,\&\, h) < P_o(h)$. Suppose, then, that we learn that $h \vdash e$, so that we alter P_o in such a way that $P_1(e/h) = 1$. That is, we now have $P_1(e \,\&\, h) = P_1(h)$. But it is clear that this result can be arrived at in any of three ways: we can lower $P_o(h)$, leave $P_o(h)$ unchanged and raise $P_o(e \,\&\, h)$, or we can raise both $P_o(h)$ and $P_o(h \,\&\, e)$ to the same level. Each of these ways of altering one's beliefs is consistent with the evolving probability model. Consequently the evolving probabil-

ity model can tell us nothing general about the effect that learning that
h ⊢ e may have on the rest of one's beliefs. The effect it has is determined
by the *way* in which one changes from P(e/h) < 1 to P(e/h) = 1, and the
evolving probability model says nothing about this.[16]

There is a second, more philosophical difficulty connected with the
evolving probability model. Although the evolving probability model gives
the Bayesian a way of dealing with logical learning, something of the
original asymmetry between logical and empirical learning still remains.
Upon learning an empirical truth, one (presumably) changes one's beliefs
through conditionalization, whereas upon learning a logical truth, one
changes one's beliefs through evolving probabilities. This continuing
asymmetry should make us feel somewhat uncomfortable. The asymmetry
could be eliminated, of course. We could declare that the evolving
probability scheme is the way to change one's beliefs whether we learn
empirical truths or logical ones, and give up conditionalization altogether,
even for empirical learning. One might say, for instance, when S learns that
e, he should simply change his beliefs in such a way that $P_1(e) = 1$, along
with whatever other changes are necessary to restore coherence. But this is
not very satisfactory. It would subject empirical learning to the same kind
of indeterminacy that logical learning has on the evolving probability
model, and prevent our saying anything interesting of a general nature
about empirical learning as well.

These two problems are serious. But there is a third problem even more
serious than the previous two. Although the evolving probability model
may give us a way of thinking about logical learning within the Bayesian
framework, it is utterly incapable of dealing with the problem of old
evidence. I argued that in the circumstances that give rise to the problem it
is learning that our new hypothesis entails some piece of old evidence (or is
related to it in some appropriate logical or mathematical way) that raises
our degree of belief in h. But if we adopt the evolving probability model,
learning that h ⊢ e in those circumstances will not change our beliefs at all!
The evolving probability model tells us that when we learn that h ⊢e, we
should alter our beliefs in such a way that P(e) = 1. But in the cases at hand,
where e is old evidence, and thus P (e) = 1, P(e/h) *already* equals 1 (as does
"P(h ⊃ e)"). So, in the cases at hand, the evolving probability model will
counsel *no change at all* in our degrees of belief. Thus learning that
h ⊢ e can have no effect at all on our degree of belief in h, if e is previously
known.

I have offered three reasons for being somewhat cautious about adopting the evolving probability model of logical learning. These arguments suggest that we turn to the conditionalization model. We must of course subject the conditionalization model to the same tests to which we subjected the evolving probability model. We must examine how well it determines the new probability function, how well it deals with the problem of asymmetry, and most important of all, how well it deals with the problem of old evidence. But first we must deal with the most basic and evident difficulty confronting the conditionalization model: can any sense be made of a probability function in which P(h ⊢ e) is anything but 0 or 1? Will allowing probability functions in which 0 < P(h ⊢ e) < 1 force us into incoherence in both the technical and nontechnical senses of that word?

3. Coherence and Logical Truth: An Informal Account

As I noted earlier, the standard definition of coherence, (D1), seems to require that all logical truths get probability 1. For surely, if h entails e, it entails e in every possible state of the world, it would seem. And if we were to assign probability less than one to a sentence like "h ⊢ e," then we would be allowed to bet that "h ⊢ e" is false, a bet that we would lose, no matter what state of the world we were in. Thus if we require P to be coherent, logical omniscience seems inescapable, and the conditionalization model of logical learning seems untenable.

One way out of this problem might be to eliminate coherence as a necessary condition of rational belief. But this is not very satisfying. If we were to eliminate coherence, we would have no synchronic conditions on rational belief at all; the Bayesian framework would reduce to an idealized psychology. It might help to reintroduce coherence as an ultimate goal of inquiry, as the evolving probabilist implicitly does. But, as I suggested in the course of our examination of the evolving probability model, this is not very attractive. This ploy has the unfortunate consequence of allowing us to say nothing of interest about the characteristics that a rational person's beliefs would have to exhibit at any given time. Explicitly relativizing coherence to an individual's state of knowledge with respect to logical truth might seem attractive, and has actually been proposed.[17] But this will give us little of the mathematical structure that we want. Moreover, it has the extra problem of introducing the philosophically problematic notion of knowledge explicitly into the Bayesian framework.

But all is not lost. Although it does not seem advisable to eliminate or weaken coherence, perhaps a more careful examination of the coherence condition itself may give us a way of weakening the requirement of logical omniscience. The definition of coherence is obviously relativized to another notion, that of a *possible state of the world*. How we understand that notion should have important consequences for the constraints that the coherence condition imposes on an individual's beliefs. And how we understand the notion of a possible state of the world, it turns out, depends on what we think the Bayesian learning model is supposed to do.

One popular conception of the Bayesian enterprise is what I shall call *global Bayesianism*.[18] On this conception, what the Bayesian is trying to do is build a global learning machine, a scientific robot that will digest all of the information we feed it and churn out appropriate degrees of belief. On this model, the choice of a language over which to define one's probability function is as important as the constraints that one imposes on that function and its evolution. On this model, the appropriate language to building into the scientific robot is the *ideal language of science,* a maximally fine-grained language L, capable of expressing all possible hypotheses, all possible evidence, capable of doing logic, mathematics, etc. In short, L must be capable, in principle, of saying anything we might ever find a need to say in science.

Now, given this global framework, there is a natural candidate for what the possible states of the world are: they are the maximal consistent sets of sentences in L. But if *these* are what we take to be the possible states of the world, then logical omniscience of the very strongest sort seems to be demanded, and the conditionalization model of logical learning goes out the window. For if the possible states of the world are the maximal consistent sets of sentences in the most fine-grained, ideal language of science, then they are, in essence, the *logically* possible states of the world. And if I am coherent with respect to these states, i.e., if I am not allowed to enter into bets that I would lose in every such logically possible state of the world, then I must have degree of belief one in all logical truths.

But there are reasons for thinking twice before accepting this conclusion. Although global Bayesianism is a position often advanced, it is a very implausible one to take. For one thing, it does not seem reasonable to suppose that there is any one language that we can be sure can do everything, an immutable language of science of the sort that the Vienna Positivists sought to construct. Without such a language, the scientific robot model of Bayesianism is untenable, as is the idea that there is some one unique set of logically possible states of the world to which we are obligated to appeal in establishing coherence. But even if it were possible to find a cannonical and complete language for science, it would not be of much use. One of the goals of the Bayesian enterprise is to reconstruct scientific practice, even if in an idealized or rationalized form. Typically when scientists or decision makers apply Bayesian methods to the clarification of inferential problems, they do so in a much more restricted scope than global Bayesianism suggests, dealing only with the sentences and degrees of belief that they are actually concerned with, those that pertain to the problem at hand.

This suggests a different way of thinking about the Bayesian learning model, what one might call *local Bayesianism*.[19] On this model, the Bayesian does not see himself as trying to build a global learning machine, or a scientific robot. Rather, the goal is to build a hand-held calculator, as it were, a tool to help the scientist or decision maker with particular inferential problems. On this view, the Bayesian framework provides a general formal structure in which one can set up a wide variety of different inferential problems. In order to apply it in some particular situation, we enter in only what we need to deal with in the context of the problem at hand, i.e., the particular sentences with which we are concerned, and the beliefs (prior probabilities) we have with respect to those sentences.

So, for example, if we are interested in a particular group of hypotheses h_i, and what we could learn about them if we were to acquire some evidence e_i, then our *problem relative language* L' would naturally enough be just the truth-functional closure of the h_i and the e_i. Our probability functions would then, for the duration of our interest in this problem, be defined *not* over the maximally specific language of science L, but over the considerably more modest problem-relative language L'.

In working only with the problem relative L', we are in effect treating each of the h_i and e_i as *atomic sentences*. This is not to say that h_i and e_i *don't* have any structure. Of course they do. It is by virtue of that structure, which determines their *meanings*, that we can tell in a given observational circumstance whether or not a given e_i is true, and it is by virtue of that structure that we know what it is that our degrees of belief are degrees of belief about! But none of this extra content is entered into our Bayesian hand-held calculator. Whatever structure h_i and e_i might have in some language *richer* than L' is submerged, so to speak, and the h_i and e_i treated as unanalyzed wholes from the point of view of the problem at hand. This extra structure is not *lost*, of course. But it only enters in *extrasystematically*, so to speak, when, for example, we are assigning priors, or when we are deciding whether or not a particular observational sentence is true in a particular circumstance.

This seems to open the door to a Bayesian treatment of logical truth. In some investigations we are interested only in sentences like "h_i" and "e_i." But in others, like those in which the problem of old evidence comes up, we are interested in other sentences, like "$h_i \vdash e_i$." Sentences like "$h_i \vdash e_i$" certainly have structure. Depending on the context of investigation, "$\vdash$" may be understood as truth-functional implication, or implication in L, the global language of science. We can even read "$h_i \vdash e_i$" as "e_i is a positive instance of h_i," or as "e_i bootstrap confirms h_i with respect to some appropriate theory," as Glymour demands.[20] But whatever *extrasystematic* content we give sentences like "$h_i \vdash e_i$," in the context of our problem-relative investigation we can throw such sentences into our problem-relative language as *atomic sentences*, unanalyzed and unanalyzable wholes, and submerge whatever content and structure they might have, exactly as we did for the h_i and e_i.

Suppose now that we are in a circumstance in which logical relations between sentences are of concern to us. Say we are interested in some implicative relations between hypotheses and evidence, sentences of the form "$h_i \vdash e_i$." The problem-relative language will be the truth-functional closure of all the h_i, e_i, and sentences of the form "$h_i \vdash e_i$," where *each* of these sentences, *including* those of the form "$h_i \vdash e_i$" is treated as an *atomic sentence* of the problem relative language. Now the crucial question is this: what constraints does coherence impose on probability functions defined over this language? In particular, does coherence require that all sentences of the form "$h_i \vdash e_i$" get 0 or 1? If not, then we are out of the woods and on our way to an account of logical learning through conditionalization.

As I argued, in order to decide what follows from coherence, we must determine what is to count as a possible state of the world. Now in giving up global Bayesianism and any attempt to formulate a maximally fine-grained language of science, we give up in effect the idea that there is some one set of logically possible states of the world that stands behind every inferential problem. But how then *are* we to understand states of the world? The obvious suggestion is this. In the context of a particular investigation, we are interested in some list of atomic sentences and their truth-functional compounds: hypotheses, possible evidence, and statements of the logical relations between the two. Insofar as we are uncertain of the truth or falsity of any of these atomic sentences, we should regard each of them as true in some states of the world, and false in others, at least in the context of our investigation. And since, in the context of investigation, we are interested in no other sentences, our problem relative states of the world are easily specified: *they are determined by every possible distribution of truth values to the atomic sentences of the local language L'*. This amounts to replacing the *logically* possible worlds of the global language with more modest *epistemically* possible worlds, specified in accordance with our immediate interests.

Now if the possible states of the world are those determined by all possible assignments of truth values to the atomic sentences of the local language L', then coherence imposes one obvious constraint on the scientist's degrees of belief: if sentence T in L' is true on all possible assignments of truth values to the atomic sentences of L', then $P(T) = 1$. That is, if T is a *tautology* of L' then $P(T) = 1$. Coherence understood in this way, however, relativized to the problem-relative states of the world, does *not* impose *any* constraints on the *atomic sentences* of L'. Since for any atomic sentence of L' there are states of the world in which it is false, we can clearly assign *whatever* degree of belief we like to *any* of the atomic sentences without violating coherence, i.e., without being caught in the position of accepting bets that we would lose in every (problem-relative) state of the world. *And this holds even if one of those atomic sentences is extrasystematically interpreted as "h logically entails e."*

This seems to get us exactly what we want. It seems to allow us to talk about uncertainty with respect to at least *some* logical truths, and in fact, it allows us to do this without even violating coherence! This is an interesting and slightly paradoxical result. In order to see better what is going on, and make sure that there is no contradiction lurking beneath the surface of the exposition, I shall try to set the result out more formally.

4. Coherence and Logical Truth: a Formal Account

In the previous section we dealt informally with relatively modest local languages, a few hypotheses, a few evidential sentences, a few logical relations. But the coherence result I argued for can be shown formally to hold for much larger languages as well. Let us consider first the language L, the truth-functional closure of a countably infinite collection of atomic sentences, {a_i}. Let us build the larger language L* by adding to L some new atomic sentences, those of the form "A ⊢ B," where A and B are in L, and again, closing under truth-functional operations. L* is a truth-functional language that allows us to talk about truth-functional combinations of an infinite set of atomic sentences {a_i}, and relations of implication between any truth-functional combination of these sentences.[21] So it is clearly adequate to handle any of the problem situations that we had been discussing earlier.

Now, L* is just a truth-functional language generated by a countably infinite number of atomic sentences, i.e., those of the form "a_i" or "A ⊢ B." So, if the possible states of the world are identified with possible assignments of truth values to the atomic sentences of L*, on analogy with what I argued above with respect to the more modest local languages, then imposing coherence will fix no degrees of belief with respect to the atomic sentences of L*. There will be coherent P-functions that will allow us to assign *whatever* values we like in [0, 1] to the atomic sentences of the form "A ⊢ B," however these may be interpreted extrasystematically. The only specific values fixed by the requirement of coherence will be those of the tautologies and truth-functional contradictions in L*, i.e., the tautological and contradictory *combinations* of atomic sentences of L*.

This almost trivial result follows directly from the fact that, from the point of view of the probability function, sentences like "A ⊢ B" are uninterpreted and treated on a par with the a_i, treated like structureless wholes. But, interestingly enough, a similar result can be obtained without such a strong assumption. That is, we can introduce a certain amount of structure on the atomic sentences of the form "A ⊢ B" without restricting our freedom to assign them probabilities strictly between 0 and 1.

In introducing the atomic sentences of the form "A ⊢ B" into our local problem-relative languages, I emphasized that "A ⊢ B" could be interpreted extrasystematically in a variety of different ways, as "A truth-functionally entails B," that is, as "'A ⊃ B' is valid in L," as "A entails B in some richer language" (e.g., in the maximally fine-grained ideal language of science), or as some logical or mathematical relation other than implication, e.g., as "B is a positive instance of A," or as "B bootstrap confirms A with respect to some appropriate theory," in the sense in which Glymour understands this relation. For the purpose of adding some additional structure, though, let us assume that we are dealing with some variety of implication or other. Now if "A ⊢ B" is to be read as "A implies

B," we may want to require that our Bayesian investigator S recognize that atomic sentences of the form "A ⊢ B" have *some* special properties, however implication is understood. Although we do not want to demand that S recognize all true and false ones, it does seem reasonable to demand that S recognize that *modus ponens* is applicable to these particular atomic sentences of L*. That is, we might require that if "A ⊢ B" is to be properly read as "A implies B," then at very least, if S knows that A, and S knows that A ⊢ B, he must also know that B as well. Put probabilistically, this amounts to adopting the following constraint over reasonable degree of belief functions on L*:

(K) P(B/A & A ⊢ B) = 1, when defined.

But since, Renyi and Popper aside, this conditional probability is undefined when P(A & A ⊢ B) = 0, we might replace (K) with the following slightly stronger condition:

(K*) P(A & B & A ⊢ B) = P(A & A ⊢ B).

(K*) clearly reduces to (K) when the conditional probability in (K) is defined.

(K*) is a stronger condition than it may appear on the surface. If, in addition to coherence, we impose (K*) on all "reasonable" probability functions defined on L*, then we get a number of interesting and desirable properties, as outlined in the following theorem:

(T1) If P is a probability function on L* and P satisfies (K*), then:
   (i) If P(A ⊢ B) = 1, then P(A ⊃ B) = 1 and P(B/A) = 1, when defined.
   (ii) P(~A/~B & A ⊢ B) = 1 when defined.
   (iii) If A and B are truth-functionally inconsistent in L, then P(A & A ⊢ B) = 0.
   (iv) P(B/(A ⊢ B) & (~A ⊢ B)) = 1, when defined.
   (v) If P(A & A ⊢ B) = 1, then P(A ⊢ ~B) = 0.
   (vi) If B and C are truth-functionally inconsistent in L, then P(A/(A ⊢ B) & (A ⊢ C)) = 0, when defined.
   (vii) As P(B) →0, P(A/A ⊢ B) →0 and P(A ⊢ B/A) →0.
   (viii) If A and ~B are both tautologies in L, then P(A ⊢ B) = 0.
   Proof: All of the arguments are trivial and left to the reader.

These properties are attractive, and seem appropriate when " ⊢ " is interpreted as a variety of implication.[22] Imposing (K*) guarantees that when we learn that A ⊢ B, our degrees of belief in "A ⊃ B" and our conditional degrees of belief in B given A will behave appropriately, by clause (i). It gives us a probabilistic version of *modus tollens* (clauses (ii) and (vii)). It also guarantees that S will be certain of the truth of anything that follows both from A and from ~A (clause (iv)), and that S will be certain of the falsity of anything that has truth-functionally inconsistent consequences (clause (vi)).

Now (K*) seems to be an appropriate constraint to impose on any probability function defined over L*, if "⊢" is to be interpreted as a variety of logical implication. Although it does not *guarantee* that we are dealing with a variety of implication,[23] it is certainly reasonable to require that any variety of implication should satisfy (K*). But now matters are not so trivial. Might adding (K*) as an extra constraint take away *all* of the freedom we had in assigning probabilities to sentences of the form "A ⊢ B" in L*? The coherence condition imposes no constraints on assigning probabilities to the atomic sentences of L*, I have argued. Most importantly, it does not force us to logical omniscience, to the position in which all sentences of the form "A ⊢ B" are forced to take on probabilities of 0 or 1. But might coherence in conjunction with (K*)? The surprising answer is that with one small exception (already given in (T1) (viii)), no! This result is set out in the following theorem:

(T2) There exists at least one probability function P on L* such that P satisfies (K*) and such that every atomic sentence in L* of the form "A ⊢ B" where not both A and ~B are tautologies gets a value strictly between 0 and 1.

Proof: Consider L and L* as above. Let P be any strictly positive probability on L. That is, for A in L, $P(A) = 0$ iff A is truth-functionally inconsistent in L. Then extend P to L* as follows:

(i) Suppose that A in L is not a tautology. Then let C be any sentence in L which is nontautologous, noncontradictory, and inconsistent with A. If A is not truth-functionally inconsistent in L, then ~A will do; otherwise let C be any atomic sentence a_i in L. Then, for any B in L, let $P(A ⊢ B) = P(C)$; and for any D in L*, let $P([A ⊢ B] \& D) = P(C \& D)$; $P([A ⊢ B] \lor D) = P(C \lor D)$; etc.

(ii) Suppose that A in L is a tautology and B is not. Then let

$P(A \vdash B) = P(B); P([A \vdash B] \& D) = P(B \& D); P([A \vdash B]vD)$
$= P(BvD)$; etc.

(iii) Suppose that A and B in L are both tautologies. Then let
$P(A \vdash B) = P(a_i)$, where "a_i" is an arbitrary atomic sentence
in L; $P([A \vdash B] \& D) = P(a_i \& D)$; $P([A \vdash B]vD) = P(a_ivD)$;
etc.

P so extended is clearly a probability on L*. Further, it can easily
be shown that P so extended satisfies (K*). And finally, since P on
L is strictly positive, $P(A \vdash B)$ will never have a value of either 0
or 1, except when both A and ~B are tautologies, in which case it
will get a value of 0 by clause (ii).[24]

So it turns out that even if we add more structure, as we do when (K*) is
introduced, we are not forced to logical omniscience. Even with (K*) *and*
coherence, we are permitted to be uncertain of logical implications.[25]

These technical conclusions call for some reflection. How can I say that I
have gotten rid of logical omniscience if S is *still* required to know all
tautologies of L*? And if S is required to know all tautologies of L*, mustn't
the freedom he is given with respect to the sentences of the form "A ⊢ B"
inevitably lead to contradiction? As regards logical omniscience, that has
been eliminated. Coherence still requires that we have some logical
knowledge. But knowing the tautologies of L* is a far cry from logical
omniscience, since there are many logical truths that are not tautologies of
L*. The threat of internal contradiction is more subtle, though. Formally
speaking, there is no contradiction. The key to seeing this lies in
understanding the distinction between those logical truths that S is
required to know and those that he is not. Let A, B, A ⊢ B be sentences in
our local problem relative language L*, where A and B are truth-functional
combinations of atomic sentences of L, and "A ⊢ B" is an atomic sentence
of L* interpreted (extrasystematically) as "A entails B." For the purposes of
discussion, it does not matter whether the turnstile is interpreted as truth-
functional entailment in L, or something weaker. Now suppose that, as a
matter of fact, A *does* truth-functionally entail B. What precisely does
coherence require? It clearly requires that $P(B/A) = 1$ and $P(A \supset B) = 1$.
That is, it requires that S be certain of B conditional on A, and certain of the
tautology "A ⊃ B." But if my argument is correct, S is not required to be
certain of the *atomic* sentence "A ⊢ B," which can get a degree of belief
strictly between 0 and 1. That is, in requiring that S be certain of "A ⊃ B,"

coherence requires that S be certain that a particular truth-functional combination of atomic sentences of L is *true*. But at the same time, in allowing uncertainty with respect to "A ⊢ B," coherence allows that S might be uncertain as to whether or not that truth-functional combination of atomic sentences is *valid*. And insofar as truth and validity are distinct, there is no formal contradiction in asserting that S may be certain that "A ⊃ B" is *true* without necessarily being certain that it is *valid*, i.e., without being certain that "A ⊢ B" is true.

But even if there is no formal contradiction, there does appear to be a kind of *informal* contradiction in requiring that S be certain of A ⊃ B when A truth-functionally entails B in L, while at the same time allowing him to be uncertain of A ⊢ B. But this informal contradiction can be resolved easily enough by adopting a new constraint on reasonable probability functions on L*:

(*)   If "A ⊃ B" is a tautology in L, then $P(A \vdash B) = 1$.

This would require that S know not only the *truth* of all tautologies of L, but also their *validity*.[26] Although I see no particular reason to adopt (*), doing so would resolve the informal appearance of contradiction without doing much damage to the formalism or its applicability to scientific reasoning. For truth-functional implication is not the only variety of implication. In fact, when we are interested in the logical relations between hypotheses and evidentiary sentences, the kind of implicatory relations we are interested in will most likely be not truth-functional implication, but quantification-theoretic implication in some background language richer than L* in which the hypotheses and evidence receive their (extrasystematic) interpretation. So, in any realistic application of the formalism developed in this section, adding (*) as a constraint will fix only a small number of sentences of the form "A ⊢ B," and leave all of the rest unaffected. (*) will fix *all* such sentences *only* in the case in which "A ⊢ B" is interpreted rather narrowly as "A truth-functionally entails B in L," a case that is not likely to prove of much use in the analysis of scientific reasoning.

5. The Conditionalization Model and Old Evidence Redux

After this rather lengthy argument, it might help to review where we have been and gauge how much farther we have to go. Starting with the problem of old evidence, I argued that a fully adequate Bayesian account of scientific reasoning must include some account of the learning of logical truths; in particular, it must allow for the fact that the logical and mathematical relations between hypotheses and evidence must be discovered, just as the empirical evidence itself must be. I then presented two Bayesian models of logical learning, the evolving probability model and the conditionalization model, argued that the evolving probability model has serious weaknesses, and suggested that we explore the conditionalization model. In the previous two sections I showed that the central problem with the conditionalization model, the widely held conviction that coherence requires that all logical truths get probability one, turns out not to be a problem at all. I showed that if we think of the Bayesian framework as problem-relative, a hand-held calculator rather than a scientific robot, then we can make perfectly good sense of assigning probabilities of less than one to the logical truths we are interested in, without even violating coherence! This conclusion enables us to return to the conditionalization model for learning logical truth, and discuss its adequacy, particularly in regard to the problem of old evidence.

On the conditionalization model, when S learns a logical truth, like "h ⊢ e," he should change his beliefs as follows:

$$P_1(-) = P_0(- / h \vdash e)$$

The investigations of the previous sections have shown that this does not necessarily reduce to triviality, nor does it force us to give up the requirement of coherence. But is it an otherwise attractive way to think about the consequences of learning a logical truth? In discussing the evolving probability model, I noted three problems: (a) the evolving probability model does not uniquely determine a new probability function upon learning that h ⊢ e; (b) the evolving probability model maintains an asymmetry between logical and empirical learning; and (c) the evolving probability model offers no solution to the (historical) problem of old evidence. It is clear that the conditionalization model deals admirably with the first two of these problems. Since "$P(-/ h \vdash e)$" is uniquely determined for all sentences in the language over which P is defined, the conditionalization model gives us a unique new value for all sentences of that language, upon learning that h ⊢ e. And there is obviously no asymmetry between logical and empirical learning: *both* can proceed by conditionalization. The third question, then, remains: how does the conditionalization model do with respect to the problem of old evidence? Unlike the previous two questions, the answer to this one is not obvious at all.

Earlier I argued that the (historical) problem of old evidence derives from the assumption of logical omniscience. For the logically omniscient S, old evidence can never be appealed to in order to *increase* his degree of belief because, as soon as h is proposed, S can immediately see all of the logical consequences of h, and thus his *initial* probability for h will be *based on* a complete knowledge of what it entails. If old evidence can be used to *raise* the probability of a new hypothesis, then, it must be by way of the *discovery* of previously unknown logical relations. In the cases that give rise to the problem of old evidence, we are thus dealing with circumstances in which hypotheses are confirmed not by the empirical evidence itself, but by the discovery of some logical relation between hypothesis and evidence, by the discovery that h ⊢ e. Now the evolving probability model of logical learning failed to deal with the problem of old evidence because on that model, when $P(e) = 1$, learning that h ⊢ e has no effects on S's degrees of belief. The evolving probability model thus breaks down in precisely the cases that are of interest to us here. But, one might ask, does the conditionalization model do any better? That is, is it possible on the conditionalization model for the discovery that h ⊢ e to change S's beliefs when e is previously known, for $P(h/h ⊢ e)$ to be greater than $P(h)$ when $P(e) = 1$? Unfortunately, (T2) will not help us very much here. (T2) does have the consequence that $P(h ⊢ e)$ can be less than one when $P(e) = 1$, which is certainly *necessary* if $P(h/h ⊢ e)$ is to be greater than $P(h)$. But because of the assumption of a strictly positive probability on L in the proof of (T2), the probability function constructed there, in which (almost) all implications get probability strictly between 0 and 1 will be such that for any e, $P(e) = 1$ if and only if e is a tautology. Thus (T2) does not assure us that $P(h ⊢ e)$ can be less than one when S is certain of a *non*tautologous e. This is not very convenient, since the old evidence we are interested in is not likely to be tautologous! Furthermore, although (T2) assures us that (K*) does not require extreme values on all logical implications, it does not assure us that that strong constraint ever allows for probability functions in which $P(h/h ⊢ e) > P(h)$ for *any* e at all, tautologous or not. But luckily it is fairly easy to show that under appropriate circumstances, there is always a probability function on L* (in fact, an infinite number of them) that satisfies (K*) in which, for *any* noncontradictory e, and for *any* nonextreme values that might be assigned to $P(h)$ and $P(h ⊢ e)$, $P(e) = 1$ and $P(h/h ⊢ e) > P(h)$. This is the content of the following theorem:

(T3) For L and L* constructed as above, for any atomic sentence of L* of the form "A ⊢ B" where B is not a truth-functional contradiction in L and where A does not truth-functionally entail ~B in L and B does not truth-functionally entail A in L, and for any r, s in (0, 1), there exist an infinite number of probability functions on L* that satisfy (K*) and are such that $P(B) = 1$, $P(A ⊢ B) = r$, $P(A) = s$, and $P(A/A ⊢ B) > P(A)$.

Proof: Consider all sentences s_i in L* of the following form (Carnapian state descriptions):

$$(\pm)a_1\&. \quad . \quad . \quad \&(\pm)a_n\&(\pm)[A ⊢ B]$$

where $a_1, . . . , a_n$ are the atomic sentences of L that appear in every sentence of L equivalent to either A or B, if B is not a tautology, or those that appear in every equivalent of A, if it is, and "$(\pm)$" is replaced by either a negation sign or a blank. Define a function P over the s_i as follows. First of all, assign a P-value of 0 to any s_i that truth-functionally entails ~B in L*. Since B is not truth-functionally inconsistent, there will be some s_i that remain after the initial assignment. Divide the remaining s_i into the following classes:

Class 1: s_i that truth-functionally entail A&[A ⊢ B]
Class 2: s_i that truth-functionally entail A&~[A ⊢ B]
Class 3: s_i that truth-functionally entail ~A&[A ⊢ B]
Class 4: s_i that truth-functionally entail ~A& ~[A ⊢ B]

Each s_i truth-functionally entails either [A ⊢ B] or ~[A ⊢ B], but not both, and since each s_i fixes the truth values of all of the atomic sentences in A, each s_i truth-functionally entails either A or ~A, but not both. Thus every remaining s_i fits into one and only one of these classes. Also, since A does not truth-functionally entail ~B, there will be some s_i that remain which entail A. And while every remaining s_i truth-functionally entails B, since B does not truth-functionally entail A, there will be some that remain which entail ~A. Thus, it is obvious that none of these classes will be empty. Now, let $\delta = \min(r(1-s), s(1-r))$, and let ε be an arbitrarily chosen number in $(0, \delta]$. Because of the constraints imposed on r and s, $\delta > 0$ and $(0, \delta]$ is nontrivial. Given the constraints imposed on r, s, and ε it can be shown that each of the following quantities is in [0, 1]:

$$rs + \varepsilon, \; s(1 - r) - \varepsilon, \; r(1 - s) - \varepsilon, \; (1 - r) (1 - s) + \varepsilon$$

So, we can extend P to the remaining s_i, those that do not truth-functionally entail ~B, as follows:

Class 1: Let P assign any values in [0, 1] to the s_i in class 1 that sum to
rs + ε

Class 2: Let P assign any values in [0, 1] to the s_i in class 2 that sum to
s(1 − r) − ε

Class 3: Let P assign any values in [0, 1] to the s_i in class 3 that sum to
r(1 − s) − ε

Class 4: Let P assign any values in [0, 1] to the s_i in class 4 that sum to
(1 − r) (1 − s) + ε

This completes the definition of P on the s_i. Since the values assigned sum to 1, P defines a unique probability function on the sublanguage of L* generated by the s_i. This can be further extended to the whole of L* by assigning a P-value of 0 to all atomic sentences of L* that do not appear in the s_i. P so defined clearly satisfies (K*), and is such that P(B) = 1. Also:

$$P(A \vdash B) = rs + ε + r(1 − s) − ε = r$$
$$P(A) = rs + ε + s(1 − r) − ε = s$$

Furthermore, $P(A\& [A \vdash B]) = rs + ε > rs$, so, $P(A \& [A \vdash B]) > P(A)P(A \vdash B)$ and thus $P(A/A \vdash B) > P(A)$. Since ε was arbitrarily chosen from (0, δ], there are an infinite number of probability functions on L* that have the required properties.[27]

To take a simple numerical example as an illustration of (T3), let us suppose that h and e are both atomic sentences of L, say a_1 and a_2, and let us suppose that we want to build a probability function on L* in which $P(a_1) = .4$, $P(a_2) = 1$, and $P(a_1 \vdash a_2) = .4$, and in which $P(a_1/a_1 \vdash a_2) > P(a_1)$. One such function can be constructed by assigning the following probabilities to the appropriate state descriptions, and extending the function to L* as in the proof of (T3):

$$P(a_1 \& a_2 \& [a_1 \vdash a_2]) = .3 \qquad P(a_1 \& a_2 \& \sim[a_1 \vdash a_2]) = .1$$
$$P(\sim a_1 \& a_2 \& [a_1 \vdash a_2]) = .1 \qquad P(\sim a_1 \& a_2 \sim[a_1 \vdash a_2]) = .5$$
$$P(a_1 \& \sim a_2 \& [a_1 \vdash a_2]) = 0 \qquad P(a_1 \& \sim a_2 \& \sim[a_1 \vdash a_2]) = 0$$
$$P(\sim a_1 \& \sim a_2 \& [a_1 \vdash a_2]) = 0 \qquad P(\sim a_1 \& \sim a_2 \& \sim[a_1 \vdash a_2]) = 0$$

(Using the notation of the proof of (T3), r = s = .4, and δ = .24, allowing ε to be any number in (0, .24]. The ε chosen in the example is .14). The extension of these probabilities on the state descriptions clearly satisfies (K^*), and clearly assigns the specified values to $P(a_1)$, $P(a_2)$, and $P(a_1 \vdash a_2)$. Furthermore, one can easily calculate that $P(a_1/a_1 \vdash a_2) = .3/.4 = .75$, which is clearly greater than $P(a_1)$. Thus, on my construction, it is *not* trivially the case that $P(h/h \vdash e) = P(h)$ when $P(e) = 1$, and the discovery that $h \vdash e$ *can* raise S's confidence in h. That is to say, unlike the evolving probability model, the conditionalization model of logical learning does not break down over the case of the problem of old evidence, even when (K^*) is assumed to hold.

With this last feature of the conditionalization model in place, we have completed our solution to the problem of old evidence. I have shown how old evidence e can contribute to the confirmation of a more recently proposed h through the discovery that $h \vdash e$, and I have shown how this can be done in a way consistent with Bayesian first principles. Or, perhaps more accurately, I have shown *one* way in which the Bayesian can explain how, on his view of things, old evidence can confirm new hypotheses. This takes the sting out of Glymour's critique. With a bit of ingenuity the Bayesian *can* accommodate the kinds of cases that Glymour finds so damaging. But work remains before one can make a final judgment on the particular proposal that I have advanced, the particular way in which I have proposed to deal with the problem of old evidence. In particular, one must examine with great care the cases that Glymour cites—the case of Copernican astronomy and the ancient evidence on which it rested, Newton's theory of gravitation and Kepler's laws, and Einstein's field equations and the perihelion of Mercury—along with other cases like them, in order to determine whether or not my analysis of the reasoning fits the cases at hand. We must show that the scientists in question were initially uncertain that $h \vdash e$ for the appropriate h and e, that their prior degrees of belief were such that $P(h/h \vdash e) > P(h)$[28], and that it was, indeed, the discovery that $h \vdash e$ that was, as a matter of fact, instrumental in increasing their confidence in h. Such investigations go far beyond the scope of this paper. My *intuition* is that when we look carefully at such cases, the details will work out in favor of the account that I propose.[29] But this is just an intuition.

6. Postscript: Bayesianizing the Bootstrap

I should point out that Clark Glymour was fully aware of the general lines of the solution to the problem of old evidence offered here at the time

Theory and Evidence was published. I proposed it to him while his book was still in manuscript, and we discussed it at some length. In the published version, Glymour gives a crude and early version of this line of argument, along with some remarks on why he does not believe it saves the Bayesian position. Glymour says:

> Now, in a sense, I believe this solution to the old evidence/new theory problem to be the correct one; what matters is the discovery of a certain logical or structural connection between a piece of evidence and piece of theory . . . . [The] suggestion is at least correct in sensing that our judgement of the relevance of evidence to theory depends on the perception of a structural connection between the two, and that the degree of belief is, at best, epiphenomenal. In the determination of the bearing of evidence on theory, there seem to be mechanisms and strategems that have no apparent connection with degrees of belief, which are shared alike by people advocating different theories . . . . But if this is correct, what is really important and really interesting is what these structural features may be. The condition of positive relevance [i.e., q confirms p iff $P(p/q) > P(p)$], even if it were correct, would simply be the least interesting part of what makes evidence relevant to theory.[30]

As I understand it, Glymour's point is that what should be of interest to confirmation theory is not degrees of belief and their relations, but the precise nature of the structural or logical or mathematical relations between hypothesis and evidence by virtue of which the evidence confirms the hypothesis. Put in terms I used earlier, Glymour is arguing that what confirmation theory should interest itself in is the precise nature of the "$\vdash$" necessary to make the above given formalism applicable to the analysis of scientific contexts, rather than in the fine details of how the discovery that $h \vdash e$ may, in some particular situation, raise (or lower) some scientist's degree of belief in h. Now the most difficult kind of criticism to answer is the one that says that a certain project is just not very interesting or important. I shall not attempt to defend the interest of my investigations; but I shall argue that they should be of some importance even to Glymour's own program by showing that the account of confirmation through the discovery of logical truth that I offered in the body of this paper can be used to fill in a large gap in Glymour's theory of confirmation.

The structural relation, which, Glymour argues, should be what is of interest to the confirmation theorist, is the main focus of *Theory and Evidence*. What he offers is a version of instance confirmation, but with an important and novel twist. Unlike previous writers, Glymour allows the use of *auxiliary theories* in the arguments used to establish that a given piece of evidence is a positive instance of a given hypotheses. Glymour summarizes his account as follows:

> [N]eglecting anomalous cases, hypotheses are supported by positive instances, disconfirmed by negative ones; instances of a hypothesis in a theory, whether positive or negative, are obtained by "bootstrapping," that is, by using the hypotheses of that theory itself (or, conceivably, some other) to make computations from values obtained from experiment, observation, or independent theoretical considerations; the computations must be carried out in such a way as to admit the possibility that the resulting instance of the hypothesis tested will be negative. Hypotheses, on this account, are not generally tested or supported or confirmed absolutely, but only *relative to a theory*.[31]

Glymour's intuitive sketch could be filled out in a number of ways. But since the idea is clear enough, I shall pass over the details here. With Glymour's bootstrap analogy in mind, I shall say that e BS confirms h with respect to T when the structural relation in question holds, and will symbolize it by "$[h \vdash e]_T$."

Glymour tells us a great deal about BS confirmation. But one thing that he doesn't say very much about is how we can compare different BS confirmations. The discovery that $[h \vdash e]_T$ is supposed to confirm h; it is supposed to support h and give us some reason for believing h. But when does one BS confirmation support h better than another? This is a general question, one that could be asked in the context of any confirmation theory. But it has special importance for Glymour. A distinctive feature of Glymour's theory of confirmation, one that he takes great pains to emphasize, is the fact that BS confirmations are *explicitly* relativized to auxiliary theories or hypotheses. By itself, this feature is unobjectionable. But it leads to a bit of a problem when we realize that for virtually any hypothesis h and any evidence e, there will be *some* auxiliary T such that $[h \vdash e]_T$. I shall not give a general argument for this, but the grounds for such a claim are evident enough when we examine how Glymour's BS method applies to systems of equations relating observational and theoretical quantities.[32] Let the hypothesis h be the following equation:

$$X(q_1, \ldots, q_j) = 0$$

where $q_1, \ldots, q_j$ are taken to be theoretical quantities; and let our evidence e consist of an n-tuple $e_1, \ldots, e_n$ of data points. The hypothesis h and evidence e may be entirely unrelated intuitively; h might be some equation relating physical magnitudes to one another, and e might be some quantities derived from a sociological study. Yet, as long as h is not itself a mathematical identity (i.e., not every j-tuple of numbers is a positive instance of h), we can always construct an auxiliary hypothesis with respect to which e BS confirms h. Let $c_1, \ldots, c_j$ be a j-tuple of numbers that satisfies h, and $d_1, \ldots, d_j$ be one that does not. The auxiliary appropriate to the data points $e = \{e_1, \ldots, e_n\}$ can then be constructed as follows. Let F be a function which takes e onto $c_1, \ldots, c_j$ and all other n-tuples onto $d_1 \ldots, d_j$. Then consider the auxiliary T:

$$F(p) = q$$

where "p" is an n-tuple of "observational" quantities, and $q = \{q_1, \ldots, q_j\}$ the j-tuple of theoretical terms appearing in h. Clearly, e BS confirms h with respect to T, since, on the assumption of T, e constitutes a positive instance of h.

Given the ease with which we can come by BS confirmations, the question of comparative confirmation becomes quite crucial: why is it that some BS confirmations count for more than others? Why is it that we take BS confirmations with respect to some auxiliaries as seriously reflecting on the acceptability of the hypothesis, whereas we ignore the great mass of trivial BS confirmations, those relativized to ad-hoc auxiliaries? Glymour attempts to offer something of an answer:

> The distinctions that the strategy of testing makes with regard to what is tested by what with respect to what else are of use despite the fact that if a hypothesis is not tested by a piece of evidence with respect to a theory, there is always some *other* theory with respect to which the evidence confirms or disconfirms the hypothesis. It is important that the bearing of evidence is sensitive to the changes of theory, but the significance of that fact is not that the distinctions regarding evidential relevance are unimportant. For in considering the relevance of evidence to hypothesis, one is ordinarily concerned either with how the evidence bears on a hypothesis with respect to some *accepted* theory or theories, or else one is concerned with the bearing of the evidence on a hypothesis with respect to definite theory containing that hypothesis.[33]

Glymour is surely correct in his intuitions about what we ordinarily do. But this just rephrases the problem. Why *should* we do what we ordinarily do?

Why *should* we take some BS confirmations, those that use the "appropriate" auxiliaries more seriously than we take others? If it is permissible to take seriously a BS confirmation relative to an untested auxiliary or relative to the hypothesis itself being tested, as Glymour often insists, how can he disregard *any* BS confirmations?

What is missing from Glymour's theory of confirmation seems obvious. Glymour gives us *no* way of mediating the gap between any one BS confirmation of h, and our increased confidence in h; he gives us no way to gauge how *much* any one BS confirmation supports, h, and the factors that go into that determination. Although there may be a number of different ways of filling in this gap in Glymour's program, the earlier sections of this paper suggest one attractive solution. Earlier I offered a Bayesian response to the problem of old evidence, in which the problem is resolved by showing how confirmation in the cases at hand can be understood as proceeding by conditionalization on the discovery of some logical relation between the hypothesis and the evidence in question. Now the logical relation I talked about most explicitly was logical implication. But almost everything I said holds good for whatever conception of the logical relation we like: *and this includes the logical relation that Glymour explicates*, $[h \vdash e]_T$. This framework is ready-made to fill in the gap in Glymour's program. Within this framework, we can show how the discovery that a given e BS confirms h with respect to T may increase our confidence in h, given one group of priors, and how, given other priors, the discovery that e BS confirms h with respect to T may have little or no effect on our confidence in H. The Bayesian framework, as interpreted above, thus gives us the tools needed to distinguish between the effects that different BS confirmations may have on our confidence in h, and gives us a way of resolving the problem of the ad-hoc auxiliary. To those of us of the Bayesian persuasion, the conclusion is obvious: Glymour's theory of confirmation can be fully adequate only if it is integrated into a Bayesian theory of reasoning.[34]

Notes

1. The criticisms are widespread, but the following are representative of the literature: Henry Kyburg, "Subjective Probability: Criticisms, Reflections and Problems," *Journal of Philosophical Logic* 7 (1978): 157-180; Isaac Levi, "Indeterminate Probabilities," *Journal of Philosophy* 71 (1974); 391-418; and Glenn Shafer, A *Mathematical Theory of Evidence* (Princeton: Princeton University Press, 1976).

2. Clark Glymour, *Theory and Evidence* (Princeton: Princeton University Press, 1980), hereafter referred to as *T & E*.

3. See I. J. Good, "46656 Varieties of Bayesians." *The American Statistician* 25, 5 (Dec. 1971): 62-63.

4. Following Kolmogorov's influential systematization, *Foundations of the Theory of Probability* (New York: Chelsea Publishing Co., 1950), most mathematical treatments of the theory of probability take probability functions to be defined over structured collections of sets, (σ-) rings or (σ-) fields, or over Boolean algebras. For obvious reasons philosophers of the Bayesian persuasian have often chosen to define probabilities over sentences in formal languages. I shall follow this practice. Because of the structural similarities among the different approaches, though, many of the theorems carry over from one domain to another, and in what follows, I shall not make use of the mathematically special features of probability functions defined on languages. Although I talk of probability functions defined on sentences rather than propositions or statements, no philosophical point is intended. Any of these objects would do as well.

5. For a fuller treatment of the coherence theorem, originally due to de Finetti, see Abner Shimony, "Coherence and the Axioms of Probability." *Journal of Symbolic Logic* 20 (1955); 1-28 or John Kemeny, "Fair Bets and Inductive Probabilities." *Journal of Symbolic Logic* 20 (1955): 263-273. The coherence theorem is not the only argument Bayesians appeal to to argue that degrees of belief ought to be probabilities. See, e.g., the arguments given in L. J. Savage, *The Foundations of Statistics* (New York: John Wiley and Dover, 1954 and 1972), chapter 3, and R. T. Cox, *The Algebra of Probable Inference* (Baltimore: Johns Hopkins University Press, 1961). However, the coherence argument is often cited and well accepted by Bayesians. Moreover, the coherence condition is closely connected with the requirement of logical omniscience, which will be one of the central foci of this paper.

6. For justifications of conditionalization, see Bruno de Finetti, *Theory of Probability* vol. 1 (New York: John Wiley, 1974), section 4.5; and Paul Teller, "Conditionalization, Observation, and Change of Preference," in W. Harper and C. Hooker, *Foundations and Philosophy of Epistemic Applications of Probability Theory* (Dordrecht: Reidel, 1976), pp. 205-259. Among other rational ways of changing one's beliefs I would include the extension of conditionalization proposed by Richard Jeffrey in chapter 11 of his *The Logic of Decision* (New York: McGraw-Hill, 1965), and the sorts of changes that one makes upon discovering an incoherence in one's beliefs. The former is appropriate when changing one's beliefs on the basis of uncertain observation, and the latter when one discovers, e.g., that one attributes probability .5 to heads on a given coin, yet attributes probability .25 to a run of three heads on the same coin. There may be other alternatives to conditionalization, but we shall not consider them here.

7. Both of these conceptions of Bayesianism are widespread. For a statement of the thought-police model, see L. J. Savage, *The Foundations of Statistics* (New York: John Wiley and Dover, 1954 and 1972), p. 57, and for a statement of the ideal learning machine model, see Rudolph Carnap, The Aim of Inductive Logic, in E. Nagel, P. Suppes, and A. Tarski, *Logic, Methodology and Philosophy of Science* (Stanford: Stanford University Press, 1962), pp. 303-318.

8. *T & E*, pp. 85-6.

9. The historical problem, as posed, appears to presuppose that evidence for an hypothesis must somehow serve to increase the scientist's degree of belief in that hypothesis. This *may* not hold for *everything* that we want to call evidence. Peter Achinstein argues that the evidence for an hypothesis may not only fail to raise the scientist's degree of belief in that hypothesis, but might actually *lower* it! See his "Concepts of Evidence," *Mind* 87 (1978); 22-45, and "On Evidence: A Reply to Bar-Hillel and Margalit," *Mind* 90 (1981): 108-112. But be that as it may, it seems clear to me that in the sorts of cases Glymour cites in this connection, we are dealing with circumstances in which considerations relating to the evidence *do* increase the scientist's degree of belief in his hypothesis. Whatever more general account of the notion of evidence we might want to adopt, there is an important question as to how the Bayesian can account for that. Closely related to what I call the historical problem of old

evidence, the question as to how old evidence can increase the scientist's degree of belief in a new hypothesis, is the question of how the Bayesian is to deal with the introduction of new theories at all. This is especially difficult for what I shall later call global Bayesianism, where the enterprise is to trace out the changes that would occur in an ideally rational individual's degrees of belief as he acquires more and more experience, and where it is assumed that the degree of belief function is defined over some maximally rich global language capable of expressing all possible evidence and hypotheses. Since I shall reject global Bayesianism, I won't speculate on how a global Bayesian might respond. I shall assume that at any time, a new hypothesis can be introduced into the collection of sentences over which S's degree of belief function is defined, and his previous degree of belief function extended to include that new hypothesis, as well as all truth-functional combinations of that hypothesis with elements already in the domain of S's beliefs. The new degrees of belief will, of course, reflect S's confidence in the new hypothesis. Although these new degrees of belief will be prior probabilities in the strictest sense of the term, they will not be without ground, so to speak, since they may be based on the relations that the new hypothesis is known to bear to past evidence, other hypotheses already considered, and so on.

10. I am indebted to Brian Skyrms for pointing out the ambiguity in Glymour's problem.

11. See *T & E*, p. 87-91 for a development of this line of argument, along with Glymour's criticisms.

12. The logical probabilist, like Carnap, does not have to go to counterfactual degrees of belief to solve the ahistorical problem of old evidence. Since a logical c-function is taken to measure the degree of logical overlap between its arguments, we can always appeal to the value of "c(h/e)" as a measure of the extent to which e confirms h, regardless of whether or not we, as a mater of fact, happen to believe that e. But, as far as I can see, the logical probabilist will be in no better shape than his subjectivist comrade is with respect to the *historical* problem of old evidence. Even for Carnap's logically perfect learning machine, once e has been acquired as evidence, it is difficult to see how it could be used to increase the degree of confirmation of a new hypothesis. I would like to thank James Hawthorne for this observation.

13. e confirms h in the relevance sense iff learning that e would increase S's confidence or degree of belief in h. On the relations among the various senses of confirmation and the importance of the notion of relevance, see Wesley Salmon, Confirmation and Relevance, in Maxwell and Anderson, eds., *Minnesota Studies in Philosophy of Science*, vol. 6 (Minneapolis: University of Minnesota Press, 1975).

14. These worries are eloquently pressed by Ian Hacking in "Slightly More Realistic Personal Probability," *Philosophy of Science* 34 (1967): 311-325. Much of my own solution to the problem of logical omniscience is very much in the spirit of Hacking's, although the details of our two accounts differ significantly.

15. The evolving probability model is suggested by I. J. Good in a number of places, though I know of no place where he develops it systematically. See, e.g., "Corroboration, Explanation, Evolving Probability, Simplicity and a Sharpened Razor," *British Journal for the Philosophy of Science* 19 (1968): 123-143, esp. 125, 129; "Explicativity, Corroboration, and the Relative Odds of Hypotheses", *Synthèse* 30 (1975): 39-73, esp. 46, 57; and "Dynamic Probability, Computer Chess, and the Measurement of Knowledge," *Machine Intelligence* 8 (1977), pp. 139-150. Good's preferred name for the position is now "dynamic probability." A similar position is expressed by Richard Jeffrey in a short note, "Why I am a Born-Again Bayesian," dated 5 Feb. 1978 and circulated through the Bayesian samizdat. To the best of my knowledge, the conditionalization model does not appear in the literature, although it is consistent with the sort of approach taken in I. Hacking, "Slightly More Realistic Personal Probability."

16. My own intuition is that in any *actual* case, the *way* we change our beliefs upon discovering that h ⊢ e will be determined by the strength of our prior belief that h ⊢ e, and that it is because the evolving probability model leaves out any considerations of these prior beliefs that it suffers from radical indeterminacy. This, it seems to me, is where the evolving probability model differs most clearly from the conditionalization model, which does, of course, take into acount the relevant prior beliefs as prior probabilities.

17. See, e.g., Hacking, "Slightly More Realistic Personal Probability," and I. J. Good, *Probability and the Weighing of Evidence* (London: C. Griffin, 1950).

18. Carnap, in The Aim of Inductive Logic, is an example of such an approach.

19. The local approach is by far the dominant one among practicing Bayesian statisticians and decision theorists, although it is often ignored by philosophers. One exception to this is Abner Shimony, who takes locality to be of central importance to his own version of the Bayesian program. See his Scientific Inference, in R. C. Colodny, ed., *The Nature and Function of Scientific Theories* (Pittsburgh: University of Pittsburgh Press, 1970), pp. 79-172, esp. pp. 99-101.

20. Glymour's bootstrap theory of confirmation will be discussed below in section 6.

21. What this formalism does *not* allow is the embedding of the turnstile. So sentences like "[A ⊢ B] ⊢ C" and "A ⊢ [B ⊢ C]" will not be well formed. An extension of the language to include such sentences may be needed if we want to talk about the confirmation of sentences of the form "A ⊢ B" and the problem of old evidence as it arises at that level.

22. Not everything of interest can be derived from (K*). The following interesting properties are *not* derivable from (K*) and the axioms of the probability calculus alone:

(a) P(A ⊢ B & A ⊢ C) = P(A ⊢ B & C)
(b) If A truth functionally entails B in L, P(A ⊢ B) = 1
(c) If A and B are truth functionally inconsistent in L, then P(A ⊢ B) = 0
(d) P(A ⊢ B v A ⊢ C) = P(A ⊢ B v C)
(e) P(A ⊢ B & B ⊢ C) ≤ P(A ⊢ C)

Later we shall discuss adding (b). But any of these properties could be added as additional constraints. The more constraints we add, however, the less freedom S has in assigning probabilities, and the closer we get to the specter of logical omniscience.

23. (K*) will be satisfied if "A ⊢ B" is interpreted as "A & B," say.

24. I would like to thank William Tait for pointing out a mistake in an earlier and stronger but, unfortunately, false version of (T2), and for suggesting the method of proof used here. On the existence of strictly positive probabilities, see, e.g., A. Horn and A. Tarski, "Measures in Boolean Algebras," *Transactions of the American Mathematical Society* 64 (1948): 467-497; or J. L. Kelley, "Measures on Boolean Algebras," *Pacific Journal of Mathematics* 9 (1959): 1165-1177. The proof of Theorem 2.5 in Horn and Tarski suggests a simple way of actually constructing an infinite number of different strictly positive probabilities on L, one corresponding to each countably infinite ordered set of numbers in (0, 1) that sum to 1. Consequently, there are an infinite number of probabilities on L* having the properties specified in (T2).

25. Although the recent literature on probability and conditionals, both indicative and subjunctive, is vast, something should be said about the relation between my results here and what others have done on conditionals. Two constraints on probabilities of conditionals have been toyed with in the literature, Stalnaker's thesis and Harper's constraint:

(C1) P(h → e) = P(e/h)
(C2) P(h → e) = 1 iff P(e/h) = 1

Unfortunately both constraints seem too strong, and lead to triviality results. David Lewis has shown that if (C1) is satisfied, then P can take on at most four different values. See his Probabilities of Conditionals and Conditional Probabilities. *Philosophical Review* 85 (1976): 297-315. Similarly, Stalnaker has shown that if (C2) is satisfied, then P(h → e) = P(h ⊃ e). See Letter by Robert Stalnaker to W. L. Harper, in Harper and Hooker, *Foundations and Philosophy*, pp. 113-115. Neither of these arguments has gone without challenge. See, e.g., Bas van Fraassen's answer to Lewis, Probabilities of Conditionals, in Harper and Hooker, *Foundations and Philosophy*, pp. 261-308 and Harper's answer to Stalnaker in Ramsey Test Conditionals and Iterated Belief Change, in Harper and Hooker, *Foundations and Philosophy*, pp. 117-135. But (C1) and (C2) are obviously strong conditions that introduce substantial complications. Luckily I don't have to worry about the complications or the triviality proofs. (C1) and (C2) fail in my formalism when "→" is replaced by " ⊢ ." Instead, I am committed only to the following more modest constraint:

(C3) If P(h ⊢ e) = 1 then P(e/h) = 1.

26. It would be unwise to adopt the slightly stronger constraint:

(**) If $P(A \supset B) = 1$, then $P(A \vdash B) = 1$.

(**) is certainly unnatural if " $\vdash$ " is interpreted as implication, since S could be certain of "$A \supset B$" because he was certain that A is false, say. Adopting (**) would also block our ability to use the formalism in the solution of the problem of old evidence, since (**) has the consequence that if $P(e) = 1$, then $P(h \vdash e) = 1$, no matter what h or e we are dealing with.

27. The same basic technique can be used to construct other probability functions of interest. If the conditions of the theorem are satisfied, and $\varepsilon = 0$, then $P(A)$, $P(B)$, and $P(A \vdash B)$ will all have the required values and $P(A/A \vdash B) = P(A)$. If ε is chosen to be in the interval $[-\delta', 0)$, where $\delta' = \min(rs, (1 - r)(1 - s))$, then $P(A/A \vdash B) < P(A)$.

28. It certainly will not be the case that *every* configuration of priors is such that the discovery that $h \vdash e$ will increase S's degree of belief that h. It can easily be shown that $P(h/h \vdash e) > P(h)$ if and only if $P(h \vdash e/h) > P(h \vdash e/{\sim}h)$. That is, the discovery that $h \vdash e$ will increase S's degree of belief in h if and only if S believes that it is more likely that h entails e if h is true than if it is false. (This has an obvious parallel in the case of e confirming h: the discovery that e confirms h if and only if e is more likely given h than it is given $\sim$h.) It is obvious that this condition will not always be satisfied. For example, when e is known to be false it is clear that $P(h \vdash e/h)$ ought to be 0. Even when $P(e) = 1$, one would not always expect $P(h \vdash e/h)$ to be greater than $P(h \vdash e/{\sim}h)$ (let h be an arbitrary hypothesis in biology and e be Kepler's laws). I have found it impossible to specify in any illuminating way a set of circumstances in which it is always reasonable to expect that $P(h \vdash e/h) > P(h \vdash e/{\sim}h)$.

29. In the discussion period following this paper when it was presented at the Minnesota Center for Philosophy of Science, Clark Glymour suggested that the historical facts of the Einstein case do indeed agree with my analysis.

30. *T & E*, pp. 92-3.

31. *T & E*, p. 122.

32. I am appealing here to the formulation of bootstrap confirmation that Glymour outlines in *T & E*, pp. 116-117

33. *T & E*, pp. 120-121. Glymour elsewhere discusses how his method can distinguish between the confirmation afforded to whole theories, i.e., *collections* of hypotheses. See *T & E* pp. 152-155, 182, 352-353. But nothing Glymour says there touches on the problem that concerns me here, so far as I can see.

34. For a very different attempt to combine the bootstrap idea with Bayesian probability, see a paper that Glymour wrote after publishing *T & E*, "Bootstraps and Probabilities," *Journal of Philosophy* 77 (1980): 691-699. In that essay, Glymour uses the tools of subjective probability directly in the explication of the relation, "e BS confirms h with respect to T," rather than considering the probability function defined over instances of that relation, itself defined independently of probabilistic notions. I am inclined to agree with Paul Horwich in thinking that "Glymour's proposal may reduce under pressure to a trivial modification of probabilistic confirmation theories" ("The Dispensability of Bootstrap Conditions," *Journal of Philosophy* 77 (1980): 699-702, esp. 700), and I am inclined to think that my way of combining bootstraps with probability yields a much richer and more palatable mixture than does Glymour's.

Bayesianism with a Human Face

What's a Bayesian?

Well, I'm one, for example. But not according to Clark Glymour (1980, pp. 68-69) and some other definers of Bayesianism and personalism, such as Ian Hacking (1967, p. 314) and Isaac Levi (1980, p. xiv). Thus it behooves me to give an explicit account of the species of Bayesianism I espouse (sections 1 and 2) before adding my bit (section 3, with lots of help from my friends) to Daniel Garber's treatment in this volume of the problem of new explanation of common knowledge: the so-called problem of old evidence.

With Clark Glymour, I take there to be identifiable canons of good thinking that get used on a large scale in scientific inquiry at its best; but unlike him, I take Bayesianism (what *I* call "Bayesianism") to do a splendid job of validating the valid ones and appropriately restricting the invalid ones among the commonly cited methodological rules. With Daniel Garber, I think that bootstrapping does well, too—when applied with a tact of which Bayesianism can give an account. But my aim here is to elaborate and defend Bayesianism (of a certain sort), not to attack bootstrapping. Perhaps the main novelty is the further rounding-out in section 3 (by John Etchemendy, David Lewis, Calvin Normore, and me) of Daniel Garber's treatment of what I have always seen as the really troubling one of Clark Glymour's strictures against Bayesianism. After that there is a coda (section 4) in which I try to display and explain how probability logic does so much more than truth-value logic.

1. Response to New Evidence

In Clark Glymour's book, you aren't a Bayesian unless you update your personal probabilities by conditioning (a.k.a. "conditionalization"), i.e., like this:

> As new evidence accumulates, the probability of a proposition changes according to Bayes' rule: the posterior probability of a hypothesis on the new evidence is equal to the prior conditional probability of the hypothesis on the evidence. (p. 69)

That's one way to use the term "Bayesian," but on that usage I'm no Bayesian. My sort of Bayesianism gets its name from another sense of the term "Bayes's rule," equally apt, but stemming from decision theory, not probability theory proper. Whereas Bayes's rule in Glymour's sense prescribes conditioning as the way to update personal probabilities, Bayes's rule in my sense prescribes what Wald (1950) called "Bayes solutions" to decision problems, i.e., solutions that maximize expected utility relative to some underlying probability assignment to the states of nature. (No Bayesian himself, Wald contributed to the credentials of decision-theoretic Bayesianism by proving that the Bayes solutions form a complete class.) The Reverend Thomas Bayes was both kinds of Bayesian. And of course, he was a third kind of Bayesian, too: a believer in a third sort of Bayes's rule, according to which the right probability function to start with is m* (as Carnap (1945) was to call it).

Why am I not a Bayesian in Glymour's sense? This question is best answered by way of another: What is the "new evidence" on which we are to condition? (Remember: the senses are not telegraph lines on which the external world sends observation sentences for us to condition upon.) Not just any proposition that newly has probability one will do, for there may well be many of these, relative to which conditioning will yield various posterior probability distributions when applied to the prior.

All right, then: what about the conjunction of all propositions that newly have probability one? That will be the total new evidence, won't it? Why not take the kinematical version of Bayes's rule to prescribe conditioning on that total?

I answer this question in Chapter 11 of my book (1965, 1983), and in a few other places (1968, 1970, 1975). In a nutshell, the answer is that much of the time we are unable to formulate any sentence upon which we are prepared to condition, and in particular, the conjunction of all the sentences that newly have probability one will be found to leave too much out for it to serve as the Archimedean point about which we can move our probabilities in a satisfactory way. Some of the cases in which conditioning won't do are characterized by Ramsey (1931, "Truth and Probability," end of section 5) as follows:

> I think I perceive or remember something but am not sure; this would seem to give me some ground for believing it, contrary to Mr. Keynes' theory, by which the degree of belief in it which it would be rational for me to have is that given by the probability relation between the

proposition in question and the things I know for certain.

Another sort of example is suggested by Diaconis and Zabell (1982): a record of someone reading Shakespeare is about to be played. Since you are sure that the reader is either Olivier or Gielgud, but uncertain which, your prior probabilities for the two hypotheses are nearly equal. But now comes fresh evidence, i.e., the sound of the reader's voice when the record is played. As soon as your hear that, you are pretty sure it's Gielgud, and the prior value ca. .5 is replaced by a posterior value ca. .9, say. But, although it was definite features of what you heard that rightly made you think it very likely to have been Gielgud, you cannot describe those features in an observation sentence in which you now have full belief, nor would you be able to recognize such a sentence (immensely long) if someone else were to produce it.

Perhaps it is the fact that there surely is definite evidence that prompts and justifies the probability shift in the Olivier/Gielgud case, that makes some people think there must be an evidence *sentence* (observation sentence) that will yield the new belief function via conditionalization. Surely it is all to the good to be able to say just what it was about what you heard, that made you pretty sure it was Gielgud. But few would be able to do that; nor is such inability a mark of irrationality; nor need one be able to do that in order to count as having had good reason to be pretty sure it was Gielgud. The Olivier/Gielgud case is typical of our most familiar sorts of updating, as when we recognize friends' faces or voices or handwritings pretty surely, and when we recognize familiar foods pretty surely by their look, smell, taste, feel, and heft.

Of course conditioning is sometimes appropriate. When? I mean, if your old and new belief functions are p and q, respectively, when is q of form p_E for some E to which p assigns a positive value? (Definition: $p_E(H)$ is the conditional probability of H on E, i.e., $p(H/E)$, i.e., $p(HE)/p(E)$.)

Here is an answer to the question:

(C) If $p(E)$ and $q(E)$ are both positive, then the conditions (a) $q_E = p_E$ and (b) $q(E) = 1$ are jointly necessary and sufficient for (c) $q = p_E$.

You can prove that assertion on the back of an envelope, via the Kolmogorov axioms and the defition of conditional probability. Here is a rough-and-ready verbal summary of (C):

Conditioning is the right way to update your probability judgments iff the proposition conditioned upon is not only (b) one you now fully believe, but is also (a) one whose relevance to each proposition is unchanged by the updating.

The point of view is one in which we take as given the old and new probability functions, p and q, and then ask whether the condition (c) $q = p_E$ is consistent with static coherence, i.e., the Kolmogorov axioms together with the definition of conditional probability applied to p and q separately. In (C), (a) is the ghost of the defunct condition of total evidence.

In the Olivier/Gielgud example, and others of that ilk, fresh evidence justifies a change from p to q even though $q \neq p_E$ for all E in the domain of p. What is the change, and when is it justified? Here is the answer, which you can verify on the back of the same envelope you used for (C):

(K) If "E" ranges over some partitioning of a proposition of p-measure 1 into propositions of positive p-measure, then the ("rigidity") condition
(r) $q_E = p_E$ for all E in the partitioning
is necessary and sufficient for q to be related to p by the following ("kinematical") formula:
(k) $q = \Sigma_E q(E)p_E$.

There is no more question of justifying (k) in (K) than there was of justifying (c) in (C): neither is always right. But just as (C) gives necessary and sufficient conditions (a) and (b) for (c) to be right, so (K) gives (r), i.e., the holding of (a) for each E in the partitioning, as necessary and sufficient for (k) to be correct—where in each case, correctness is just a matter of static coherence of p and q separately. We know when (k) is right:

The kinematical scheme (k) yields the correct updating iff the relevance of each member E of the partitioning to each proposition H is the same after the updating as it was before.

It is an important discovery (see May and Harper 1976; Williams 1980; Diaconis and Zabell 1982) that in one or another sense of "close," (k) yields a measure q that is closest to p among those that satisfy the rigidity condition (r) and assign the new probabilities q(E) to the Es, and that (c) yields a measure that is closest to p among those that satisfy the conditions (a) and (b) in (C). But what we thereby discover is that (so far, anyway) we have adequate concepts of closeness: we already knew that (k) was equivalent to (r), and that (c) was equivalent to (a) and (b) in (C). This is not to deny the interest of such minimum-change principles, but rather to emphasize that their importance lies not in their serving to justify (c) and (k)—for they don't—but in the further kinematical principles they suggest in cases where (k) holds for no interesting partitioning. To repeat: (c) and (k) are justified by considerations of mere coherence, where their proper conditions of applicability are met, i.e., (a) and (b) for (c), and (r) for (k). And where those conditions fail, the corresponding rules are unjustifiable.

Observe that in a purely formal sense, condition (r) is very weak; e.g., it holds whenever the Boolean algebra on which p and q are defined has atoms whose p-values sum to 1. (Proof: with "E" in (r) ranging over the atoms that have positive p-measure, $p(H/E)$ and $q(H/E)$ will both be 1 or both be 0, depending on whether E implies H or –H.) Then in particular, (k) is always applicable in a finite probability space, formally. But if (k) is to be useful to a human probability assessor, the E partitioning must be coarser than the atomistic one. To use the atomistic partitioning is simply to start over from scratch.

The Olivier/Gielgud example is one in which the partitioning is quite manageable: {O, G}, say, with O as the proposition that the reader is Olivier, and G for Gielgud. The hypothesis H that the reader (whoever he may be) loved Vivien Leigh serves to illustrate the rigidity conditions. Applied to H, (r) yields

$$q(H/O) \; = \; p(H/O), \quad q(H/G) \; = \; p(H/G).$$

Presumably these conditions both hold: before hearing the reader's voice you attributed certain subjective probabilities to Olivier's having loved Leigh (high), and to Gielgud's having done so (low). Nothing in what you heard tended to change those judgments: your judgment about H changed only incidentally to the change in your judgment about O and G. Thus, by (k),

$$q(H) \; = \; q(O)p(H/O) \; + \; q(G)p(H/G).$$

$q(H)$ is low because it is a weighted average of $p(H/O)$, which was high, and $p(H/G)$, which was low, with the low value getting the lion's share of the weight: $q(G) \; = \; .9$.

2. Representation of Belief

In Clark Glymour's book, Bayesianism is identical with personalism, and requires not only updating by conditioning, but also a certain superhuman completeness:

> There is a class of sentences that express all hypotheses and all actual
> or possible evidence of interest; the class is closed under Boolean
> operations. For each ideally rational agent, there is a function defined
> on all sentences such that, under the relation of logical equivalence,
> the function is a probability measure on the collection of equivalence
> classes. (pp. 68-69)

The thought is that Bayesian personalism must represent one's state of
belief at any time by a definite probability measure on some rather rich
language. And indeed the two most prominent personalists seem to
espouse just that doctrine: de Finetti (1937) was at pains to deny the very
meaningfulness of the notion of unknown probabilities, and Savage (1954)
presented an axiomatization of preference according to which the agent's
beliefs must be represented by a unique probability.

But de Finetti was far from saying that personal probabilities cannot fail
to *exist*. (It is a separate question, whether one can be unaware of one's
existent partial beliefs. I don't see why not. See Mellor (1980) and Skyrms
(1980) for extensive discussions of the matter.) And Savage was far from
regarding his 1954 axiomatization as the last word on the matter. In
particular, he viewed as a live alternative the system of Bolker (1965) and
Jeffrey (1965), in which even a (humanly unattainable) complete preference ranking of the propositions expressible in a rich language normally
determines no unique probability function, but rather an infinite set of
them. The various members of the set will assign various values throughout
intervals of positive length to propositions about which the agent is not
indifferent: see Jeffrey (1965, section 6.6) for details.

Surely the Bolker-Jeffrey system is not the last word, either. But it does
give one clear version of Bayesianism in which belief states—even
superhumanly definite ones—are naturally identified with infinite sets of
probability functions, so that degrees of belief in particular propositions
will normally be determined only up to an appropriate quantization, i.e.,
they will be interval-valued (so to speak). Put it in terms of the thesis of *the
primacy of practical reason*, i.e., a certain sort of pragmatism, according to
which belief states that correspond to identical preference rankings of
propositions are in fact one and the same. (I do not insist on that thesis, but
I suggest that it is an intelligible one, and a clearly Bayesian one; e.g., it
conforms to Frank Ramsey's (1931) dictum (in "Truth and Probability,"
section 3): "the kind of measurement of belief with which probability is
concerned . . . is a measurement of belief *qua* basis of action.") Applied to

the Bolker-Jeffrey theory of preference, the thesis of the primacy of practical reason yields the characterization of belief states as sets of probability functions (Jeffrey 1965, section 6.6). Isaac Levi (1974) adopts what looks to me like the same characterization, but labels it "un-Bayesian."

But of course I do not take belief states to be determined by full preference rankings of rich Boolean algebras of propositions, for our actual preference rankings are fragmentary, i.e., they are rankings of various subsets of the full algebras. Then even if my theory were like Savage's in that full rankings of whole algebras always determine unique probability functions, the actual, partial rankings that characterize real people would determine belief states that are infinite sets of probability functions on the full algebras. Here is the sort of thing I have in mind, where higher means better:

A, B	C
	D
W	W
	–C
–A, –B	–D
(1)	(2)

This is a miniature model of the situation in which the full Boolean algebra is infinite. Here the full algebra may be thought of as consisting of the propositions A, B, C, D and their truth-functional compounds. W is the necessary proposition, i.e., $W = Av - A = Cv - C$, etc. Here is a case in which the agent is indifferent between A and B, which he prefers to W, which in turn he prefers to –A and to –B, between which he is indifferent. But he has no idea where AB, Av – B, etc. come in this ranking: his preferences about them remain indeterminate. That is what ranking (1) tells us. And ranking (2) gives similar sorts of information about C, D, and their denials: the agent's preferences regarding CD, Cv – D, etc. are also indeterminate. But the two rankings are related only by their common member, W. Thus C and D are preferred to –A and –B, but there is no information given about preference between (say) A and C.

That is the sort of thing that can happen. According to (1), we must have $p(A) = p(B)$ for any probability function p in the belief state determined by preferences (1) and (2): see Example 3 in chapter 7 of Jeffrey (1965). And according to (2), we must have $p(C) < p(D)$ for any such p: see problem 1 in section 7.7. Then the belief state that corresponds to this mini-ranking (or this pair of connecting mini-rankings) would correspond to the set {p:p(A) = p(B) and p(C) < p(D)}.

The role of definite probability measures in probability logic as I see it is the same as the role of maximal consistent sets of sentences in deductive logic. Where deductive logic is applied to belief states conceived unprobabilistically as *holdings true* of sets of sentences, maximal consistent sets of sentences play the role of unattainable completions of consistent human belief states. The relevant fact in deductive logic is

> *Lindenbaum's Lemma:* a truth value assignment to a set of sentences is consistent iff consistently extendible to the full set of sentences of the language.

(There is a one-to-one correspondence between consistent truth-value assignments to the full set of sentences of the language and maximal consistent sets of sentences of the language: the truth value assigned is t or f depending on whether the sentence is or is not a member of the maximal consistent set.) The corresponding fact about probability logic is what one might call

> *De Finetti's Lemma:* an assignment of real numbers to a set of sentences is coherent (= immune to Dutch books) iff extendible to a probability function on the full set of sentences of the language.

(See de Finetti 1972, section 5.9; 1974, section 3.10.)

It is a mistake to suppose that someone who assigns definite probabilities to A and to B (say, .3 and .8 respectively) is thereby committed in Bayesian eyes to some definite probability assignment to the conjunction AB, if de Finetti (1975, (2) on p. 343, and pp. 368-370) is to be counted as a Bayesian. On the other hand, probability logic in the form of the Kolmogorov axioms, say, requires that any assignment to that conjunction lie in the interval from .1 to .3 if the assignments p(A) = .3 and p(B) = .8 are to be maintained: see Boole (1854, Chapter 19), Hailperin (1965), or Figure 1 here. Thus probability logic requires that one or both of the latter assignments be abandoned in case it is discovered that A and B are logically incompatible, since then p(AB) = 0 < .1.

Clearly indeterminacies need not arise as that of p(AB) did in the

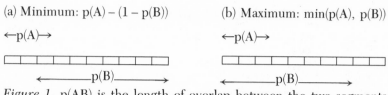

Figure 1. p(AB) is the length of overlap between the two segments.

foregoing example, i.e., out of an underlying determinate assignment to the separate components of the conjunction. See Williams (1976) for an extension of de Finetti's lemma to the case where the initial assignment of real numbers $p(S_i) = r_i$ is replaced by a set of conditions of form $r_i \leq p(S_i) \leq s_i$. And note that indeterminacies need not be defined by such inequalities as these. They might equally well be defined by conditions (perhaps inequalities) on the mathematical expectations $E(X_i)$ of random variables—conditions that impose conditions on the underlying probability measures p via the relation $E(X_i) = \int_W X_i \, dp$. More complex special cases arise when "X_i" is replaced by "$(X_i - EX_i)^2$", etc., so that belief states are defined by conditions on the variances etc. of random variables.

Such definitions might be thought of as generalizing the old identification of an all-or-none belief state with the proposition believed. For propositions can be identified with sets of two-valued probability measures. Each such measure, in which the two values must be 0 and 1, can be identified with the possible world in which the true statements are the ones of probability 1. Then a set of such measures works like a set of possible worlds: a proposition. Now Levi and I take belief states to be sets of probability measures, omitting the requirement that they be two-valued. Call such sets "probasitions." The necessary probasition is the set P of all probability measures on the big Boolean algebra in question. P is the *logical space* of probability logic. My current belief state is to be represented by a probasition: a region R in this space. If I now condition upon a proposition E, my belief state changes from R to

$$R/E = {}_{Df} \{p_E : p \; \varepsilon \; R \text{ and } p(E) \neq 0\}.$$

Perhaps R/E is a proper subset of R, and perhaps it is disjoint from R, but for the most part one would expect the change from R to R/E to represent a new belief state that *partly* overlaps the old one. And in some cases one would expect the operation of conditioning to *shrink* probasitions, e.g., perhaps in the sense that the diameter of R/E is less than that of R when diameter is defined

$$\text{diam}(R) = \sup_{p,q \, \epsilon \, R} \| p - q \|$$

and the norm $\| p - q \|$ is defined

$$\| p - q \| = \sup_A | p(A) - q(A) |$$

where "A" ranges over all propositions in the Boolean algebra.

That is how I would ride that hobby-horse. But I would also concede I. J. Good's (1952, 1962) point, that probasitions are only rough characterizations of belief states, in which boundaries are drawn with artificial sharpness, and variations in the acceptability of different members of probasitions and in the unacceptability of various nonmembers go unmarked. In place of probasitions, one might represent belief states by probability measures μ on suitable Boolean algebras of subsets of the space P. Good himself rejects that move because he thinks that μ would then be equivalent to some point μ^* in P, i.e., the point that assigns to each proposition A the definite probability

$$\mu^*(A) = \int_{p \epsilon P} p(A) \, d\mu(p).$$

In our current terminology, the thought is that a definite probability measure μ on P must correspond to a sharp belief state, viz., the probasition $\{\mu^*\}$. To avoid this reduction, Good proposes that μ be replaced by a nondegenerate probasition of type 2, i.e., a nonunit set of probability measures on P; that in principle, anyway, that probasition of type 2 be replaced by a nondegenerate probasition of type 3; and so on. "It may be objected that the higher the type the woolier the probabilities. It will be found, however, that the higher the type the less wooliness matters, provided the calculations do not become too complicated." (Good 1952, p. 114)

But I do not see the need for all that. It strikes me that here, Good is being misled by a false analogy with de Finetti's way of avoiding talk of unknown probabilities (i.e., the easy converse of his representation theorem for symmetric probability functions). De Finetti's point was that where objectivists would speak of (say) coin-tossing as a binominal process with unknown probability of success on each toss, and might allow that their subjective probability distribution for the unknown objective probability x of success is uniform throughout the unit interval, an uncompromising subjectivist can simply have as his belief function the subjectively weighted average of the various putatively objective possibilities, so that, e.g., his subjective probability for heads on the first n tosses would be

$$p(H_1H_2 \ldots H_n) = \int_0^1 x^n dx = \frac{1}{n + 1}.$$

In the analogy that Good is drawing, the probasition R is the set all binomial probability functions p_x where $p_x(H_i) = x$, and μ is the probability measure on P that assigns measure 1 to R and assigns measure $b - a$ to any nonempty subset $\{p_x : a \leq x < b\}$ of R. But whereas for de Finetti the members of R play the role of (to him, unintelligible) hypotheses about what the objective probability function might be, for Good the members of R play the role of hypotheses about what might be satisfactory as a subjective probability function. But if only the members of R are candidates for the role of satisfactory belief function, their subjectively weighted average, i.e., p as above, is not a candidate. (*That* p is not binomial: $p(H_1/H_2) = 2/3 \neq p(H_1) = 1/2$, whereas for each p_x in R, $p_x(H_1/H_2) = p(H_1) = x$.)

The point is that the normalized measure μ over P is not being used as a subjective probability distribution that indicates one's degrees of belief in such propositions as that the true value of x lies between .1 and .3. On the contrary, the uniformity of the μ distribution within R is meant to indicate that one would be indifferent between having to behave in accordance with p_x and having to behave in accordance with p_y for any x and y in the unit interval (where such behavior is determined as well by his utility function); and the fact that $\mu(R) = 1$ is meant to indicate that one would prefer having to behave in accordance with any member of R to having to behave in accordance with any member of $P - R$. (These are rough characterizations because μ assigns measure 0 to each unit subset of P. A precise formulation would have to talk about having to behave in accordance with randomly selected members of intervals, $\{p_x : a \leq x \leq b\}$.)

Then I think Good's apprehension unfounded: I think one can replace probasitional belief states R by probability distributions μ over P that assign most of their mass to R, without thereby committing oneself to a belief state that is in effect a singleton probasition, $\{\mu^*\}$. But this is not to say that one must always have a sharp probability distribution over P: perhaps Good's probasitions of types 2 and higher are needed in order to do justice to the complexities of our belief states.

On the other hand, I think that in practice, even the relatively simple transition from probasitional belief states to belief states that are sharp probability measures on P is an idle complexity: the probasitional representation suffices, anyway, for the applications of probability logic that are considered in the remainder of this paper.

An important class of such examples is treated in Chapter 2 of de Finetti (1937), i.e., applications of what I shall call

> *de Finetti's Law of Small Numbers:* the estimated number of truths among the propositions $A_1, \ldots, A_n$ must equal the sum of their probabilities.

That follows from the additivity of the expectation operator and the fact that the probability you attribute to A is always equal to your estimate of the number of truths in the set {A}: as de Finetti insists, the thing is as trivial as Bayes's theorem. (He scrupulously avoids applying any such grand term as "law" to it.) Dividing both sides of the equation by n, the law of small numbers takes this form:

> The estimated relative frequency of truths among the propositions is the average $(p(A_1) + \ldots + p(A_n))/n$ of their probabilities.

Suppose, then, that you regard the A's as equiprobable but have no view about what their common probability is (i.e., you have no definite degree of belief in the A's), and suppose that tomorrow you expect to learn the relative frequency of truths among them, without learning anything that will disturb your sense of their equiprobability. Thus you might represent your belief state tomorrow by the probasition $\{p:p(A_1) = \ldots = p(A_n)\}$, or by a measure on P that assigns a value near 1 to that probasition. But what's the point? If you don't need to do anything on which tomorrow's belief state bears until tomorrow, you may as well wait until you learn the relative frequency of truths among the A's, say, r. At that point, your estimate of the relative frequency of truths will be r (with variance o), and by mere coherence your degree of belief in each of the A's will also be r. You know all that today.

Note that in the law of small numbers, the A's need not be independent, or exchangeable, or even distinct! The "law" is quite general: as general and as trivial as Bayes's theorem, and as useful.

A mistake that is easy to make about subjectivism is that anything goes, according to that doctrine: any weird belief function will do, as long as it is coherent.

The corresponding mistake about dress would go like this: any weird getup will do, if there are no sumptuary laws, or other laws prohibiting inappropriate dress. That's wrong, because in the absence of legislation about the matter, people will generally dress *as they see fit*, i.e., largely in a manner that they think appropriate to the occasion and comfortable for them on that occasion. The fact that it is legal to wear chain mail in city buses has not filled them with clanking multitudes.

Then have no fear: the fact that subjectivism does not prohibit people from having two-valued belief functions cannot be expected to produce excessive opinionation in people who are not so inclined, any more than the fact that belief functions of high entropy are equally allowable need be expected to have just the opposite effect. For the most part we make the judgments we make because it would be unthinkable not to. Example: the foregoing application of de Finetti's law of small numbers, which explains to the Bayesian why knowledge of frequencies can have such powerful effects on our belief states.

The other side of the coin is that we generally suspend judgment when it is eminently thinkable to do so. For example, if I expect to learn the frequency tomorrow, and I have no need for probabilistic belief about the A's today, then I am not likely to spend my time on the pointless project of eliciting my current degrees of belief in the A's. The thought is that we humans are not capable of adopting opinions gratuitously, even if we cared to do so: we are generally at pains to come to opinions that strike us as right, or reasonable for us to have under the circumstances. The laws of probability logic are not designed to prevent people from yielding to luscious doxastic temptations—running riot through the truth values. They *are* designed to help us explore the ramifications of various actual and potential states of belief—our own or other people's, now or in the past or the future. And they are meant to provide a Bayesian basis for methodology. Let us now turn to that—focussing especially on the problem ("of old evidence") that Clark Glymour (1980, Chapter 3) identifies as a great Bayesian sore point.

3. The Problem of New Explanation

Probability logic is typically used to reason in terms of partially specified probability measures meant to represent states of opinion that it would be fairly reasonable for people to be in, who have the sort of information we take ourselves to have, i.e., we who are trying to decide how to proceed in some practical or (as here) theoretical inquiry. Reasonableness is assessed by us, the inquirers, so that what none of us is inclined to believe can be summarily ruled out, but wherever there is a real issue between two of us, or whenever one of us is of two minds, both sides are ruled reasonable. Of course, if our opinions are too richly varied, we shall get nowhere; but such radical incommensurability is less common in real inquiry, even in revolutionary times, than romantics would have us think.

It is natural to speak of "the unknown" probability measure p that it would be reasonable for us to have. This is just a substitute for more intelligible speech in terms of a variable "p" that ranges over the probasition (dimly specified, no doubt) comprising the probability measures that we count as reasonable. Suppose now that in the light of evidence that has come to our attention, we agree that p should be modified in a certain way: replaced by another probability measure, p'. If $p'(H)$ exceeds $p(H)$ for each allowable value of "p," we regard the evidence as supporting or confirming H, or as positive for H. The degree of support or confirmation is

$$p'(H) - p(H)$$

In the simplest cases, where $p' = p_E$, this amounts to

$$p(H/E) - p(H),$$

and in somewhat more complicated cases, where p' comes from p by kinematics relative to the partitioning $\{E_1, \ldots, E_n\}$, it amounts to

$$\Sigma_i p(H/E_i)\, (p'(E_i) - p(E_i)).$$

But what if the evidence is the fresh demonstration that H implies some known fact E? In his contribution to this volume, Daniel Garber shows that—contrary to what one might have thought—it is not out of the question to represent the effect of such evidence in the simplest way, i.e., by conditioning on the proposition $H \vdash E$ that H implies E, so that H's implying E supports H if and only if

(1) $p(H/H \vdash E) > p(H).$

And as he points out in his footnote 28, this inequality is equivalent to either of the following two (by Bayes' theorem, etc.):

(2) $p(H \vdash E/H) > p(H \vdash E)$

(3) $p(H \vdash E/H) > p(H \vdash E/ -H)$

This equivalence can be put in words as follows:

(I) A hypothesis is supported by its ability to explain facts in its *explanatory domain*, i.e., facts that it was antecedently thought likelier to be able to explain if true than if false.

(This idea was suggested by Garber some years ago, and got more play in an early version of his paper than in the one in this volume.) This makes sense intuitively. Example: Newton saw the tidal phenomena as the sorts of things that ought to be explicable in terms of the hypothesis H of universal gravitation (with his laws of motion and suitable background data) if H was true, but quite probably not if H was false. That is why explanation of those phenomena by H was counted as support for H. On the other hand, a purported theory of acupuncture that implies the true value of the gravitational red shift would be undermined thereby: its implying *that* is likely testimony to its implying everything, i.e., to its inconsistency.

But something is missing here, namely the supportive effect of belief in E. Nothing in the equivalence of (1) with (2) and (3) depends on the supposition that E is a "known fact," or on the supposition that $p(E)$ is 1, or close to 1. It is such suppositions that make it appropriate to speak of "explanation" of E by H instead of mere implication of E by H. And it is exactly here that the peculiar problem arises, of old knowledge newly explained. As E is common knowledge, its probability for all of us is 1, or close to it, and therefore the probability of H cannot be increased much by conditioning on E before conditioning on $H \vdash E$ (see 4a)—or after (see 4b), unless somehow the information that $H \vdash E$ robs E of its status as "knowledge."

(4a) $p(H/E) \approx p(H)$ if $p(E) \approx 1$

(4b) $p_{H \vdash E}(H/E) \approx p_{H \vdash E}(H)$ if $p_{H \vdash E}(E) \approx 1$

As (4b) is what (4a) becomes when "p" is replaced by "$p_{H \vdash E}$" throughout, we shall have proved (4b) as soon as we have proved (4a) for arbitrary probability functions p. Observe that (4b) comes to the same thing as this:

$$p(H/E \ \& \ H \vdash E) \approx p(H/H \vdash E) \text{ if } p(E/H \vdash E) \approx 1.$$

There and in (4), statements of form $x \approx y$ are to be interpreted as saying that x and y differ by less than a ("small") unspecified positive quantity, say ε.

Proof of (4a). The claim is that for all positive ε,

if $p(-E) < \varepsilon$ then $-\varepsilon < p(H/E) - p(H) < \varepsilon$.

To prove the "$< \varepsilon$" part of the consequent, observe that

$p(H/E) - p(H) \leq p(H/E) - p(HE)$ since $p(HE) \leq p(H)$

$$= \frac{p(HE)}{p(E)} - p(HE) = \frac{p(HE)p(-E)}{p(E)}$$

$$\leq p(-E) \text{ since } p(HE) \leq p(E).$$

Then $p(H/E) - p(H) < \varepsilon$. To prove the "$-\varepsilon <$" part, note that it is true iff

$$p(H) - \frac{p(HE)}{p(E)} < \varepsilon, \text{ i.e., iff}$$

$$p(H - E) + (\frac{p(HE)}{1} - \frac{p(HE)}{p(E)}) < \varepsilon, \text{ i.e., iff}$$

$$\frac{p(HE)}{p(E)}(p(E) - 1) < \varepsilon - p(H - E)$$

where the left-hand side is 0 or negative since $p(E) \leq 1$, and where the right-hand side is positive since $p(H-E) \leq p(-E) < \varepsilon$. Then the "$-\varepsilon <$" part of the consequent is also proved.

Yet, in spite of (4), where E reports the facts about the tides that Newton explained, it seems correct to say that his explanation gave them the status of evidence supporting his explanatory hypotheses, H—a status they are not deprived of by the very fact of being antecedently known.

But what does it mean to say that Newton regarded H as the sort of hypothesis that, if true, ought to imply the truth about the tides? I conjecture that Newton thought his theory ought to explain the truth about the tides, *whatever that might be*. I mean that I doubt whether Newton knew such facts as these (explained in *The System of the World*) at the time he formulated his theory:

> [39.] The tide is greatest in the syzygies of the luminaries and least in their quadratures, and at the third hour after the moon reaches the meridian; outside of the syzygies and quadratures the tide deviates somewhat from that third hour towards the third hour after the solar culmination.

Rather, I suppose he hoped to be able to show that

(T) H implies the true member of $\mathscr{E}$

where H was his theory (together with auxiliary data) and $\mathscr{E}$ was a set of mutually exclusive propositions, the members of which make various claims about the tides, and one of which is true. I don't mean that he was able to specify $\mathscr{E}$ by writing out sentences that express its various members. Still less do I mean that he was able to identify the true member of $\mathscr{E}$ by way of such a sentence, to begin with. But he knew where to go to find people who could do that to his satisfaction: people who could assure him of such facts as [39.] above, and the others that he explains at the end of his *Principia* and in *The System of the World*. Thus you can believe T (or doubt T, or hope that T, etc.) without having any views about which member of $\mathscr{E}$ is the true one, and, indeed, without being able to give an account of the makeup of $\mathscr{E}$ of the sort you would need in order to start trying to deduce members of $\mathscr{E}$ from H. (Nor do I suppose it was clear, to begin with, what auxiliary hypotheses would be needed as conjuncts of H to make that possible, until the true member of $\mathscr{E}$ was identified.)

David Lewis points out that in these terms, Garber's equivalence between (1) and (2) gives way to this:

(5)   p(H/T) > p(H) iff p(T/H) > p(T).

Lewis's thought is that someone in the position I take Newton to have been in, i.e., setting out to see whether T is true, is in a position of being pretty sure that

(S)   H implies *some* member of $\mathscr{E}$

without knowing which, and without being sure or pretty sure that (T) *the member of $\mathscr{E}$ that H implies is the true one*. But in exactly these circumstances, one will take truth of T to support H. Here I put it weakly (with "sure" instead of "pretty sure," to make it easy to prove):

(II) If you are sure that H implies *some* member of $\mathscr{E}$, then you take H to be supported by implying the *true* member of $\mathscr{E}$ unless you were already sure it did.

Proof. The claim is that

if p(S) = 1 ≠ p(T) then p(H/T) > p(H).

Now if p(S) = 1 then p(S/H) = 1 and therefore p(T/H) = 1 since if H is true it cannot imply any falsehoods. Thus, if 1 ≠ p(T), i.e., if 1 > p(T), then p(T/H) > p(T), and the claim follows via (5).

Notice one way in which you could be sure that H implies the true member of $\mathscr{E}$: you could have known which member that was, and cooked H up to imply it, e.g., by setting H = EG where E is the true member of $\mathscr{E}$

and G is some hypothesis you hope to make look good by association with a known truth.

Now (II) is fine as far as it goes, but (John Etchemendy points out) it fails to bear on the case in which it comes as a surprise that H implies *anything* about (say) the tides. The requirement in (II) that p(S) be 1 is not then satisfied, but H may still be supported by implying the true member of $\mathscr{E}$. It needn't be, as the acupuncture example shows, but it may be. For example, if Newton had not realized that H ought to imply the truth about the tides, but had stumbled on the fact that H ⊢ E where E was in $\mathscr{E}$ and known to be true, then H would have been supported by its ability to explain E.

Etchemendy's idea involves the propositions S, T, and

(F)   H implies some false member of $\mathscr{E}$.

Evidently F = S – T, so that –F is the material conditional, –F = –SvT ("If H implies any member of $\mathscr{E}$ then it implies the true one"), and so the condition p(F) = 0 indicates full belief in that conditional. Etchemendy points out that Lewis's conditions in (II) can be weakened to p(F) ≠ 0 and p(HS) = p(H)p(S); i.e., you are not antecedently sure that H implies nothing false about X (about the tides, say), and you take truth of H to be independent of implying anything about X. Now Calvin Normore points out that Etchemendy's second condition can be weakened by replacing "=" by "≥", so that it becomes: your confidence in H would not be weakened by discovering that it implies something about X. Then the explanation theorem takes the following form:

(III) Unless you are antecedently sure that H implies nothing false about X, you will regard H as supported by implying the truth about X if learning that H implies something about X would not make you more doubtful of H.

The proof uses Garber's principle

(K*)   p(A & A ⊢ B) = p(A & B & A ⊢ B).

This principle will hold if "⊢" represents (say) truth-functional entailment and if the person whose belief function is p is alive to the validity of *modus ponens*; but it will also hold under other readings of "⊢," as Garber points out. Thus it will also hold if A ⊢ B means that p(A – B) = 0, on any adequate interpretation of probabilities of probabilities. The proof also uses the following clarifications of the definitions of T and S:

(T)   For some E, E ε 𝒞 and H ⊢ E and E is true
(S)   For some E, E ε 𝒞 and H ⊢ E.

Proof of (III). The claim is this:

If p(S − T) ≠ 0 and p(HS) ≥ p(H)p(S) then p(HT) > p(H)p(T).

By (K*), p(HS) = p(HT), so that the second conjunct becomes p(HT) ≥ p(H)p(S). With the first conjunct, that implies p(HT) > p(H)p(T) because (since T implies S) p(S − T) ≠ 0 implies p(S) > p(T).

Note that (III) implies (II), for they have the same conclusion, and the hypotheses of (II) imply those of (III):

(6)   If p(S) = 1 ≠ p(T) then p(S − T) ≠ 0 and p(HS) ≥ p(H)p(S)

Proof: p(S − T) ≠ 0 follows from p(S) = 1 ≠ p(T) since T implies S, and p(S) = 1 implies that p(HS) = p(H) = p(H)p(S).

The explanation theorem (III) goes part way toward addressing the original question, "How are we to explain the supportive effect of belief in E, over and above belief in H ⊢ E, where H is a hypothesis initially thought especially likely to imply E if true?" Here is a way of getting a bit closer:

(IV)   Unless you are antecedently sure that H implies nothing false about X, you take H to be supported more strongly by implying the truth about X than by simply implying *something* about X.

Proof: the claim is that

if p(S − T) ≠ 0 then p(H/T) > p(H/S),

i.e., since T implies S, that

if p(S) > p(T) then p(HT)p(S) > p(HS)p(T),

i.e., by (K*), that

if p(S) > p(T) then p(HT)p(S) > p(HT)p(T).

But the original question was addressed to belief in a particular member E of 𝒞: a particular truth about X, identified (say) by writing out a sentence that expresses it. The remaining gap is easy to close (as David Lewis points out), e.g., as follows.

(7)   For any E, if you are sure that E is about X, implied by H, and true, then you are sure that T is true.

Proof. The claim has the form

For any E, if p(Φ) = 1 then p(for some E, Φ) = 1 where Φ is this:

E ε $\mathscr{E}$ and H $\vdash$ E and E is true.

Now the claim follows from this law of the probability calculus

p(X) $\leq$ p(Y) if X implies Y

in view of the fact that Φ implies its existential generalization.

Here is an application of (III):

> Since Newton was not antecedently sure that H implied no falsehoods about the tides, and since its implying anything about the tides would not have made it more doubtful in his eyes, he took it to be supported by implying the truth about the tides.

And here is a corresponding application of (7):

> Newton came to believe that H implied the truth about the tides when he came to believe that H implied E, for he already regarded E as a truth about the tides.

To couple this with (III), we need not suppose that Newton was antecedently *sure* that H implied something or other about the tides, as in (II). In (III), the condition p(S) = 1 is weakened to p(S) > p(T), which is equivalent to p(S – T) $\neq$ 0, i.e., to p(F) $\neq$ 0.

Observe that in coming to believe T, one also comes to believe S. But if it is appropriate to conditionalize on T in such circumstances, it is not thereby appropriate to conditionalize on S, unless p(S) = p(T), contrary to the hypotheses of (III).

Observe also that although we have been reading "H $\vdash$ E" as "H implies E," we could equally well have read it as "p(E/H) = 1" or as "p(H – E) = 0": (K*) would still hold, and so (III) would still be provable.

4. Probability Logic

Let us focus on the probabilistic counterpart of truth-functional logic. (See Gaifman 1964 and Gaifman and Snir 1982 for the first-order case.)

With de Finetti (1970, 1974) I take expectation to be the basic notion, and I identify propositions with their indicator functions, i.e., instead of taking propositions to be subsets of the set W of all possible "worlds," I take them to be functions that assign the value 1 to worlds where the propositions are true, and 0 where they are false.

Axioms: the expectation operator is
   linear: $E(af + bg) = aEf + bEg$
   positive: $Ef \geq 0$ if $f \geq 0$
   normalized; $E1 = 1$

("f > 0" means that f(w) > 0 for all w in W, and 1 is the constant function that assigns the value 1 to all w in W.)

Definition: the probability of a proposition A is its expectation, **EA**, which is also written more familiarly as p(A). De Finetti (1974, section 3.10) proves what he calls "The Fundamental Theorem of Probability":

> Given a coherent assignment of probabilities to a finite number of propositions, the probability of any proposition is either determined or can coherently be assigned any value in some closed interval.

(Cf. de Finetti's Lemma, in section 2 above.)

A remarkably tight connection between probability and frequency has already been remarked upon. It is provided by the law of small numbers, i.e., in the present notation,

$$E(A_1 + \ldots + A_n) = p(A_1) + \ldots + p(A_n).$$

That is an immediate consequence of the linearity of **E** and the definition of "$p(A_i)$" as another name for EA_i. But what has not yet been remarked is the connection between observed and expected frequencies that the law of small numbers provides.

Example: "Singular Predictive Inference," so to speak. You know that there have been s successes in n past trials that you regard as like each other and the upcoming trial in all relevant respects, but you have no information about which particular trials produced the successes. In this textbook case, you are likely to be of a mind to set $p(A_1) = \ldots = p(A_n) = p(A_{n+1}) = x$, say. As $E(A_1 + \ldots + A_n) = s$ because you *know* there were s successes, the law of small numbers yields $s = nx$. Thus your degree of belief in success on the next trial will equal the observed relative frequency of successes on the past n trials: $p(A_{n+1}) = s/n$.

In the foregoing example, no particular prior probability function was posited. Rather, what was posited was a condition $p(A_i) = x$ for $i = 1, \ldots,$ n + 1, on the posterior probability function p: what was posited was a certain probasition, i.e., the domain of the variable "p." The law of small numbers then showed us that for all p in that domain, $p(A_i) = s/n$ for all i = 1, . . . ,n + 1. But of course, p is otherwise undetermined by the condition of the problem, e.g., there is no telling whether the A_i are independent, or exchangeable, etc., relative to p, if all we know is that p belongs to the probasition $\{p : p(A_1) = \ldots = p(A_n) = p(A_{n+1})\}$.

A further example: *your expectation of the relative frequency of success on the next m trials will equal the observed relative frequency s/n of success on the past n trials in case*

$$(8) \quad p(A_1) = \ldots = p(A_n) = x = p(A_{n+1}) = \ldots = p(A_{n+m}).$$

Proof: as we have just seen, the first part of (8) assures us that $x = s/n$, and by the second part of (8), the law of small numbers yields an expected number of successes on the next m trials of $\mathbf{E}(A_{n+1} + \ldots + A_{n+m}) = mx$. Then by linearity of $\mathbf{E}$, the expected relative frequency of success on the next m trials is

$$\mathbf{E}(\frac{A_{n+1} + \ldots + A_{n+m}}{m}) = \frac{ms/n}{m} = \frac{s}{n},$$

i.e., the observed relative frequency of success in the first n trials.

What if you happen to have noticed which particular s of the first n trials yielded success? Then the first part of (8) will not hold: $p(A_i)$ will be 0 or 1 for each $i = 1, \ldots, n$. Still, your judgment *might* be that

$$(9) \quad \frac{s}{n} = p(A_{n+1}) = \ldots = p(A_{n+m}),$$

in which case the expected relative frequency of success on the next m trials will again be s/n, the observed relative frequency on the first n. But maybe the pattern of successes on the first n trials rules (9) out, e.g., perhaps your observations have been that $p(A_1) = \ldots = p(A_s) = 1$ but $p(A_{s+1}) = \ldots = p(A_n) = 0$, so that you guess there will be no more successes, or that successes will be rarer now, etc. The cases in which (9) will seem reasonable are likely to be ones in which the pattern of successes on the first n trials exhibits no obvious order.

These applications of the law of small numbers are strikingly un-Bayesian in Clark Glymour's sense of "Bayesian": the results $p(A_{n+1}) = s/n = \mathbf{E}(A_{n+1} + \ldots + A_{n+m})/m$ are not arrived at via conditioning (via "Bayes's theorem"), but by other theorems of the calculus of probabilities and expectations, no less Bayesian in my sense of the term.

The emergence of probability in the mid-seventeenth century was part of a general emergence of concepts and theories that made essential use of (what came to be recognized as) real variables. These theories and concepts were quite alien to ancient thought, in a way in which two-valued logic was not: witness Stoic logic. And today that sort of mathematical probabilistic thinking remains less homely and natural than realistic reasoning from definite hypotheses ("about the outside world") to conclusions that must hold if the hypotheses do. Perhaps "Bayesian" is a misnomer—perhaps one should simply speak of *probability logic* instead. (Certainly "Bayesian *inference*" is a misnomer from my point of view, no less than from de Finetti's and from Carnap's.) But whatever you call it, it is a matter of thinking in terms of estimates (means, expectations) as well as, or often instead of, the items estimated. Thus one reasons about estimates of truth values, i.e., probabilities, in many situations in which the obvious reasoning, in terms of truth values themselves, is unproductive. The steps from two-valued functions (= 0 or 1) to probability functions, and thence to estimates of functions that need not be two-valued brings with it an absurd increase in range and subtlety. To take full advantage of that scope, I think, one must resist the temptation to suppose that a probasition that is not a unit set must be a blurry representation of a sharp state of belief, i.e., one of the probability measures that make up the probasition: an imprecise measurement (specified only within a certain interval) of some precise psychological state. On the contrary, I take the examples of "prevision" via the law of small numbers to illustrate clearly the benefits of the probasitional point of view, in which we reason in terms of a variable "p" that ranges over a probasition R without imagining that there is an unknown true answer to the question, "Which member of R *is* p?"

References

Bolker, Ethan. 1965. *Functions Resembling Quotients of Measures*. Harvard University Ph.D. dissertation (April).

Boole, George. 1854. *The Laws of Thought*. London: Walton and Maberley. Cambridge: Macmillan. Reprinted Open Court, 1940.

Carnap, Rudolf. 1945. On inductive logic. *Philosophy of Science* 12: 72-97.

de Finetti, Bruno. 1937. La prévision: ses lois logiques, ses sources subjectives. *Annales de l'Institut Henri Poincaré* 7. Translated in Kyburg and Smokler (1980).

—. 1972. *Probability, Induction, and Statistics*. New York: Wiley.

—. 1970. *Teoria delle Probabilità*, Torino: Giulio Einaudi editore s.p.a. Translated: *Theory of Probability*. New York: Wiley, 1974 (vol. 1), 1975 (vol. 2).

Diaconis, Persi and Sandy Zabell. 1982. Updating Subjective Probability. *Journal of the American Statistical Association* 77: 822-30.

Gaifman, Haim. 1964. Concerning Measures on First Order Calculi. *Israel Journal of Mathematics* 2: 1-18.

—, and Snir, Mark. 1982. Probabilities over Rich Languages, Testing, and Randomness. *The Journal of Symbolic Logic* 47: 495-548.

Glymour, Clark. 1980. *Theory and Evidence*, Princeton, Princeton University Press.

Good, I.J. 1952. Rational decisions. *Journal of the Royal Statistical Assn.*, Series B, 14: 107-114.

—. 1962. Subjective Probability as the Measure of a Non-Measurable Set. In *Logic, Methodology and Philosophy of Science*, Ernest Nagel, Patrick Suppes, and Alfred Tarski, Stanford: Stanford University Press. Reprinted in Kyburg and Smokler, 1980, pp. 133-146.

Hacking, Ian. 1967. Slightly More Realistic Personal Probability. *Philosophy of Science* 34: 311-325.

Hailperin, Theodore. 1965. Best Possible Inequalities for the Probability of a Logical Function of Events. *The American Mathematical Monthly* 72: 343-359.

Jeffrey, Richard. 1965. *The Logic of Decision*. New York: McGraw-Hill. University of Chicago Press, 1983.

—. 1968. Probable knowledge. In *The Problem of Inductive Logic*, ed. Imre Lakatos. Amsterdam: North-Holland, 1968, pp. 166-180. Reprinted in Kyburg and Smokler (1980), pp. 225-238.

—. 1970. Dracula Meets Wolfman: Acceptance vs. Partial Belief. In *Induction, Acceptance, and Rational Belief*, ed. Marshall Swain, pp. 157-185. Dordrecht: Reidel.

—. 1975. Carnap's empiricism. In *Induction, Probability, and Confirmation*, ed. Grover Maxwell and Robert M. Anderson, pp. 37-49. Minneapolis: University of Minnesota Press.

Keynes, J.M. 1921. *A Treatise on Probability*, Macmillan.

Koopman, B.O. 1940. The Axioms and Algebra of Intuitive Probability. *Annals of Mathematics* 41: 269-292.

Kyburg, Henry E., Jr. and Smokler, Howard E., eds. *Studies in Subjective Probability*, 2nd edition, Huntington, N.Y.: Krieger.

Levi, Isaac. 1974. On Indeterminate Probabilities. *The Journal of Philosophy* 71: 391-418.

—. 1980. *The Enterpirse of Knowledge*. Cambridge, Mass.: MIT Press.

May, Sherry and William Harper. 1976. Toward an Optimization Procedure for Applying Minimum Change Principles in Probability Kinematics. In *Foundations of Probability Theory, Statistical Inference, and Statistical Theories of Science*, ed. W.L. Harper and C. A. Hooker, volume 1. Dordrecht: Reidel.

Mellor, D.H., ed. 1980. *Prospects for Pragmatism*. Cambridge: Cambridge University Press.

Newton, Isaac. 1934. *Principia* (Volume 2, including *The System of the World*). Motte/Cajori translation. Berkeley: University of California Press.

Ramsey, Frank. 1931. *The Foundations of Mathematics and Other Logical Essays*. London: Kegan Paul. Also: *Foundations*, Cambridge U.P., 1978.

Savage, Leonard J. 1954. *The Foundations of Statistics*, New York: Wiley. Dover reprint, 1972.

Skyrms, Brian. 1980. Higher Order Degrees Of Belief. In Mellor (1980).

Wald, Abraham. 1950. *Statistical Decision Functions*, New York: Wiley.

Williams, P.M. 1980. Bayesian Conditionalization and the Principle of Minimum Information. *British Journal for the Philosophy of Science* 31.

—. 1976. Indeterminate Probabilities. In *Formal Methods in the Methodology of Empirical Sciences*, Proceedings of the Conference for Formal Methods in the Methodology of Empirical Sciences, Warsaw, June 17-21, 1974, ed. Marian Przetecki, Klemena Szaniawski, and Ryszand Wojcick.

Three Ways to Give a
Probability Assignment a Memory

Consider a model of learning in which we update our probability assignments by conditionalization; i.e., upon learning S, the probability of not-S is set at zero and the probabilities of statements entailing S are increased by a factor of one over the initial probability of S. In such a model, there is a certain peculiar sense in which we lose information every time we learn something. That is, we lose information concerning the initial relative probabilities of statements not entailing S.

The loss makes itself felt in various ways. Suppose that learning is meant to be corrigible. After conditionalizing on S, one might wish to be able to decide that this was an error and "deconditionalize." This is impossible if the requisite information has been lost. The missing information may also have other theoretical uses; e.g., in giving an account of the warranted assertability of subjunctive conditionals (Adams 1975, 1976; Skyrms 1980, 1981) or in giving an explication of "evidence E supports hypothesis H" (see the "paradox of old evidence" in Glymour 1980).

It is therefore of some interest to consider the ways in which probability assignments can be given a memory. Here are three of them.

I. *Make Like an Ordinal* (Tait's Suggestion)[1]: A probability assignment will now assign each proposition (measurable set) an ordered pair instead of a single number. The second member of the ordered pair will be the probability; the first member will be the memory. To make the memory work properly, we augment the rule of conditionalization. Upon learning P, we put the current assignment into memory, and put the result of conditionalizing on P as the second component of the ordered pairs in the new distribution. That is, if the pair assigned to a proposition by the initial distribution is (x, y), then the pair assigned by the final distribution is $((x, y), z)$, where z is the final probability of that proposition gotten by conditionalizing on P. (If P has initial probability zero, we go to the closest state in which it has positive probability to determine the ratios of final probabilities for propositions that entail P.)

This suggestion gives probability assignments a perfect memory. From a practical viewpoint, the price that is paid consists in the enormous amount of detail that is built into an assignment for a relatively old learning system of this kind, and the consequent costs in terms of capacity of the system.

II. *Don't Quite Conditionalize* (Probability Kinematics with or without Infinitesimals): Upon learning P, one might not quite give P probability one, but instead retain an itty-bitty portion of probability for its negation, distributing that portion among the propositions that entail not-P in proportion to their prior probabilities. There are two versions of this strategy, depending on whether the itty-bitty portion is a positive real magnitude or an infinitesimal one. Let us consider the first alternative. It should be noted that this may simply be a more realistic model of learning for some circumstances. But I am concerned here only with its value as a memory device. As such it has certain drawbacks. In the first place, the itty-bitty probabilities used as memory might be hard to distinguish from genuinely small current probabilities. In the second place, we get at best short-term memory. After a few learning episodes where we learn P_1 . . . P_n, the information as to the relative initial values of propositions that entailed not-P_1 & . . . & not-P_n is hopelessly lost. On the other hand, we have used no machinery over and above the probability assignment. This gives us a cheap, dirty, short-term memory.

The drawbacks disappear if we make the itty-bitty portion infinitesimal. The development of nonstandard analysis allows us to pursue this possibility in good mathematical conscience.[2] There is no danger of confusing an infinitesimal with a standard number. Furthermore, we can utilize *orders* of infinitesimals to implement long term-memory. (Two nonstandard reals are of the same *order* if their quotient is finite.) There is some arbitrariness about how to proceed, because there is no largest order of infinitesimals. (But, of course, arbitrariness is already present in the choice of a nonstandard model of analysis). Pick some order of infinitesimals to function as the largest working order. Pick some infinitesimal i of that order. On learning P, we update by probability kinematics on P; not-P, giving not-P final probability i. Successive updatings do not destroy information, but instead push it down to smaller orders of infinitesimals. For instance, if we now learn Q, the information as to the relative magnitude of the initial probabilities of propositions that entail not-P & not-Q lives in infinitesimals of the order i^2.

This strategy of probability kinematics with infinitesimals gives probability distributions a memory that is almost as good as that supplied by Tait's suggestion.[3] It has the advantage of a certain theoretical simplicity. Again it is only the probability assignment that is doing the work. Memory is implicit rather than something tacked on. This theoretical simplicity is bought at the price of taking the range of the probability function to be non-Archimedian in a way that reduces consideration of the practical exemplification of the model to the status of a joke.

III. *Keep a Diary:* Our system could start with a given probability distribution, and instead of continually updating, simply keep track of what it has learned. At any stage of the game its current probability distribution will be encoded as a pair whose first member is the original prior distribution, and whose second member is the total evidence to date. If it needs a current probability, it computes it by conditionalization on its total evidence. Such a *Carnapian* system has its memory structured in a way that makes error correction a particularly simple process—one simply deletes from the total evidence.

Information storage capacity is still a problem for an old Carnapian robot, although the problem is certainly no worse than on the two preceeding suggestions. Another problem is choice of the appropriate prior, providing we do not believe that rationality dictates a unique choice.

In certain tractible cases, however, this storage problem is greatly simplified by the existence of *sufficient statistics*.[4] Suppose I am observing a Bernoulli process, e.g., a series of independent flips of a coin with unknown bias. Each experiment or "observation" will consist of recording the outcome of some finite number of flips. Now instead of writing down the whole outcome sequence for each experiment, I can summarize the experiment by writing down (1) the number of trials and (2) the number of heads observed. The ordered pair of (1) and (2) is a sufficient statistic for the experiment. Conditioning on this summary of the experiment is guaranteed to give you the same results as conditioning on the full description of the experiment. Where we have sufficient statistics, we can save on memory capacity by relying on statistical summaries of experiments rather than exhaustive descriptions of them.

In our example, we can do even better. Instead of writing down an ordered pair (x, y) for each trial, we can summarize a totality of n trials by writing down a single ordered pair.

$$\left(\sum_{i=1}^{i=n} x_i, \ \sum_{i=1}^{i=n} y_i \right).$$

We have here a *sufficient statistic of fixed dimension*, which can be gotten by component-by-component addition from sufficient statistics for the individual experiments. (Such sufficient statistics of fixed dimension can be shown—under certain regularity conditions—to exist if and only if the common density of the individual outcomes is of exponential form.) Where such sufficient statistics exist, we need only store one vector of fixed dimension as a summary of our evidence.

The existence of sufficient statistics of fixed dimension can also throw some light on the other problem, the choice of an appropriate prior. In our example, the prior can be represented as a probability distribution over the bias of the coin; the actual physical probability of heads. Denote this parameter by "w." Suppose that its prior distribution is a beta distribution; i.e., for some α and β greater than zero, the prior probability density is proportional to $w^{\alpha-1}(1 - w)^{\beta-1}$. Then the posterior distribution of w will also be a beta distribution. Furthermore, the posterior distribution of w depends on the prior distribution and the summary of the evidence in an exceptionally simple way. Remembering that x is the number of trials and y is the number of heads, we see that the posterior beta distribution has parameters α' and β', where $\alpha' = \alpha + y$ and $\beta' = \beta + x - y$. The family of beta distributions is called a *conjugate* family of priors for random samples from a Bernoulli distribution. It can be shown that whenever the observations are drawn from a family of distributions for which there is a sufficient statistic of fixed dimension, there exists a corresponding family of conjugate priors. (There are conjugate priors for familiar and ubiquitous distributions such as Poisson, Normal, etc.) Random sampling, where the observation is drawn from an exponential family and where the prior is a member of the conjugate family, offers the ultimate simplification in data storage and data processing. The diary need only include the family of priors, the parameters of the prior, and the current value of the sufficient statistic of fixed dimension. For these reasons, Raiffa and Schlaifer (1961) recommend, in the case of vague knowledge of priors, to *choose* a member of the relevant family of conjugate priors that fits reasonably well.

We are not always in such a nice situation, where sufficient statistics do so much work for us; but the range of cases covered or approximated by the exponential families is not inconsiderable. In these cases, keeping a diary (in shorthand) is not as hopeless a strategy for a quasi-Carnapian robot as it might first appear.

One might wonder whether these techniques of diary-keeping have some application to an Austinian robot which, on principled grounds, never learns anything with probability one. I think that they do, but this question goes outside the scope of this note. (See Field 1978, Skyrms 1980b, and Skyrms forthcoming).

Notes

1. Proposed by Bill Tait in conversation with myself and Brian Ellis.
2. For a thumbnail sketch see appendix 4 of Skyrms (1980a). For details see the references listed there.
3. There is this difference. Suppose we think that we learn P, but then decide it was a mistake and in fact learn not-P. On the infinitesimal approach traces of the "mistake" are wiped out, while on Tait's suggestion they remain on the record.
4. On the Bayesian conception of sufficiency, sufficient statistics of fixed dimension, and conjugate priors, see Raiffa and Schlaifer (1961).

References

Adams, E. 1975. *The Logic of Conditionals*. Dordrecht, Reidel.
—. 1976. Prior Probabilities and Counterfactual Conditionals. In *Foundations of Probability Theory, Statistical Inference and Statistical Theories of Science*, ed. W. Harper and C. Hooker. Dordrecht: D. Reidel.
Field H. 1978. A Note on Jeffrey Conditionalization. *Philosophy of Science* 45: 171-85.
Glymour C. 1980. *Theory and Evidence*. Princeton: Princeton University Press.
Raiffa, H. and Schlaifer, R. 1961. *Applied Statistical Decision Theory*. Cambridge, Mass.: Harvard Business School. Paperback ed. Cambridge, Mass.: MIT Press, 1968.
Skyrms, B. 1980a. *Causal Necessity*. New Haven, Conn.: Yale University Press.
Skyrms, B. 1980b. Higher Order Degrees of Belief. In *Prospects for Pragmatism*, ed. D. H. Mellor. Cambridge, England: Cambridge University Press.
Skyrms, B. 1981. The Prior Propensity Account of Subjunctive Conditionals. In *Ifs*, ed. W. Harper, R. Stalnaker, and G. Pearce. Dordrecht: Reidel.
Skyrms, B. Forthcoming. "Maximum Entropy Inference as a Special Case of Conditionalization." *Synthèse*.

III. EVIDENCE AND EXPLANATION

Glymour on Evidence and Explanation

In Chapter VI of *Theory and Evidence* (specifically pages 199-203) and in a subsequent paper, Clark Glymour develops an account of scientific explanation to go with his theory of relevant evidence.[1] Especially significant for me is his use of these ideas in support of his contention that we can have more reason to believe one theory to be true than another even in cases in which the two theories are empirically equivalent. For that contention poses a challenge to the empiricist account of science that I have proposed.[2] Although I do not conceive of the empiricist-realist debate concerning science as addressed directly to an epistemological issue, I am acutely conscious of the fact that the empiricist view of what science is, will not be ultimately tenable in the absence of a tenable epistemology of a certain sort. And so, in response to Glymour's views on explanation and its relation to confirmation, I shall sketch here a preliminary version of such an epistemological position.

Glymour has put forward an account of three topics relevant to our present concern: hypothesis testing and confirmation, theory comparison, and explanation. The first, his theory of testing and relevant evidence, I admire greatly. In a companion paper to this one, I discussed that first topic in more detail, and below I shall draw a little on results of that discussion.[3] Here I shall sketch an alternative account of theory comparison and acceptance, as a case of decision making in the face of conflicting desiderata, and use that to introduce some objections to Glymour's account of explanation as well.

1. Theoretical Virtues; a Story of Conflict

The virtues that may be attributed to a scientific theory in order to support its acceptance, are diverse. I cannot attempt a complete typology, but I shall point to two important sorts. The first I shall call *confirmational* virtues; they are features that give us more reason to believe this theory (or

Research for this paper was supported by an NSF grant which is hereby gratefully acknowledged. The paper was first presented at a symposium with Clark Glymour and Wesley Salmon at the American Philosophical Association, Western Division, Detroit 1980.

part of it) to be true. This is equivalent, I take it, to the assertion that they are features that make the theory (or part of it) more likely to be true. The second sort I shall call *informational* virtues: one theory may be able to tell us more about what the world is like than another, or it may be able to tell us about parts of the world concerning which the other theory is quiet.

Before putting this division to use, some preliminary remarks. I have described both sorts of virtue in the comparative; for it may be that we can never do better than compare theories with each other. Let us be careful to assume none of the theories philosophers and logicians have proposed concerning these comparisons. There are, for example, theories of confirmation and of information that imply the applicability of quantitative measures, or at least linear orderings of degrees of confirmation and amounts of information. When I speak of more or less confirmation or information, my usage will be at odds with that. For if there are quantitative measures of these concepts, then one theory may have greater confirmation (overall!) or contain more information (in toto) even though some specific evidence confirms it less than another, or even though it is quiet about some phenomena that the other tells us much about. In placing some emphasis on this, I place myself on the side of Glymour on at least one important issue.

Second, I identified the two sorts of virtues by means of very different criteria. For the first is identified by what it does for the theory—making it more likely to be true—and the second by what it consists in—information the theory gives about this subject or that. Hence there is prima facie no reason to deny that the two sorts may overlap.

Without saying very much about what theory acceptance is, at this point, I take it for granted that a virtue of a theory is a feature that, if we are told that the theory has it, gives us more reason to *accept* that theory. Virtues are attributed in order to provide reasons for acceptance. One question Glymour and I differ on is whether acceptance of a theory is belief that the theory is true. If it is then all virtues are automatically confirmational virtues. But there is a strong tradition in philosophy of science, so prevalent that it does not even have a name, which speaks against this. Almost all writers on confirmation or evidential support make explicit some such principle as the following:

> (1.1) If theory T provides information that T' does not provide, and not conversely, then T is no more likely to be true than T'.

There are of course measures of how well the theory accounts for the evidence, or is corroborated by it, or whatever, which may be increased when a theory provides more information. But this makes them suspect as reasons for belief. Belief is always belief to be true; I take that as part of the logic of these words:

> (1.2) A feature of T cannot provide more reason to believe that T unless it makes T more likely to be true.

Perhaps we should emphasize at this point that assessment of whether a theory provides more information or whether we have more reason to believe it, is always made in the light of background assumptions. These are meant to be kept fixed in principles 1 and 2. Thus a theory may well provide more information from a logical point of view (i.e., ignoring our present background theories) and yet be more likely to be true (in the light of that background); but such equivocation must be avoided. From principles 1 and 2 we must of course deduce

> (1.3) Informational virtues are not confirmational.

This is a general conclusion; it does not imply that if the *overall* amount of information is greater in one theory than in another, then the former must be less likely to be true. For it may be that the theories contradict one another, so that the former does not represent *merely* an increase in *overall* information. If one theory provides all the information that another gives, however, and some more in addition, then the second cannot be less likely to be true. A corollary to this is:

> (1.4) A theory cannot be more likely to be true than any of its parts; and the relation between the whole theory and the evidence cannot provide better reason to believe it, than the relation between a part of the theory and the same evidence, provides to believe that part.

This is a controversial conclusion from not very controversial-looking premises. It may be sugared a little by the reflections that for the whole theory we can provide a much larger list of relevant considerations, such as tests that it has passed. The point is only that *whatever procedure we then adopt to reach an overall comparison* of how likely it is to be true in view of the evidence must, to satisfy principle 1, give at least as good a grade to any part of that theory.

Finally, in view of the intimate link between virtues and acceptance, we derive the corollary:

(1.5) Some reasons to accept a theory are not reasons to believe it to be true; hence acceptance is not the same as belief; any reason for belief is a fortiori a reason for acceptance, but not conversely.

The situation is in fact a bit worse than I formulated explicitly in principle 1: there is in general a conflict between the desire for information and the desire for truth. (This was the basic theme of Isaac Levi's *Gambling with Truth* and is explored there at length, though by means of quantitative measures that would, in the present context of theory evaluation, not be used by either Glymour or myself.)[4] Thus the problem of theory acceptance has the structure of a practical decision-making problem: conflicting desiderata must be weighed against one another.

Indeed, this fits well with my own view on theory acceptance, which is that it consists in (a) belief that the theory is empirically adequate, and (b) commitment to use of the theory's conceptual scheme as guide to further research. Since we are never so lucky as to get a theory of truly universal scope and completeness, point (a) does not make point (b) inevitable or even automatically advisable. In addition, it may be more rational to pursue a theory that is capable of testing in the short run, or capable of combination with other theories already being pursued. And finally, the commitment may be made while the belief is still tentative or qualified, in which case the acceptance is also called tentative or qualified.

2. Explanation: What Sort of Virtue Is It?

Wesley Salmon and I have theories of explanation that are in some ways as opposite as can be. In explaining explanation, he looks to objective relations among facts correctly depicted by theory, whereas I look to context-dependent relations among questions, answers, and accepted theories.[5] But on one central point we are in complete agreement: we agree that to give an explanation is to give relevant information. For him, the criteria of relevance and informativeness are combined in relations of objective statistical correlation and spatio-temporal connection. For me, the relevance is initially determined contextually, but then statistical relations (implied by accepted background theories) may serve to evaluate relevant answers, which count as explanations, as good or comparatively better. In my account, these statistical criteria of evaluation are offered only tentatively, and have in any case not a very central role. But the main agreement stands: for both of us, explanation is an informational virtue.

Although I do not know precisely what role Salmon would assign to explanation in the theory of evidence, my own position will be clear from the foregoing. That a theory provides explanations for phenomena in its intended scope is indeed a virtue, and gives us reason to accept the theory; but being informational, it generally gives no added reason for belief. Glymour's paper is devoted to the contrary thesis that explanation does give reason for belief. But then, his theory of explanation is quite different: for he characterizes explanation not in terms of information, but in terms of the elimination of contingency and chaos.

The first question I want to ask is whether explanation, as characterized by Glymour, is perhaps not after all an informational virtue. Since his theory of explanation is new, it does not yet provide explicit answers to some questions of a formal or procedural sort. For example, if asked what *is* an explanation, Hempel would say, an argument, and I would say, an answer to a why-question. The locution "theory T explains fact E" Hempel would gloss as "there is an explanation with the conclusion that E is the case, whose theoretical premises are furnished by T." I would gloss it as "once you accept theory T, you are in a good position to answer the question *why E?* (understood as asked in the present context)." Thus Newton's theory explains the tides in the sense that, once we accept Newton's theory we can say (correctly by our lights) that the sea is subject to tides because of the gravitational influence of the moon, and (still relative to our thus enlarged background) this is a good answer; holders of other theories are not in a position to say that, or at least are not able to give an equally good answer by their lights. I do not quite know how to regiment such terminology to accord with Glymour's theory of explanation. But I think I do know quite well how to handle examples in accordance with it in the way he does, and this should suffice.

In the remainder of this chapter I shall therefore examine the account of explanation via identifications, and also the way in which a theory may unify our description of hitherto unconnected phenomena, both of which Glymour gives as instances of theoretical explanation. The first specific idea I mean to explore is this: perhaps Glymour agrees by and large with section 1 above, and disagrees with the thesis that giving an explanation must always (or ever) consist at least in part in giving information (or more specifically with the thesis that if we amend a theory so as to make it more explanatory, we must be making it more informative). I shall argue that this cannot be so, for explanation, elimination of contingency, and unification, in his sense, are all bought at the expense of making the theory more informative.

3. Explanation from Identity and Necessity

We may imagine the following process, which is not too farfetched as a historical account: first a theory is developed in which temperature, pressure, volume, and all the mechanical parameters appear independently, and which contains the molecular hypothesis identifying bodies with aggregates of molecules. Call this theory T. Then a theoretical improvement is introduced. We add the identity:

(3.1) if X is aggregate A of molecules, then the temperature of X is identical with the mean kinetic energy of the molecules in A.

Since every body is such an aggregate (a single molecule being a small aggregate), this identifies the temperature of every body. Thus (3.1) can be rephrased as

(3.2) Temperature is identical with mean molecular kinetic energy.

I do not use the identity sign, for that has in science the typical meaning of mere equality of values. That the values of the two quantities are always the same is already implied by old theory T. That is exactly the fact which we wanted to have explained. Accepting new theory T', which is T plus (3.2), we can explain this universal equality by pointing to an identity that implies it.

At this point Glymour says that if the identity (3.2) were contingent, then it would need an explanation as much as the equality itself did. But it needs no explanation; the requests for explanation stop here. From these two considerations we can deduce that the identity is not contingent. Although this conclusion would have been controversial some years ago, it is much less now. Of course, the identity is not a priori true or false; we cannot possibly give a proof to demonstrate or to refute it. But this is merely an epistemic fact. Glymour can appeal to the theories of Kripke and Putnam to support his view that a priori truth and necessity do not coincide; and that all such identities are necessary if true. Which explains nicely why requests for explanation should stop there, at least within the confines of physics.

But the very distinction between necessity and the a priori establishes that if it is necessary that A, it may still be informative to assert that A.

Indeed, the only information we have before we are told by theory T′ that (3.2) is the case, is that (3.2) is either necessary or impossible. We do not know which. To be told that it is true, and hence necessary, conveys a great deal of information. Therefore I would say that, explanatory though T′ is, it is less likely to be true than its part T. And the fact that T′ is more explanatory than T, does not constitute a reason to believe it, a fortiori.

I have tried to think of this in terms of Glymour's theory of testing. I understand that theory best if it is applied to a theory like the present one, which is stated initially as a set of functional relationships among physical parameters, some of which are directly measurable and some not. But if we add (3.2), then any Glymourian test I can think of for that hypothesis is no more than a test of the equality

(3.3) For all bodies X, the temperature of X is (has the same value as) the mean kinetic energy of the molecular aggregate which is X.

It may be as unfair, however, to ask for a test of a necessary identity as it is to ask for an explanation.

Perhaps we should look instead at the inevitable simplification that acceptance of such identity brings in its train. One term, the simpler one (in this case, "temperature") is deleted from the primitive vocabulary in the description of the theory and reintroduced by definition. Does the new theory—call it T′′—contain more or less information than the original? Here we face a dilemma, it seems to me, depending on how we construe what the definitional extension says. The definition is used to deduce that procedures heretofore called measurements of temperature are measurements of mean kinetic energy. The original theory said only that they yielded numbers equal to the value of that quantity. Thus viewed, the definitional extension is informative, and indeed, explains this further equality. But alternatively we can say that all information about temperature has been dropped from the theory since a definition is merely a stipulation as to the use of the word hence forth. In that case T′′ has less information than T, but it also does not explain anything that T did not explain. If the concept of temperature is truly dropped and disappears, then fewer why-questions can be asked; and this may be an advantage, but the identification to which explanatory power was attributed has also disappeared.

To sum up, then, it appears that the explanation of a universal equality of values by appeal to a theoretical identification is possible only when that identification is an informative statement.

4. Explanatory Unification

Newton's theory covers the laws of fall, planetary motion, pendulums, tides, comets, and much else. He brought unity to the field of physical science. We also say that his theory *explains* the phenomena, the motions, that I have just listed. Glymour suggests that the two assertions are not unconnected; that the unification constitutes the explanation. Thus one might say: Newton's theory explains the tides, because Newton's theory exhibits the tides as one instance of a general pattern that has as instances a number of phenomena heretofore considered diverse.

I find it very hard to evaluate this claim, since it is not supported by a claim that all explanations are thus (explanations by theoretical identification are not, for example), and I don't know how to search systematically for a random sample of theories that unify diverse phenomena to see whether the connection persists in general. But it does seem to me that unification requires the introduction of additional information.

Let us again use a fictional history. Suppose that in one country, people develop theories of fall, planetary motion, tides, comets, and pendulums, and that the laws of motion they state for these are *exactly* those implied by Newton's theory. Then a quite different person, call him Newton*, is born into this country, sees all those diverse theories, and proposes Newton's theory. All the claims about unification and explanation presumably still hold. Yet Newton's theory entails a great deal that the sum of the diverse theories does not. It entails, for example, laws of planetary motion for solar systems other than this one, and tides for planets with moons different from ours. Indeed, just because he does not merely unify the phenomena we know already, but because he unifies them together with a large range of unexplored and unobserved phenomena, do we say he truly unifies them. Thus the addition of information appears to be crucial to the process of unification.

5. Testing and Comparision of Theories

I have so far examined the one alternative: that Glymour will accept the basic epistemological principles I stated at the outset, while denying that explanation is an informational virtue. It appears that this position would not be tenable. Hence it is high time to examine the second alternative, that Glymour will deny those principles. The crucial ones are (1.1) and (1.2). It is possible to deny both, or else to deny the second while affirming the first, provided one *denies*:

(5-1) It is inconsistent to say that you have more (respectively, less) reason to believe T while saying that T is less (respectively, more) likely to be true.

(with each part relativized, as always, to the light of accepted background evidence and theories). In order to explore this second alternative, we must of course look to Glymour's writings, and especially his book, to see what he says about this. The matter is, however, not clear there.

To begin, Glymour does not hold with rules for acceptance, except for the trivial one that we ought not to accept a theory if we know of a better one (Chapter V, pages 152-155, "Comparing Theories"). This reduces the problem of acceptance, which I take him to equate with that of belief, to the problem of comparing theories. Glymour's account of testing and relevant evidence focuses on the notion:

(5-2) E confirms hypothesis A relative to Theory T

where E is a body of evidence. (In another paper I examine, or rather reconstruct, that account, and give warrant there for the claims I shall make about it here.[3])

In order to say anything at all about having better reasons to believe one theory rather than another on a given occasion, Glymour *must* go beyond the purely relative notion (5.2). He must "derelativize" if we are to have a comparison of theories with each other. A naive suggestion would be to compare the hypothesis of two given theories with respect to confirmation received relative to previously accepted theory. This would make nonsense of Glymour's general account. For his central argument is that when a theory is advocated, it is on the basis of successful tests of hypotheses of that theory relative to that theory. The wealth of examples, and the sophisticated analysis of the logical complexities of this procedure, that Glymour gives us, convince me that he is right. The bootstrap account of testing is a major achievement.

Glymour makes various suggestions for the correct "derelativization," both explicit and implicit, and I shall examine what I take to be the main one. In all cases, the burning question for us here is how the suggested comparison relates to increase or decrease in information. I shall quote, to begin, his fifth explicit suggestion:

(5-3) It is better that a body of evidence test a set of hypotheses sufficient to entail the whole of the theory than that it test a

> logically weaker set, for if the logically sufficient set can be
> established the rest will follow. (*Theory and Evidence*, pp. 153-
> 154)

The content and general theme of the book make it clear that the testing in
question is to be confirmation of those hypotheses relative to the theory
itself. And it appears, in this passage, that confirmation is inherited by the
theorems from the axioms. But this is not so in the strict sense in which
Glymour explicates confirmation. The following two facts are crucial
features of the account, which are used to defuse philosophical problems
elsewhere in the book:

(5-4) If E confirms A relative to T, and A logically implies B, it does
       not follow that E confirms B relative to T.

(5-5) If E confirms A relative to T, and also confirms B relative to T, it
       does *not* follow that E confirms (A & B) relative to T.

(See my other paper for examples; it is shown there also that various
obvious weakenings of the denied principles fare no better.)

It looked for a moment as if Glymour were prepared to accept some such
principles as I stated at the outset; now it seems that this appearance was
deceptive. But then, what warrant can he claim for the quoted passage
(5.3)? It is clear that "the rest will follow" in the sense that the theorems are
as likely to be true, at least, as the axioms. But the crucial question is
whether the extent to which they are confirmed by the evidence is as great.

At this point I am tempted to suggest the introduction of an ancestral to
the confirmation relation. We could say that although the theorems need
not be confirmed, we nevertheless have as much reason to believe them,
because of the support they *inherit* from the axioms. I find it difficult to
conceive of any other sort of warrant for the quoted passage; but if Glymour
says this, he will accept the epistemological principles I set out to begin, it
seems, and our second alternative has run out.

A later passage appears to rule out this maneuver anyway. One recurring
theme for Glymour has been the assertion that the evidence may provide
better support for a theory than for the observational consequences of that
theory taken by themselves (Chapter V, pages 161-167, "The Theoreti-
cian's Dilemma"). I am champing at the bit here to argue that nothing like
what Glymour and others call the observational consequences constitutes
anything like the empirical import of the theory. But that is irrelevant for
the present dispute; it suffices to read Glymour as saying here that a certain

subtheory of a theory comes out the loser in the comparision between the two. It is clear that the theory entails its subtheory; hence we have here, it appears, a clear denial of the principle that we have at least as much reason to believe the theorems as we have to believe the axioms. The property that the whole theory has and the subtheory lacks is this:

> (5-6) There is a body of evidence E, and a complete set A of axioms for T such that E confirms each member of A relative to T itself.

The principle we can apparently glean from what Glymour has said so far, then, is this:

> (5-7) If we have no evidence that disconfirms any theorem of T relative to T, while T but not T′ has property (5.6), then T is better confirmed than T′.

Now I would deny this, because in the case in which T′ is a subtheory of T, hence consists of theorems of T, I would say that as far as any nonrelative evidential support we have for theories on that occasion, is concerned, logical consequences inherit support from confirmed hypotheses. We can read (5.7) as saying that there is no inherited support. If a theory does very well in testing, we cannot cite that as evidence for the proposition that any of its consequences are true, at least not until we take the great leap and decide to believe that the theory itself is true. (For belief, presumably, is inherited!)

I find this very surprising. I should think that if a theory is doing well, that is good evidence that certain relevant parts are true, the evidence being the better when the part is more circumscribed. But a clash of intuitions won't get us anywhere. Instead I must ask Glymour to confront the problem I have with (5.3) and its apparent consequence (5.7). The only sort of reason I can think of to warrant the former is at odds with the latter. In the nonrelative or derelativized sense of evidential support (which will be needed if we are to compare theories with each other), it seems to me sensible to say that evidential support for the axioms is especially important, *only if* we also say that this (nonrelative) support is inherited by their consequences. But in *that* case, I can see no way in which we shall ever have less reason to believe a subtheory than we have to believe the whole. All I can see—and I urge this on Glymour as the correct diagnosis—is that we might have less reason to *accept* the subtheory (questions of belief left aside), because, taken as a theory, it has considerably fewer virtues than its parent does.

6. The Real Significance of a Test

If theory comparison and acceptance are a matter of striking a balance between competing desiderata, we should expect a procedure designed to elicit simultaneously the strengths of the theory on several different counts. That is exactly how I see the testing procedure, and Glymour's account I take as support for my view. To pass a well-designed test provides confirmation and evidential support for a theory, but to admit such a test in the first place the theory must be sufficiently informative. Hence to say of a theory that it has passed well-designed tests speaks simultaneously to the two competing criteria of confirmation and information, and the support testing provides is not to be equated with simple confirmation.[6]

Notes

1. Clark Glymour, *Theory and Evidence* (Princeton University Press, 1980) and Explanations, Tests, Unity and Necessity. *Nous, 14* (1980) 31-50.

2. *The Scientific Image* (Oxford: Oxford University Press, 1980).

3. Theory Comparison and Relevant Evidence, this volume.

4. Isaac Levi, *Gambling with Truth* (New York: Knopf, 1967; Cambridge, Mass.: MIT Press, 1973); see also his *The Enterprise of Knowledge* (Cambridge, Mass.: MIT Press, 1980).

5. See Chapter V of *The Scientific Image* where Salmon's theories are discussed in sections 2.2 and 2.6.

6. *Note on inheritance of support*. Should evidential support for a theory be inherited by its consequences? A few logical distinctions help. Let us call a judgment *diachronic* if it takes the form "T is now more X than formerly." There "X" may stand for "believed" or "supported by the available evidence" etc. Clearly there is no inheritance in the diachronic case, since the change in attitude or evidence may have to do directly with one part of T and not another. Call the judgment *synchronic* if it compares several theories from a single historical point of reference. Then there is obvious inheritance in the case of belief (if (A&B) is more strongly believed than C, then so is A), and obvious non-inheritance for the effect of *new* evidence (same reason as in the diachronic case). The final disputed case concerns therefore a synchronic judgment concerning support by the total evidence: could the total evidence on a given occasion support a theory more than one of its parts? Note that when Bayesians define "E confirms H" as "$P(H|E) > P(H)$" their concern is either with the diachronic or synchronic/new evidence case.

IV. HISTORICAL CASE STUDIES

Newton's Demonstration of Universal Gravitation and Philosophical Theories of Confirmation

Newton consistently asserted that his method was not hypothetico-deductive.

I cannot think it effectuall for determining truth to examine the severall ways by wch Phaenomena may be explained, unless there can be a perfect enumeration of all those ways. You know the proper Method for inquiring after the properties of things is to deduce them from Experiments. And I told you that the Theory wch I propounded was evinced by me, not by inferring tis this because not otherwise, but by deriving it from Experiments concluding positively & directly.

. . . what I shall tell . . . . is not an Hypothesis but most rigid consequence, not conjectured by barely inferring 'tis thus . . . because it satisfies all phaenomena (the Philosophers universall Topick), but evinced by the mediation of experiments concluding directly and without any suspicion of doubt.

In this [my experimental] philosophy particular propositions are inferred from the phenomena, and afterwards rendered general by induction. Thus it was that [among other things] the laws of . . . gravitation were discovered.[1]

Philosophers—and critical historians—have not tended to give Newton a sympathetic ear. Probably influenced by the hypothetico-deductive account (henceforth HD) and by its first cousin holism, they have tended to dismiss Newton's remarks as the necessarily ineffectual rationalizations of the paranoid who is unable to accept human limitations and honest criticism. Newton's seclusion after his early optical battles, the publication of the *Opticks* after Hooke's death, and the acrimonious debate over the calculus are all well-known and documented incidents, which lend easy credence to a dismissal of Newton's claims of demonstration. There is also the not inconsiderable point that Newtonian mechanics is, strictly speaking, false!

Given Glymour's confirmation theory, however, Newton's assertions begin to ring true, for on this theory confirmation is the coherent *deduction* of instances of hypotheses *from* observational data and other laws and theories. Glymour's application of his bootstrapping confirmation theory to Newton's argument for universal gravitation is particularly impressive given Newton's hitherto bad philosophical press.[2]

One important difference between Newton and Glymour on confirmation is Newton's insistence that hypotheses are deduced, and Glymour's insistence that only instances are deduced. But this difference depends primarily on Newton's idiosyncratic use of his Rules of Reasoning as premises in a purportedly deductive argument. Glymour correctly notes that interpreted as a bootstrapper, Newton does *not* compute instances of universal gravitation, but instead computes instances of special-case corollaries of universal gravitation. So, for example, the data about the planets and the moons of Jupiter and Saturn are used to calculate, assuming the second law, instances of the corollary that "the sun and each of the planets exert inverse square attractions on whatever satellites they may have." (TE, p. 217) These corollaries, which are confirmed by bootstrap instantiation, are then fed into Newton's *Regulae Philosophandi*, aptly described by Glymour as rules of detachment, and generalized *in stages* to universal gravitation. Glymour makes no pretense that Newton's argument is a pure case of bootstrapping. Obviously the Rules do much of the work.

My aim in this paper is not to analyze how good a mix can be made of Newton's Rules and Glymour's bootstrapping procedures. It is Newton's use of *idealized evidence* that will be of concern here: what it is, what controls its use, and what, if anything, bootstrapping has to do with the use of simplified and computationally tractable data. Before I consider these questions, allow me first to review the first four propositions of *Principia*, Book III, as seen through the eyes of a bootstrapper. These propositions will be the focus of my analysis.

The first three propositions can be seen as Glymourian computations of instances of this corollary of universal gravitation: all satellites of the sun and of the planets are attracted to their respective centers by an inverse square force. For the purpose of constructing a bootstrapping computation diagram, I shall represent this corollary as:

$$((Sx \lor Px) \& Txy) \rightarrow Ixy$$

where Sx = x is the sun
     Px = x is a planet
     Txy = y is a satellite of x
     Ixy = y is attracted to x with an inverse square force

In order to generate an instance of this corollary, Newton needs to compute—i.e., generate an instance of—the "quantity" Ixy on the basis of the observational quantities of the antecedent and other laws. In the case of the moons of Jupiter (Prop. I), Newton computes his instances by using Proposition II "or" III, and Corollary VI of Proposition IV, all of Book I. In the case of the "primary" planets (Prop. II), Newton makes up his mind and uses only Proposition II as well as Corollary VI. Propositions II and III state in modern terms that if the area swept by the radius vector of a curved motion varies directly as time, then any object in such motion is acted on by a centripetal force directed toward the origin of the radius vector. In short, satisfaction of the area law requires a centripetal force. The difference between the two propositions is that Proposition II is with respect to a center in inertial motion, whereas Proposition III is with respect to a center "howsoever moved." In the latter case, the total force on the body in motion is the vector sum of the centripetal force and those forces acting on the center. I shall discuss the importance of this difference later. Corollary VI of Proposition IV is the well-known result that if in the case of concentric circular motions the ratios of the square of the periods and the cube of the radii are constant, then the centripetal forces are inverse square. Implicit in this corollary, and explicit in its proof, is the satisfaction of the area law, which is the antecedent condition of Propositions II and III.

The instantiation of the desired corollary of universal gravitation, i.e., the confirmationally relevant content of Propositions I and II of Book III, can now be seen as a straightforward bootstrapping deduction from the "observational" data and two theorems. This deduction is represented by the right-hand fork of the following computational diagram.

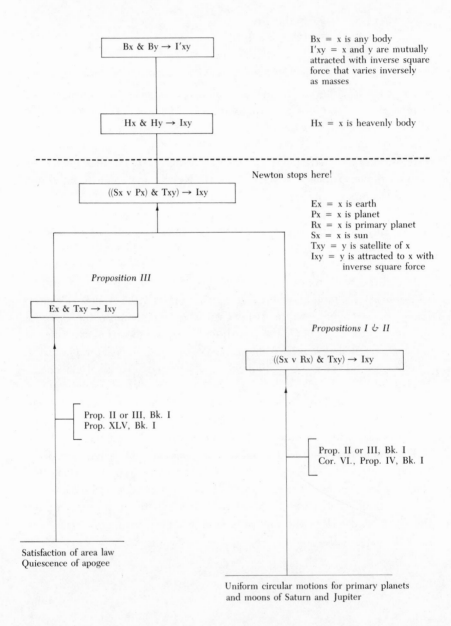

Figure 1. Newton's bootstrapping computations in Propositions I, II, and III.

Proposition III, that the moon is attracted to the earth with an inverse-square force, receives a similar bootstrapping interpretation. Again, Proposition II "or" III is used to show that the moon is acted on by a centripetal force. That this force is inverse-square is shown by using the near absence of the motion of the moon's apogee and Proposition XXV, Corollary 1 of Book I. These computations are represented by the left-hand fork of the diagram.

Proposition IV is the famous moon test. The centripetal force acting on the moon is calculated and compared with that calculated to be acting on bodies on the earth's surface. There is equality of magnitude. Therefore, "by Rule 1 and 2," there is identity of force. Although no bootstrapping per se occurs here, this proposition will be important for the purposes of this paper given Newton's use, as we shall see, of idealized evidence.[3]

As already noted, because of Newton's use of his Rules of Reasoning, he cannot be interpreted as having computed instances of universal gravitation. So bootstrapping for Newton stops at the location indicated in the diagram.[4] It must be admitted that the propositions leading to this stage of the argument appear to be no more than extremely straightforward, indeed trivial, bootstrapping computations. They do not in fact satisfy Glymour's conditions for successful bootstrapping, however. Specifically, the computations violate his fourth condition, which states, in essence, that it must be possible for observational data to lead to a negative-instance of the hypothesis in question. (TE, pp. 130-132) But this is *not* possible if we use as resources only those theorems and corollaries that Newton explicitly mentions. The reason is that on this basis alone an input of nonsatisfaction of the area law or failure of the constancy of $period^2/radius^3$ will not allow for a computation of *not*-Ixy. If we interpret Newton as having intended the corollaries of Propositions II and III, and Proposition IV itself, then our problem is solved, since these additions allow for the deduction of negative instances of inverse-square attraction. Newton's normal practice, though, was to distinguish explicitly between propositions and corollaries used. So some creative reconstruction is required to make good Glymour's claim of having captured Newton's practice.[5] But since the possibility of a negative instance is, as Glymour stresses, the very heart of the bootstrapping requirement (TE, pp. 114-117), Newton's apparent carelessness is a cause for worry if we view bootstrapping as an historical account. And Glymour's claim is that bootstrapping receives some of its normative support from the fact that it is historically instantiated. (TE, pp. 64, 176)

More serious problems for a bootstrapping interpretation arise when we consider Newton's use of Propositions II and III of Book I. As already noted, they state that satisfaction of the area law entails a centrally directed force with respect to a center in inertial motion (Prop. II) or to a center "howsoever moved" (Prop. III). With respect to Jupiter's moons and that of the earth, Newton uses Proposition II "or" III. The reason for the apparent indecision is clear. Proposition II is inappropriate, since Jupiter and the earth are presumably not in inertial motion. But, on the other hand, Proposition III is also inappropriate, since Newton needs *real forces* for his argument and not indeterminate sums or components.[6] This dilemma is not explicitly resolved by Newton in the *Principia*. With respect to the planets, Newton makes a decision and opts for Proposition II. This choice raises the obvious question of how Newton can know at this stage of the argument that the sun is in inertial motion. Duhem claimed, as the reader may recall, that Newton could not know this, since such knowledge entailed already knowing what were real forces.[7] Therefore, a sun in inertial motion was for Duhem a *convention*. Glymour, however, makes the case that Newton can be read as having given arguments for the sun being in inertial motion that are "both cogent and powerful." (TE, p. 213)[8]

The components of these arguments occur in the context of Proposition XIV of Book III and its corollaries. The proposition states that the aphelions and nodes of the planetary orbits are fixed. The propositions given as justification—respectively, XI and I of Book I—require that there be no forces other than an inverse-square force acting at one focus and that the orbits be elliptical with respect to absolute space. Stationary apsides, therefore, are a deduction from an inverse-square force originating at a stationary sun plus the additional assumption that there are no disturbing forces. Corollary I, states that the stars are at absolute rest, since they are stationary with respect to orbits in absolute rest. Now none of this is relevant per se to showing the sun to be at absolute rest (or in inertial motion), since that is assumed from the beginning. The proof goes from one kind of absolute motion to another, from that of the sun to that of the stars.

Corollary II, though, circles back and shows the noninfluence of the stationary stellar system on our solar system:

> And since these stars are liable to no sensible parallax from the annual motion of the earth, they can have no force, because of their immense distance, to produce any sensible effect in our system. Not to mention that the fixed stars, everywhere promiscuously disposed in the

heavens, by the contrary attractions destroy their mutual actions, by Prop. LXX, Book I.[9]

In HD terms, this can be read as stating that no untoward consequences follow from stationary stars, since, being far away and randomly dispersed, they will have no influence on our system. In other words, a consequence of universal gravitation and a stationary sun—namely, stationary stars—is not disconfirmed by observation since this consequence when conjoined with distant and random placement yields a null influence on the planets. On this reading of Proposition XIV and its *corollaries*, we do not have an independent proof for a sun in inertial motion, or for that matter, for a solar system in inertial motion. Glymour, however, sees Corollary II as not really being a corollary at all, but instead as an "independent" argument for the sun's inertial motion. (TE, p. 212) Evidence will be given below that suggests strongly that this is incorrect.

An argument similar to that given in Corollary II is given by Newton in the *System of the World* (MS Add. 3990), which is an early version of Book III of *Principia*. But here the argument is clearly intended to be an independent proof for the inertial motion of the solar system. In this work, composed sometime before 1685 but not published until 1728, Newton argues directly from stars in relative rest to the conclusion that the center of gravity of the solar system (calculated to be near the surface of the sun) "will either be quiescent, or move uniformly forwards in the right line."[10] It is this move that Glymour claims can be seen, with elaboration, as "both cogent and powerful." (TE, p. 213) Since my interest here is with Newton's use of idealized evidence, I shall not dispute this claim, but shall note only that understood this way Newton is in some trouble because the planetary aphelions *do* rotate. This was known by Newton and is in fact mentioned in Proposition XIV and its scholium! There is also the complication that showing the center of gravity of the solar system to be in inertial motion does no good, since our original problem was to justify Newton's assumption that the sun was in inertial motion. This assumption was needed in order to utilize Proposition II, Book I, in the bootstrapping computation of Proposition II (and really I and III as well) of Book III. So again Newton is somewhat wide of his mark.

Pemberton, apparently thinking in terms similar to our HD account, communicated very much this sort of observation to Newton when involved in the preparation of the third edition of *Principia:*

Do not the words in which prop. 14 are [sic] expressed seem almost to be contradicted in the demonstration of it? For as in the proposition it is said, that the Aphelia and Nodes remain fixed; in the demonstration it is only shewn, that they would remain so, if they were not moved by certain causes, which, both here and more particularly in the following scholium, are allowed to take effect.[11]

That Newton was concerned by the observable rotation of the aphelions is clear upon examination of the additions made to Proposition XIV in the second edition of *Principia*. In the first edition, Newton simply dismissed these rotations with the comment that "it is true that some inequalities may arise from the mutual actions of the planets and comets in their revolutions; but these will be so small, that they may be here passed by."

In the second edition, however, Newton added a scholium to justify the claim that these inequalities could be successfully explained away:

> Since the planets near the sun (viz., Mercury, Venus, the earth, and Mars) are so small that they can act with but little force upon one another, therefore their aphelions and nodes must be fixed, except so far as they are disturbed by the actions of Jupiter and Saturn, and other higher bodies. And hence we may find, by the theory of gravity, that their aphelions move forwards a little, in respect of the fixed stars, and that as the $^3/_2$th power of their several distances from the sun. So that if the aphelion of Mars, in a space of a hundred years, is carried forwards $33'20''$, in respect of the fixed stars, the aphelions of the earth, of Venus, and of Mercury, will in a hundred years be carried forwards $17'40''$, $10'53''$, and $4'16''$, respectively.[12]

What we have in Proposition XIV, then, is this. A stationary sun and inverse-square force entail stationary orbits and (given the observational data) stationary stars. And the great distance and "promiscuous" placement of these stationary stars entail the lack of a *disturbing* force on the planetary orbits. But since the orbits *do* rotate, this prediction of stationary apsides is violated. This violation, though, can be explained away by the hypothesis that Jupiter and Saturn are sources of disturbing forces. Furthermore, a successful conditional prediction can be obtained from data on Mars's precession. So on my view the essential features of Newton's argumentation are these. First, the initial data used are idealized so that clean simple calculations can be obtained. Second, *supplementary* arguments are added to show that *better* observational fit is obtainable if certain complications are attended to. There is *not* a unified monolithic HD or bootstrapping account that accommodates the known data in a single coherent calculation. And it is this patchwork nature that was the source of Pemberton's complaint to Newton. As a last observation, the role of Corollary II is now seen to be not that of providing an independent proof for the sun's inertial motion, but of providing some justification for the identification of Jupiter and Saturn as the sole sources of significant disturbing forces.

Duhem was on a similar interpretational trail when he noted that, whereas Newton accepted as a datum the description of the planetary orbits as elliptical, it is a formal consequence of his theory that because of perturbational forces, these orbits cannot be elliptical.[13] (Actually, as we have seen, Newton starts with circular orbits!) That is, the theory is used to correct its originally supporting data. Newton's use of idealized stationary apsides and elliptical orbits raises the question, then, of what *controls* the form that data take when used as inputs to theories and as confirmational tests. Glymour, after recording Duhem's observation, suggests this analysis: scientists may describe data in a simplified way so as to readily attach them to theory *only if* that description is compatible with the data within the range of calculated observable error: "one fraction of Newton's genius was to see that empirical laws, inconsistent with a theory, could still be used to argue for that theory, provided the inconsistencies were within observational uncertainties." (TE, pp. 222-224)

So on Glymour's view, what Newton did was to pick a mathematically tractable description of data that was later corrected by theory, but that was all the while compatible with then accepted observation. There are significant problems here with the concept of "observational uncertainties," since the calculation of these uncertainties is theory-dependent. And in some important cases, these uncertainties have been differing functions of the theories to be tested. An experiment may be feasible from one theoretical point of view and yet be unfeasible from the point of view of a competing theory.[14] Leaving aside this sort of problem, however, there is this *crushing* difficulty for Glymour: *Newton's descriptions of the phenomena were typically incompatible with the then accepted observational data.*

The most striking example of using idealized data not consistent with known observational uncertainties is Newton's use in Proposition II, Book III, of circular and not elliptical orbits for the planets. Glymour accounts for this counter example to his above thesis with the plausible though ad hoc (and nonbootstrapping) suggestion that "Newton wanted elliptical orbits to be demonstrated in his system rather than to be assumed at the outset . . . ." (TE, p. 208) That is, Newton wanted to show that his system was self-correcting: idealized data would lead to laws that could then be applied to correct the data. An inverse-square force entails elliptical orbits as a more general case. What I want to show is that this sort of self-correction is not an isolated feature restricted to the move from circular to elliptical orbits. Following Duhem, we have already noted that elliptical orbits also get corrected to accommodate perturbations due to other planets. But here, unlike the circular orbit case, we have a theory suggesting to the experimenter likely corrections to be made to currently accepted observational description. Newton told Flamsteed where and what to look for.[15] In the circular orbit case, on the other hand, theory had to catch up to accepted data. The "logic" of both sorts of cases is similar in that theory is articulated in order to demonstrate how experimental fit can be improved with respect to data less idealized as compared with known or suspected data.

The correction for apsidal rotation in Proposition XIV is like the circular orbit case (Prop. II) in that the originally supporting data were known *beforehand* to be idealized and false. Again, theory is articulated to show how better observational fit is possible. There is also significant use made of idealized data in Propositions III and IV dealing with the moon. As noted above in our initial review of Glymour's analysis of the beginning of Book III, Newton's bootstrap computation for the moon depends on the assumptions that the area law is satisfied and that the moon's apogee does not rotate. As Newton admits, however, the moon's apogee rotates at the eminently observational rate of $3°3'$ per revolution. Furthermore, it is a consequence of Proposition XLV, Book I, that because of this rotation the (net) force acting on the moon cannot be inverse square, and in fact must vary "inversely as $D^{2\ 4/243}$." Therefore Newton's attempt to *deduce* universal gravitation directly from phenomena has gone awry. And so as well has Glymour's interpretation of Newton as having bootstrapped his way from phenomena to instances of universal gravitation. Newton's response to this failure was at first to dismiss it, since the fractional increase "is due to the action of the sun." But, as was the case with the unexplained motion of the aphelions in Proposition XIV, Newton felt some obligation to explain away better the factor 4/243 in the second edition of *Principia*. After several false starts the following finally appeared:

> The action of the sun, attracting the moon from the earth, is nearly as the moon's distance from the earth; and therefore (by what we have shown in Cor. II, Prop. XLV, Book I) is to the centripetal force of the

moon as 2 to 357.45 or nearly so; that is, as 1 to $178^{29/40}$. And if we neglect so inconsiderable a force of the sun, the remaining force, by which the moon is retained in its orb, will be inversely as D^2.[16]

Seen as an *isolated* piece of argumentation, the passage can be reconstructed, as Glymour notes, as being HD in form: if only the sun and the earth are exerting inverse square forces on the moon, then the moon's apogee should rotate. But, such a reconstruction by itself does not explain what this apparently isolated piece of HD argumentation is doing in the *midst* of a Glymour-like instance derivation.

Glymour proposes this two-part explanatory account. First, Prop. III is *not* to be interpreted as a piece of bootstrapping: the moon data are not being used as a premise in the deduction of an instance of some corollary of universal gravitation. Second, that "the main point of the *argument* for Theorem III . . . is to demonstrate that the motion of the moon is *consistent* with the assumption that the Earth exerts an inverse square attractive force upon it." (TE, pp. 217-218) These proposals taken together are mildly surprising, since we would expect Glymour to have made some attempt to show that this is truly a case of *bootstrapping* in the sense of assuming the very hypothesis that one wishes to confirm.[17] Furthermore, it is not specified in what sense this is a consistency proof other than the obvious HD sense discussed earlier. But if this is so then Glymour's account seems to collapse into the admittedly unsatisfactory HD account. Finally, there is no real explanation as to why a consistency proof would be needed here. Glymour says merely that "without such a demonstration it might appear that the phenomena contradict the claim." (TE, p. 218)

These problems with Glymour's account worsen when we observe (and Glymour seems to have overlooked this) that a similarly HD-looking argument is given in Prop. II! Newton argues that the *accuracy* of the claim that the planets are attracted to the sun with an inverse square force is demonstrated by "the quiescence of the aphelion points"—i.e., on the assumption of universal gravitation, what we are out to prove, and Proposition XLV of Book I. And this is the same proposition used in Proposition III of Book III.

Looking on to Proposition IV of Book III, we find more instances of Newton's use of data not within the range of experimental error. First, the discrepancy is ignored between the predicted value of the rate of fall at the earth's surface ("15 [*Paris*] feet, 1 inch, and 1 line 4/9") and the observed rate of fall ("15 *Paris* feet, 1 inch, 1 line 7/9"). Second, an admittedly inaccurate earth-to-moon distance of 60 earth radii is used instead of the more accurate value of 60½ radii.

Newton planned, for the second edition, an explanation for the discrepancy in Proposition IV between observed and predicted rates of fall. This explanation, which was supposed to appear in the scholium to Proposition IV, was based on improved values for the size of the earth, the complication that the experimental values for the rate of fall were obtained not at the equator but at "the latitude of Paris," and once again the assumption of universal inverse-square gravitation. But, as Cotes objected, these improved values and corresponding calculations did not mesh with the rest of Book III, especially Propositions XIX (dealing with variations in the moon's orbit) and XXXVII (dealing with the influence of the moon on the tides). There was a lengthy and involved correspondence between Cotes and Newton as to "how to make the numbers appear to best advantage."[18] The net result was that Newton was unable to meet his printer's deadlines, so the proposed explanation was moved to the scholia of Propositions XIX and XXXVII.[19] (Does this sound familiar?) The relevance of all of this, then, is that in Proposition IV, as in the previous two propositions, there is a discrepancy between prediction and observed result that is not explained away as being due to observational error, as required by Glymour's account. Instead there was an attempt to explain away, in HD fashion, the discrepancy as being due to a hitherto ignored complication.

The other idealized datum used in Proposition IV is that the distance between earth and moon is 60 earth radii. The consensus was that the actual value was in excess of this figure by about one-half a radius, and Newton knew and accepted this.[20] Even so, the value 60 was used in his comparison of the force at the surface of the earth with that acting on the moon. What the argument of Proposition IV in fact establishes is this *counterfactual:*[21]

If (a) the earth is stationary (it is not);
(b) the distance between earth and moon is 60 earth radii (it is not);
(c) the moon's period is $27^d7^h43^m$ (an accurate though mean value);
(d) bodies fall at the surface of the earth at 15 1/12 feet/second (not observed at Paris latitudes);
then via Rules of Reasoning I and II, the same inverse square force acts on the moon and bodies on the surface of the earth.

Although, as noted, Newton gave the justification for (d) only in Propositions XIX and XXXVII, he gave the justification for (a) and (b) within his discussion of Proposition IV. Propositions LVII and LX of Book I show how to convert systems of rotation about a stationary body to systems of rotation about the center of mass of such systems. In particular, Proposition LX shows what the separation must be in the latter sort of case, given that the period of rotation is to be preserved and given that the force of attraction is inverse square and varies directly as the product of the masses. Newton asserts that use of this proposition converts the distance of the counterfactual situation into the more accurate value of 60 and 1/2 earth radii. (The details, though, are only given in *De mundi systemate*.)[22] All of this, however, is on the *assumption* of universal gravitation, the deductive goal of the first seven propositions of Book III.

Focusing attention on just the first four propositions of Book III shows that if Newton is a bootstrapper, it is only with highly idealized data not compatible with known observational error. Furthermore, bootstrapping gives an incomplete account of the argumentation, since Newton uses supplementary arguments to justify, on the basis of universal gravitation, these idealizations. A summary of these idealizations and their justifications will perhaps be useful. In Proposition II the quiescence of the aphelion points is used to prove "the great accuracy" of the existence of an inverse-square force and not per se to correct or justify anything. But showing great accuracy here is just to say that the original description of the orbits as circular is not crucial to the result, given the immobility of the aphelions. That the planetary orbits are not really stationary is admitted and explained away later in Proposition XIV. In Proposition III, the motion of the moon's apogee is at first assumed zero for bootstrapping purposes, but later in the proposition a more accurate nonzero value is given for this motion and accounted for. A conversion algorithm is referred to in Proposition IV that will transform the idealized counterfactual into something more realistic. In addition, if it were not for a printer's deadline, Proposition IV would have contained an account explaining away the discrepancy between observed and predicted rates of fall for terrestrial bodies.

Given, then, that Newton's deductive bootstrapping is filled with inaccurate and simplified data descriptions, our problem is to give a unified explanation of *the confirmatory value* of this sort of (strictly speaking) unsound argument. Since philosophers have almost universally ignored questions of accuracy and precision, as well as complications of the sort here illustrated, I shall return to Duhem as a useful source of insight. With respect to descriptions involving physical magnitudes, Duhem distinguished between what he called *theoretical* and *practical* facts.[23] The basic idea is this: true, precise, quantitative descriptions of phenomena, practical facts, are because of the complexity of nature either unavailable or unusable. So science must do with theoretical facts, i.e., with idealizations that are usually *not* compatible with experimental error, but that are justified in part by their *logical attachability* to scientific theories. Truth in descriptions can be achieved only at the cost of vagueness, but science demands precision. Typically, scientists have large amounts of quantitative data that are individual measurements of key parameters and properties. But these numbers do not immediately attach to theory. Newton gives only a carefully selected distillation of locational data in *Principia*, but the problem is clear enough. How are these individual pieces of data to be incorporated into theory? The problem is a species of curve fitting. What prima facie justifies our curves, however, is not simplicity per se, but theoretical attachability and practically possible computation.

In the case at hand, the idealized descriptions (or summaries) used by Newton were prima facie justified because of their ready attachment to the theorems of Book I of the *Principia*. Glymour *is* right: partial instances of universal gravitation were deducible from data and the laws of motion—but only from false and simplified data. However, logical attachability and practically computable consequences are obviously not enough to insure the rationality of the procedure. What is needed, I contend, is argumentation showing that *if* more accurate descriptions were fed into the theoretical hopper, correspondingly more accurate output would be obtained. That is, one must show that there are real possibilities for accommodating more accurate descriptions of data. To show that such possibilities exist need not require actually constructing the appropriate HD (or statistical) account. All that needs be shown is that a better (in terms of experimental fit) account is possible. And since, as I shall show, not every theory can demonstrate such possibilities, this is not a trivial requirement to place on theories.

A few examples will indicate what it is I am after. In his studies on optics, when confronted with the plain observational fact that there is color separation and diffusion after the second prism of the famous *experimentum crucis*, Newton responded by noting that in his treatment of the experiment, the aperture descriptions were excessively idealized as being infinitely small. Finite aperture size would, Newton argued, lead to the observed color separation and diffusion. But Newton did not actually construct an HD account that did this; he merely argued that it *could* be done. "But why the image is in one case circular, and in others a little oblong, and how the diffusion of light lengthwise may in any case be diminished at pleasure, I leave to be determined by geometricians, and compared with experiment."[24]

When Shankland argued that Miller's carefully obtained results from numerous repetitions of the Michelson-Morely experiment did not refute the Special Theory of Relativity, he tried to show that there was no reason to think that more accurate thermal convection theories *if available* could not explain away Miller's results. Again, Shankland did not actually construct the better HD mousetrap. "It is practically impossible to carry through calculations which would predict the over-all behavior of the interferometer due to temperature anomalies, since hardly any of the necessary data for such calculations exist."[25]

A rough estimate of the temperature fluctuations needed to account for Miller's positive results required these fluctuations to be ten times greater than those recorded. Nevertheless, Shankland went on to assert and conclude that:

> There is no doubt, however, that this factor ten would be very considerably reduced if convection of the air inside the casing were taken into account and if the contribution of the cover of this casing, facing the hut, could be evaluated. . . . We conclude from the foregoing estimate that an interpretation of the systematic effects in terms of the radiation field established by the non-uniform temperatures of the roof, the walls, and the floor of the observation hut is not in quantitative contradiction with the physical conditions of the experiment.[26]

Of course, the most decisive way to show that something is possible is to actually go ahead and do it. Just as good is to give an algorithm that completely specifies a constructive procedure. And these sorts of possibility proofs certainly have occurred in science. An example of a constructive procedure is Newton's reference to Propositions LVII and LX of Book I as recipes for converting a stationary earth-moon system of 60 radii separation into a more realistic moving system with a 60 1/2 radii separation. My point is that actually constructing better accounts and providing constructive procedures do not exhaust the types of arguments used in science to show that better experimental fit is possible. Newton's elaboration of the *experimentum crucis* shows this. Another, more complex, instance is his response in Proposition XIV to the fact that the aphelions do rotate. Newton did not construct an HD account that predicted these rotations solely on the basis of masses, locations, and initial velocities. The computational problems were too severe. What he did do was to provide a conditional account that predicted aphelion rotations when that of Mars was given. This showed that the theory was on the right track and could, with better data and computational methods, provide better experimental fit.

The other side of the possibility coin is that sometimes it can be demonstrated that better input and computational methods will not lead to better results. Newton's projection of a beam of sunlight through a prism provides an example of such a case. According to the then received view, the image should have been circular, but the experiment showed it to be oblong by a factor of about five. Despite this large discrepancy, Newton did not announce at this stage of his narrative the refutation of the received view.

> But because this computation was founded on the Hypothesis of proportionality of the *sines* of Incidence, and Refraction, which through by my own & others Experience I could not imagine to be so erroneous, as to make that Angle by 31′, which in reality was 2 deg. 49′; yet by curiosity caused me again to take my Prisme. And having placed it at my window, as before, I observed that by turning it a little about its *axis* to and fro, so as to vary its obliquity to the light, more than by an angle of 4 or 5 degrees, the Colours were not thereby sensibly translated from their place on the wall, and consequently by that variation of Incidence, the quantity of Refraction was not sensibly varied.[27]

It is only after this variation on the experiment that Newton claims to have achieved refutation. Given the perspective of this paper, the point of Newton's prism rotation is easy to see. An exactly circular image would on the received view occur only in the case of a prism placed symmetrically with respect to incoming and outgoing rays. All other orientations would result in oblong images. So it appeared open to supporters of the received view to claim that Newton was in error with respect to his experimental assessment of symmetrical orientation. What the rotation of the prism shows, however, is that even if Newton were grossly in error on this point, by as much as five degrees, the experimental outcome, the oblong spectrum, would not be sensibly varied. In others words, one type of saving argument is not a real possibility for the received view. Even if more accurate data were available, they would not help.[28]

Pardies objected that Newton's theoretical prediction was based on a mathematical treatment that assumed the beam aperture to be infinitely small. Perhaps a more realistic account would save the received view. But Newton was able to show that the benefits in terms of image elongation did not come close to accommodating the observed discrepancy.[29] Lorentz made a similar response to some detractors of the Michelson-Morley experiment. They contended that more realistic and detailed treatments of the experiment, as opposed to simple single ray accounts, would show that a null result was to be anticipated on the basis of aether theory. And several such demonstrations were in fact published. But Lorentz put an end to such demonstrations by constructing a generalized proof showing that the null result would remain *despite* more realistic treatments. Hence there were mistakes in the published and very complicated detailed treatments of the experiment.[30]

What these examples suggest is that scientific reasoning is a *two-part* affair. First there is the idealization of data until they can be attached in some coherent way to theory. A theory is confirmed with respect to idealized but (strictly speaking) false data if it satisfies Glymour's bootstrapping, or the HD, or perhaps other accounts of confirmation. This, however, is just the initial step. A more severe test of theory occurs at the next stage. Here there are arguments, quite various in form, showing that *if* the idealizations were to be replaced with more realistic descriptions, then the relations between theory and data would *continue* to be confirmatory. A theory then is confirmed with respect to more accurate and less idealized data, if this latter class of arguments, which I have elsewhere called modal auxiliaries, exists.[31] A theory is *disconfirmed* if it can be shown that the introduction of realism does not lead to convergence with more accurate descriptions of the data.

I shall now apply this sort of gestalt to the case of gravitation. Newton can be reconstructed as showing how universal gravitation is deducible from the Rules and from bootstrapped instances. In particular, the first three propositions of Book III can be read as examples of simple bootstrapping. But this tidy connection is achieved only at the cost of using idealized evidence, i.e., evidence simplified for theoretical convenience and not consistent with observational error. What adds confirmatory value to this deductive but unsound story are Newton's various demonstrations that experimental fit can be improved, or, in the case of Proposition IV, that the same result is obtainable if more accurate data are used. The precession of the moon's apogee can be accounted for in terms of the sun's interference. The use of the earth-to-moon distance of 60 earth radii converts to the observationally more accurate 60 and 1/2 radii if the earth's motion around the sun is taken into account. The variation in predicted and observed rates or terrestrial acceleration can be explained away in terms of the interfering centrifugal forces at higher latitudes. The purpose of the various HD arguments sprinkled among the deductive derivation of theory from data is to provide the necessary justification for the initial use of idealized evidence. Newton is showing that things can be made better.

Finally, it should be noted that Newton's demonstration of universal gravitation is immediately followed by Proposition VIII, which shows that no harm is done in treating spherical gravitational sources as having their mass concentrated at their centers. At distances greater than their radii, gravitational attraction will continue to be inverse square. Therefore, *a more realistic treatment* of objects as spatially extended will *not* disturb Newton's basic results.[32]

Glymour admits in *Theory and Evidence* the greater historical accuracy of my account. It provides a better surface grammar than his account. He balks, however, at my "radical treatment of observation." (TE, p. 217) I have tried to show here that my approach also makes good *normative* sense, since it provides a way of living with the Duhemian dilemma of truth or theoretical tractability. To quote one of Glymour's requirements for a confirmation theory, my account does, I contend, "*explain* both methodological truism and particular judgments that have occurred within the history of science." (TE, p. 64) It is a necessary adjunct to Glymour's bootstrapping account.

Notes

1. The first two quotations are from Newton's correspondence with Oldenburg (6 July 1672, 6 February 1671/2) and are reprinted in H. W. Turnbull, ed., *The Correspondence of Isaac Newton*, I (Cambridge: Cambridge University Press, 1959), 209, 96-97. The last quotation is from the General Scholium of Newton's *Principia* and originally appeared as a late addition in Newton's correspondence with Cotes (28 March 1713), A. Rupert Hall and Laura Tilling, eds., *Correspondence*, V (1975), 397. See also Alexandre Koyré and I. Bernard Cohen, ed.,

Isaac Newton's Philosophiae Naturalis Principia Mathematica, in two volumes (Cambridge, Mass.: Harvard University Press, 1972), II, pp. 763-764. I shall use Cajori's revision of Motte's translation of Principia: *Sir Isaac Newton: Principia* (Berkeley and Los Angeles: University of California Press, 1966), p. 547.

2. Clark Glymour, *Theory and Evidence* (Princeton: Princeton University Press, 1980), pp. 203-226. For the sake of brevity I shall make all references to this work *within* the text of the paper and I shall use the notation TE.

3. Glymour's account of Proposition IV (TE 218) is somewhat abbreviated, but I do not believe he intends this to be interpreted as bootstrapping. Given some stretching of the text, however, a bootstrapping interpretation is not impossible.

4. Because of Newton's use of his Rules of Reasoning, it is impossible to determine exactly where bootstrapping stops and where use of the Rules begins. For example, Proposition II deals only with the primary planets, i.e., excluding the earth, whereas in Proposition V Newton assumes that all of the planets are attracted to the sun. Newton can be interpreted therefore as having instantiated a more general proposition or as having applied his Rules. See also *The System of the World*, translated in Cajori, *Principia*, pp. 554-559.

5. Even liberally reconstructed, Newton's computations are still only "partial" since no instance or counter-instance is computable on the basis of noncurved orbits. But Glymour, with laudable foresight, allows for such partial functions in his statement of the bootstrapping conditions. See TE, pp. 158-159.

6. Cf. Cotes to Newton, 18 March 1712/13, *Correspondence*, V, p. 392.

7. Pierre Duhem, *The Aim and Structure of Physical Theory*, trans. Philip P. Wiener (New York: Atheneum, 1962), p. 192.

8. Another option, surprisingly not noted by Glymour, would be to simply accept universal gravitation as a bootstrapping assumption of the computation and to then demonstrate that the production of negative instances is nevertheless possible. But this need not be unwelcome by a Duhemian, since it could be interpreted as specifying the consistency requirements that need be satisfied by our conventions. Cf. Hans Reichenbach, *The Philosophy of Space and Time* (New York: Dover, 1959), p. 17.

9. Koyré and Cohen, *Principia*, II, p. 590. Cajori, *Principia*, p. 422.

10. Cajori, *Principia*, pp. 574-575. For the history of this early version of Book III of *Principia*, see I. Bernard Cohen, *Introduction to Newton's 'Principia'* (Cambridge: University Press, 1971), pp. 327-335.

11. Queries on *Principia*, (February 1725), reprinted in A. Rupert Hall and Laura Tilling, eds., *Correspondence* VII, p. 306.

12. For documentation of Newton's revisions to Proposition XIV, see Koyré and Cohen, *Principia*, II, pp. 590-591. Translation in Cajori, *Principia*, p. 422. Incidently, no changes were made in response to Pemberton's query.

13. Duhem, *Aim and Structure*, p. 193.

14. For some examples see my Independent Testability: The Michelson-Morley and Kennedy-Thorndike Experiments. *Philosophy of Science* 47 (1980): pp. 1-35.

15. To Flamsteed (30 December 1684), reprinted in H. W. Turnbull, *Correspondence*, II, pp. 406-407.

16. Koyré and Cohen, *Principia*, II, p. 566; Cajori, *Principia*, p. 407. For the correspondence leading up to this addition start with Cotes to Newton (23 June 1711), *Correspondence*, V, p. 170, where Cotes complains (rightly) to Newton concerning a proposed addition: "I should be glad to understand this place, if it will not be too much trouble to make it out to me. I do not at present so much as understand what it is yt You assert."

17. Glymour, in his closing appraisal of Newton's demonstration of universal gravitation (TE, p. 225), mentions as a possibility that Proposition III be given a bootstrapping interpretation.

18. Cotes to Newton (16 February 1711/12), *Correspondence*, V, p. 226. See also Cotes to Newton (23 February 1711/12), Ibid., p. 233 where Cotes writes: "I am satisfied that these exactnesses, as well here as in other places, are inconsiderable to those who can judge rightly

of Your book: but ye generality of Your Readers must be gratified with such trifles, upon which they commonly lay ye greatest stress. . . . You have very easily dispatch'd the 32 Miles in Prop. XXXIXth, I think You have put the matter in the best method which the nature of the thing will bear."

19. See Cotes to Newton (13 March 1711/12) and Newton to Cotes (18 March 1711/12), *Correspondence*, V, pp. 246-248.

20. This, I believe, is clearly the sense of Newton's discussion of the data in Proposition IV. However, if there is any doubt about this reading, for additional evidence see Newton's draft scholium to Proposition IV, reprinted in *Correspondence*, V, pp. 216-218. See also Corollary VII of Proposition XXXVII, Book III, where some of this draft finally appeared.

21. For additional support that this is the correct reading, see Pemberton's Queries on *Principia* (? February 1725), *Correspondence*, V, p. 306, where he writes: "the whole paragraph seems to me not to express, what is intended by it, in the fullest manner: your design being to give a reason why you assumed the distance of the moon from the earth a little less than what you shew astronomical observations to make it. Would not this intent be a little more fully expressed after the following manner? 'As the computation here is based on the hypothesis that the Earth is at rest, the distance of the Moon from the Earth is taken a little less than astronomers have found it. If account is taken of the Earth's motion about the common centre of gravity of the Earth and the Moon, the distince here postulated must be increased (by Prop. 60, Book I) so that the law of gravity may remain the same; and afterwards (corol. 7, Prop. 37 of this book) if may be found to be about 60 1/2 terrestrial radii.'"

22. I.e., MS Add. 3990; Cajori, *Principia*, pp. 560-561.

23. Duhem, *Aim and Structure*, pp. 132-138.

24. To Oldenburg for Pardies (10 June 1672), *Correspondence*, I, p. 167. For more details about the *experimentum crucis* see my Newton's *Experimentum Crucis* and the Logic of Idealization and Theory Refutation. *Studies in History and Philosophy of Science* 9 (1978): 51-77.

25. R. S. Shankland, New Analysis of the Interferometer Observations of Dayton C. Miller. *Reviews of Modern Physics* 27 (1955): 175.

26. Ibid. For more details about this case and a comparison with Newton's *experimentum crucis* see my Newton's *Experimentum Crucis*, pp. 71-75.

27. To Oldenburg (6 February 1671/2) [from *Philosophical Transactions* 6 (1671/2), 3075-3087], *Correspondence*, I, pp. 93-94.

28. What Newton showed experimentally is that at orientations near symmetrical, the size of the spectrum is insensitive to changes in orientation. So large errors in initial conditions have only small effect on predicted effect. Duhem gives a sort of converse example in *Aim and Structure*, p. 139, where small errors in initial conditions yield large errors in predicted effect. Surprisingly, no commentator that I am aware of has correctly reported the logic of Newton's argument. They all give a tradtional *modus tollens* analysis whereby it is held that Newton rejected the received view simply because of the lack of experimental fit. And these commentators include the usually perceptive Thomas Kuhn in, for example, his Newton's Optical Papers, *Isaac Newton's Papers & Letters On Natural Philosophy*, ed. I. B. Cohen (Cambridge, Mass.: Harvard University Press, 1958), p. 32. For more details about the spectrum experiment see my Newton's Advertised Precision and His Refutation of the Received Laws of Refraction. *Studies in Perception: Interrelations in the History and Philosophy of Science*, ed. P. K. Machamer and R. G. Turnbull (Columbus: Ohio State University Press, 1977), pp. 231-258.

29. Pardies to Oldenburg (30 March 1672), Newton to Oldenburg (13 April 1672), *Correspondence*, I, pp. 130-133, 140-142. For an analysis of this interchange see my Newton's Advertised Precision, pp. 252-254.

30. Lorentz's proof, several references, as well as discussion, appear in Conference on the Michelson-Morley Experiment, *Astrophysical Journal* 68 (1928): 341-373.

31. In my Idealization, Explanation and Confirmation, *PSA 1980: Proceedings of the 1980 Biennial Meeting of the Philosophy of Science Association*, I, ed. Peter D. Asquith and Ronald N. Giere (East Lansing: Philosophy of Science Association, 1980), pp. 336-350. What

distinguishes my view from that of Lakatos is his use of series of (predictionally) complete theories as compared with my emphasis on possibility proofs. What distinguishes my view from that of Kuhn is my insistence on the rationality of possibility proofs. See my "Idealization," 347, fn. 6 and 8.

32. See the well-known Newton to Halley (26 June 1686), *Correspondence*, II, p. 435.

— Michael R. Gardner —

Realism and Instrumentalism in Pre-Newtonian Astronomy

1. Introduction

There is supposed to be a problem in the philosophy of science called "realism versus instrumentalism." In the version with which I am concerned, this supposed problem is whether scientific theories in general are put forward as true, or whether they are put forward as untrue but nonetheless convenient devices for the prediction (and retrodiction) of observable phenomena.

I have argued elsewhere (1979) that this problem is misconceived. Whether a theory is put forward as true or merely as a device depends on various aspects of the theory's structure and content, and on the nature of the evidence for it. I illustrated this thesis with a discussion of the nineteenth-century debates about the atomic theory. I argued that the atomic theory was initially regarded by most of the scientific community as a set of false statements useful for the deduction and systematization of the various laws regarding chemical combination, thermal phenomena, etc.; and that a gradual transition occurred in which the atomic theory came to be regarded as a literally true picture of matter. I claimed, moreover, that the historical evidence shows that this transition occurred because of increases in the theory's proven predictive power; because of new determinations of hitherto indeterminate magnitudes through the use of measurement results and well-tested hypotheses; and because of changes in some scientists' beliefs about what concepts may appear in fundamental explanations. I posed it as a problem in that paper whether the same or similar factors might be operative in other cases in the history of science; and I suggested that it might be possible, on the basis of an examination of several cases in which the issue of realistic vs. instrumental acceptance of a

This material is based upon work supported by the National Science Foundation under Grants No. SOC 77-07691 and SOC 78-26194. I am grateful for comments on an earlier draft by Ian Hacking, Geoffrey Hellman, Roger Rosenkrantz, Robert Rynasiewicz and Ferdinand Schoeman.

theory has been debated, to put forward a (normative) theory of when it is reasonable to accept a theory as literally true and when as only a convenient device.

For simplicity I shall usually speak of the acceptance-status of a theory as a whole. But sometimes it will be helpful to discuss individual hypotheses within a theory, when some are to be taken literally and others as conceptual devices.

In the present paper I would like to discuss a closely analogous case—the transition, during the Copernican revolution, in the prevailing view of the proper purpose and correlative mode of acceptance of a theory of the planetary motions. I shall briefly discuss the evidence—well known to historians—that from approximately the time of Ptolemy until Copernicus, most astronomers held that the purpose of planetary theory is to permit the calculations of the angles at which the planets appear from the earth at given times, and not to describe the planets' true orbits in physical space. For a few decades after Copernicus's death, except among a handful of astronomers, his own theory was accepted (if at all) as only the most recent and most accurate in a long series of untrue prediction-devices. But eventually the Copernican theory came to be accepted as the literal truth, or at least close to it. That this transition occurred is well known; why it occurred has not been satisfactorily explained, as I shall try to show. I shall then try to fill in this gap in the historical and philosophical literature. In the concluding section I shall also discuss the relevance of this case to theses, other than the one just defined, which go by the name "realism."

2. The Instrumentalist Tradition in Astronomy

The tradition of regarding theories of planetary motion as mere devices to determine apparent planetary angles is usually said to have originated with Plato; but it also has roots in the science of one of the first civilizations on earth, that of the Babylonians. Their reasons for being interested in the forecasting of celestial phenomena were largely practical in nature. Such events were among those used as omens in the conduct of government. Predictions of eclipses, for example, were thus needed to determine the wisdom of planning a major undertaking on a given future date. In addition, since the new month was defined as beginning with the new moon, one could know when given dates would arrive only through predictions of new moons. The Babylonians therefore appear to have developed purely arithmetical procedures for the prediction of apparent angles of celestial bodies. Their main technique was the addition and subtraction of quantities, beginning with some initial value and moving back and forth between fixed limits at a constant rate. Through such techniques they could forecast such quantities as lunar and solar velocity, the longitudes of solar-lunar conjunctions, the times of new moons, etc. Although numerous tablets have survived showing techniques for the computation of such quantities, and others showing the results of the computations, we have no evidence that these techniques were based on any underlying theory of the geometrical structure of the universe (Neugebauer 1952, Chapter 5). Moreover, the arithmetical character of the regularities in the variations of the quantities also suggests that the computations are based on purely arithmetical rather than geometrical considerations. Of course it is possible that the forecasting techniques were based upon statements about and/or pictures of the geometry of the universe that were not committed to clay or that do not survive. But it appears that in Babylonian mathematical astronomy—probably developed between about 500 and 300 BC—we have an extreme instance of instrumentalism: not even a theory to be interpreted as a prediction-device rather than as purportedly true, but only a set of prediction-techniques in the literal sense.

As I have said, another major source of the instrumentalist tradition in astronomy is Plato. I do not mean that he was a source of the instrumentalists' conception of the aims of astronomical theorizing, but rather of their strictures upon the permissible means for achieving those ends. Vlastos (1975, pp. 51-52) has assembled from various Platonic works an argument that runs as follows: celestial bodies are gods and are moved by their souls; these souls are perfectly rational; only uniform circular motions befit rationality; hence all celestial motions are uniform (in speed) and circular. To document this, we must note that in *Laws* Plato (in Hamilton and Cairns 1963) asserted that the sun's "soul guides the sun on his course," that this statement is "no less applicable to all. . . celestial bodies," and that the souls in question possess "absolute goodness"—i.e., divinity (898 C-E). Plato required in the argument the tacit assumption that goodness entails rationality. He could then conclude that since bodies "moving regularly and uniformly in one compass about one center, and in one sense" must "surely have the closest affinity and resemblance that may be to the revolution of intelligence," celestial motions must be uniform and circular (898 A-B). Plato was also in a position to argue for the rationality of a celestial soul on the grounds that it, unlike our souls, is not associated with a human body, which causes "confusion, and whenever it played any part, would not allow the soul to acquire truth and wisdom." (*Phaedo* 66A, in Bluck 1955)

Two clarifications are needed. Although Plato was not explicit on the point, he seems to have assumed that all these circular motions are concentric with the earth. This is presumably why at *Timaeus* 36D (in Cornford 1957) he says that the circles of the sun, moon, and five (visible) planets are "unequal"—so that they can form a nest centered about the earth, as Cornford suggests. (p. 79) Second, we are not to assume that these "circular" motions trace out simple circles. Rather, the complex observed paths are compounded out of a plurality of motions, each of which is uniform and circular. (*Timaeus* 36 C-D)

A story of uncertain reliability (Vlastos 1975, pp. 110-111) has it that Plato, perhaps aware of the empirical inadequacies of his own theory, proposed to astronomers the problem: "What uniform and orderly motions must be hypothesized to save the phenomenal motions of the stars?" (Simplicius, quoted in Vlastos 1975) The first astronomer to make a serious attempt to answer this question in a precise way was Plato's student Eudoxus. He postulated that each of the seven "wandering" celestial bodies was moved by a system of three or four earth-centered, nested, rotating spheres. Each inner sphere had its poles attached at some angle or other to the next larger sphere. The celestial body itself was located on the equator of the innermost sphere of the system. By adjustments of the angles of the poles and the (uniform) speeds of rotation, Eudoxus was able to obtain a fairly accurate representation of the motion in longitude (including retrogression) of three planets and the limits (but not the times) of the planets' motion in latitude. (Dreyer 1953, pp. 87-103)

In the extant sources on the Eudoxian theory (Aristotle and Simplicius), no mention occurs of any theories regarding the physical properties of these spheres—their material, thickness, or mutual distances—nor on the causes producing their motions. And even though the outermost sphere for each celestial body moves with the same period (one day), Eudoxus appears to have provided no physical connections among the various systems. (Dryer 1953, 90-91) The (admittedly inconclusive) evidence therefore suggests that Eudoxus regarded the spheres not as actual physical objects in the heavens, but as geometrical constructions suitable for computing observed positions. If this is correct, Eudoxus, the founder of quantitatively precise geometrical astronomy, was also an important early instance of some aspects of the instrumentalist tradition within that field. By this I do not mean that Eudoxus denied that his theory is true, or even that he was sceptical of its truth, but only that he appears to have limited the aim of astronomy to a geometrical representation (not including a causal explanation) of the observed phenomena. From this position it is only one step to the instrumentalist view that there *cannot* be a causal explanation of the orbits postulated by a given astronomical theory, and that the theory's descriptions of the orbits are therefore *not true*.

Whether Eudoxus held this view about the status of the spheres or not, it is quite clear that Aristotle held the opposite view. In *Metaphysics* XII, 8, he remarked that in addition to the seven spheres added to the Eudoxian system by Callipus for the sake of greater observational accuracy, additional spheres must be intercalated below the spheres of each planet, in order to counteract the motions peculiar to it and thereby cause only its daily motion to be transmitted to the next planet below. "For only thus," he said, "can all the forces at work produce the observed motion of the planets." (in McKeon 1941, 1073b-1074a) Clearly, then, Aristotle thought of the spheres as real physical objects in the heavens, and supposed that some of their motions are due to forces transmitted physically from the sphere of stars all the way to the moon.

At *Physics* II, 2 Aristotle expressed his realistic view of astronomy in a different way. He raised the question,

> Is astronomy different from physics or a department of it? It seems absurd that the physicist should be supposed to know the nature of sun or moon, but not to know any of their essential attributes, particularly as the writers on physics obviously do discuss their shape also and whether the earth and the world are spherical or not.
>
> Now the mathematician, though he too treats of these things, nevertheless does not treat of them as the limits of a physical body, nor does he consider the attributes indicated as the attributes of such bodies. That is why we separated them; for in thought they are separable from motion. . . .
>
> Similar evidence is supplied by the more physical of the branches of mathematics, such as optics, harmonics, and astronomy. These are in a way the converse of geometry. While geometry investigates physical lines, but not *qua* physical, optics [and, presumably, astronomy] investigates mathematical lines, but *qua* physical, not *qua* mathematical.

Although the message of this passage is somewhat ambiguous, its main point can be discerned. Aristotle thought that, generally, mathematicians deal with geometrical objects, such as shapes, as nonphysical entities, and thus they are presumably unconcerned with such questions as the constitution of objects and the forces upon them. Although astronomy is a branch of mathematics, it is one of the more physical branches; thus the foregoing generalization does not apply to astronomers, who must treat spheres, circles, etc. as physical objects. They cannot, then, as an astronomical instrumentalist would say they should, limit their concerns to the question of whether various geometrical constructions permit correct determinations of planetary angles.

The most influential astronomer before Copernicus was, of course, Ptolemy (fl. 127-150 AD); and it is therefore of considerable interest whether he regarded his theory as providing a true cosmological picture or as only a set of prediction-devices. This question, however, admits of no simple answer. To some extent, Ptolemy worked within traditions established by Plato and by Aristotle. Echoing (but not explicitly citing) Plato's postulate about uniform circular motions in the divine and celestial realm, he wrote in the *Almagest*, IX, 2:

> Now that we are about to demonstrate in the case of the five planets, as in the case of the sun and the moon, that all of their phenomenal irregularities result from regular and circular motions—for such befit the nature of divine beings, while disorder and anomaly are alien to their nature—it is proper that we should regard this achievement as a great feat and as the fulfillment of the philosophically grounded mathematical theory [of the heavens]. (Trans. Vlastos 1975, p. 65)

Again, he wrote, "we believe it is the necessary purpose and aim of the mathematician to show forth all the appearances of the heavens as products of regular and circular motions." (*Almagest*, III, 1)

Ptolemy followed Aristotle's division of the sciences, but diverged somewhat when it came to placing astronomy within it:

> For indeed Aristotle quite properly divides also the theoretical into three immediate genera: the physical, the mathematical, and the theological. . . .the kind of science which traces through the material and ever moving quality, and has to do with the white, the hot, the sweet, the soft, and such things, would be called physical; and such an essence [ousia], since it is only generally what it is, is to be found in corruptible things and below the lunar sphere. And the kind of science which shows up quality with respect to forms and local motions,

seeking figure, number, and magnitude, and also place, time, and similar things, would be defined as mathematical. (*Almagest*, I, 1)

He then remarked that he himself had decided to pursue mathematics (specifically, astronomy), because it is the only science that attains "certain and trustworthy knowledge." Physics does not do so because its subject (sublunar matter) is "unstable and obscure." (*Almagest*, I, 1) Ptolemy, then, separated astronomy from physics more sharply than Aristotle did: he held that it is a branch of mathematics and did not say that it is one of the "more physical" branches. Such a claim, indeed, would make no sense in view of Ptolemy's claim that physics deals with the sublunar, corruptible realm.

Since Ptolemy denied that astronomy is a branch of physics, or even one of the more physical branches of mathematics, he did not put forward any theory analogous to Aristotle's of a cosmos unified through the transmission of physical forces productive of the planetary motions. Rather, he treated each planet's motion as a separate problem, and held that the motions are not explained in terms of physical forces at all, but in terms of the individual planet's essence: "The heavenly bodies suffer no influence from without; *they have no relation to each other;* the particular motions of each particular planet follow from the essence of that planet and are like the will and understanding in men." (*Planetary Hypotheses*, Bk. II; trans. Hanson 1973, p. 132)

But were these "motions" mere geometric constructions, or were they physical orbits? A key point is that Ptolemy did not claim in the *Almagest* to know how to determine the distances of the planets from the earth or even to be entirely certain about their order: "Since there is no other way of getting at this because of the absence of any sensible parallax in these stars [planets], from which appearance alone linear distances are gotten," he said he had no choice but to rely on the order of "the earlier mathematicians." (IX, 1) Although the angles given by Ptolemy's constructions in the *Almagest* are to be taken as purportedly corresponding to physical reality, the supposed physical distances were left unspecified. But this leaves open the possibility that Ptolemy thought his constructions gave the physical orbits up to a scale factor—i.e., that they gave the orbit's shape. Dreyer argues that Ptolemy's theory of the moon, if interpreted in this way, implies that the moon's distance and therefore apparent diameter vary by a factor of 2; and Ptolemy (like nearly everyone) must have been aware that no such variation in diameter is observed. (1953, p. 196) Against this it must be said that in the *Almagest* Ptolemy claimed that the moon's distance, unlike those of the five planets, can be determined from observational data via his model and that it varies from about 33 to 64 earth radii. (V, 13) He gave no indication that these numbers were not to be taken literally, despite what he obviously knew they implied about apparent diameter. He also took the numbers seriously enough to use them (and data on the apparent diameters of the sun, moon, and the earth's shadow) to obtain a figure for absolute distance to the sun—1210 earth radii. (*Almagest*, V, 15)

A stronger piece of evidence for Dreyer's view is Ptolemy's admission in *Planetary Hypotheses:* "I do not profess to be able thus to account for all the motions at the same time; but I shall show that each by itself is well explained by its proper hypothesis." (Trans. Dreyer 1953, p. 201) This quotation certainly suggests that Ptolemy did not think the totality of his geometric constructions was a consistent whole, and thus that he could not have thought that it (that is, all of it) represented physical reality.

Again, discussing the hypothesis that a given planet moves on an epicycle and the hypothesis that it moves on a deferent eccentric to the earth, he claimed: "And it must be understood that all the appearances can be cared for interchangeably according to either hypothesis, when the same ratios are involved in each. In other words, the hypotheses are interchangeable.... (*Almagest*, III, 3) This relation of "interchangeability" is apparently logical equivalence of actual asserted content; for when confronted with the identity of the two hypotheses' consequences for the appearances—i.e., planetary angles at all times—Ptolemy did not throw up his hands and say he could not decide between them (as would have been rational had he thought them nonequivalent). He straightway declared (in the case of the sun) that he would "stick to the hypothesis of eccentricity which is simpler and completely effected by one and not two movements." (*Almagest*, III, 4) In sum, his arguments in these passages appear to make sense only on the assumption that the genuine asserted content of the orbital constructions is exhausted by what they say about observed angles.

On the other hand, his use of the lunar model to determine distances makes sense only on a realist interpretation. Moreover, his planetary models are based partly on the assumption that maximum brightness (minimum distance) coincides with retrogression because both occur on the inner part of the epicycle (Pedersen 1974, p. 283); and this reasoning also presupposes that the model gives true orbits. We can only conclude that Ptolemy's attitude was ambivalent, or at least that the evidence regarding it is ambiguous.

Whatever we say about Ptolemy's attitude towards the planetary orbits, it would clearly be an exaggeration to say that his theory had no asserted content at all beyond what it implies about observed angles. For Ptolemy explicitly subscribed, of course, to a cosmology that placed the earth at rest in the center of an incomparably larger stellar sphere; and he argued for this conception on the basis of physical considerations—supposed effects the earth's motion would have on bodies on or near the earth. (*Almagest*, I, 2 and 7)

Moreover, some of his arguments are based on physical considerations regarding the nature of the ether (*Almagest*, I, 3) and heavenly bodies (XIII, 2)—their homogeneity, eternity, constancy, irresistibility, etc. (Lloyd 1978, p. 216) Thus the theory has considerable (literally-intended) cosmological and physical as well as observational content.

By the time he wrote *Planetary Hypotheses*, Ptolemy had changed his view about the possibility of determining the distances of the five planets despite their lack of visible parallax. He explained his procedure as follows: "We began our inquiry into the arrangement of the spheres [i.e., spherical shells] with the determination, for each planet, of the ratio of its least distance to its greatest distance. We then decided to set the sphere of each planet between the furthest distance of the sphere closer to the earth, and the closest distance of the sphere further (from the earth)." (Goldstein 1967, pp. 7-8) I shall call the assumption decided upon here the "nesting-shell hypothesis": no space between shells. Using the ratios mentioned here and his values for the lunar distances—both supplied by the theory of the *Almagest*—together with this new hypothesis, Ptolemy computed least and greatest distances for all the planets. And what was the evidence for the nesting-shell hypothesis? Ptolemy said only that "it is not conceivable that there be in nature a vacuum, or any meaningless and useless thing," and that his hypothesis (after some ad hoc fiddling with parameters) meets the weak constraint that it agree with his independently measured value for solar distance. He was apparently not entirely convinced by the argument himself; for he continued: "But if there is space or emptiness between the (spheres), then it is clear that the distances cannot be smaller, at any rate, than those mentioned."

To conclude on Ptolemy, then: in his post-*Almagest* writings, Ptolemy added a further quantity that was to be taken seriously and not just as an angle-predictor—namely, earth-to-planet distance. But he could compute the values of this quantity from observational data (on lunar eclipses) only through the use of a hypothesis for which he had almost no observational evidence.

In the period between Ptolemy and Copernicus, when Ptolemy's approach dominated astronomy, it is possible to distinguish several positions concerning the sense (if any) in which Ptolemaic astronomy should be accepted. One is the view—suggested by some of Ptolemy's own remarks—that the circles associated with the planets determine only observed angles and not orbits in physical space. A prominent example of a person who held such a view quite explicitly is Proclus, a fifth-century commentator. He indicated that he was alive to the distinction between the instrumentalist and realist interpretation of Ptolemaic astronomers by asking, "what shall we say of the eccentrics and the epicycles of which they continually talk? Are these only inventions or have they a real existence in the spheres to which they are fixed?" He concluded his discussion by conceding that some Ptolemaic system was acceptable at least with the former interpretation: "these hypotheses are the most simple and most fitting ones for the divine bodies. They were invented in order to discover the mode of the planetary motions which in reality are as they appear to us, and in order to make the measure inherent in these motions apprehendable." (Proclus 1909, VII 50 (236, 10); trans. Sambursky 1962, pp. 147-149) Earlier, however, Proclus had given a variety of arguments to show that no epicyclic-eccentric theory could be accepted as genuinely true. For example, he argued that such a theory, literally interpreted, is "ridiculous" because it is inconsistent with the principles of physics he accepted—specifically, those of Aristotle: "These assumptions make void the general doctrine of physics, that every simple motion proceeds either around the center of the universe or away from or towards it." (Proclus 1903, 248c [III 146, 17]; trans. Sambursky 1962, pp. 147-149) He argued further: "But if the circles really exist, the astronomers destroy their connection with the spheres to which the circles belong. For they attribute separate motions to the circles and to the spheres and, moreover, motions that, as regards the circles, are not at all equal but in the opposite direction." The point here is that the oppositeness of the deferent and epicyclic motions (for sun and moon) is physically implausible in view of their spatial interrelations (Proclus 1909, VII 50 (236, 10); trans. Sambursky 1962, pp. 147-149)

In addition, Proclus gave two arguments of an epistemological character purporting to show that we have insufficient evidence to justify acceptance of any Ptolemaic theory on a realistic interpretation. One of these is that the heavens are necessarily beyond the reach of merely human minds: "But when any of these [heavenly] things is the subject of investigation, we, who dwell, as the saying goes, at the lowest level of the universe, must be satisfied with 'the approximate.' " (Proclus 1903, I, pp. 352-353; trans. Duhem 1969, pp. 20-21) This argument appeals to the specific character of astronomy's domain; he bolstered his conclusion with a second epistemological argument that is, in principle, applicable to other sciences as well. It is that there could be, and in fact are, alternative systems of orbits (epicycle, eccentric, and homocentric) equally compatible with the observed data; hence which of these sets comprises the true physical orbits is unknowable: "That this is the way things stand is plainly shown by the discoveries made about these heavenly things—from different hypotheses we draw the same conclusions relative to the same objects...hypotheses [about]...epicycles...eccentrics...counterturning spheres." (Proclus 1903, I, pp. 352-353; trans. Duhem 1969, pp. 20-21)

A complication in Proclus's position is that whereas he conceded that the Ptolemaic theory is useful for predictions, even though it is not physically true, he did not think one can in the long run be satisfied with astronomical theories that are only conceptual devices and fail to provide causes: "For if they are only contrived, they have unwittingly gone over from physical bodies to mathematical concepts and given the causes of physical movements from things that do not exist in nature." (Trans. Lloyd 1978, p. 205. My interpretation follows Lloyd, in disagreement with Duhem.)

Essentially the same three arguments against realistic acceptance of epicycles and eccentrics—their violations of physical principles, the impossibility of human astronomical knowledge, and the existence of alternative observationally equivalent theories—are found in many writers in the period between Ptolemy and Copernicus. (See Duhem 1969, passim.) We should also take note of one further type of argument against accepting an epicyclic theory on a realistic interpretation: it is that the theory thus understood has consequences that have been observed to be false. Along these lines, Pontano, a widely read astronomer in the early sixteenth century, argued that if the epicycles were real physical orbits, they would have been formed (like the planets) from "solidification" of the spheres carrying them, and thus would be visible—as, of course, they are not. (Quoted in Duhem 1969, pp. 54-55)

Among those who conceded that the evidence showed Ptolemy's theory agrees with the observed celestial motions—and was thus acceptable in just that sense—we can distinguish a subgroup who made it clear that they nonetheless thought that since the theory could not be accepted on a realistic interpretation, it should be replaced with one that could. Some writers also maintained that such an astronomical revolution would also open the way to the sort of logical reconciliation between astronomy and physics attempted by Aristotle: real physical orbits, unlike Ptolemaic epicycles, must obey the laws of physics. An example of such a writer is the great Islamic philosopher Averroes (1126-1198):

> The astronomer must, therefore, construct an astronomical system such that the celestial motions are yielded by it and that nothing that is from the standpoint of physics impossible is implied. . . . Ptolemy was unable to see astronomy on its true foundations. . . . The epicycle and the eccentric are impossible. We must therefore, apply ourselves to a new investigation concerning that genuine astronomy whose foundations are principles of physics. . . . Actually, in our time astronomy is nonexistent; what we have is something that fits calculation but does not agree with what is. (Quoted in Duhem 1969, p. 31)

Another cognitive attitude towards the Ptolemaic theory, held by a few pre-Copernican thinkers, is that it should be accepted not only as an angle-determiner, but also as literally true. One possible argument for such a view is that if the theory (taken literally) were false, it would certainly have some false observational consequences. The thirteenth-century scholastic Bernardus de Virduno (1961, p. 70) gave such an argument: "And up to our time these predictions have proved exact; which could not have happened if this principle [Ptolemy's theory] had been false; for in every department, a small error in the beginning becomes a big one in the end." (Trans. Duhem 1969, p. 37) A second reason why someone who accepts a Ptolemaic type of theory as at least a convenient device may also accept it as true is that he, unlike Proclus, accepts a physical theory with which the astronomical theory in question is consistent. For example, Adrastus of Aphrodisius and Theon of Smyrna, near-contemporaries of Ptolemy, interpreted an epicycle as the equator of a sphere located between two spherical surfaces concentric with the earth. The smaller sphere moves around the earth between the two surfaces to produce the deferent motion, while it rotates on its axis to produce the epicyclic motion. (Duhem 1969, pp. 13-15) These three spheres are evidently to be understood as real physical objects, since Theon argued that their existence is demanded by physical considerations, specifically, the physical impossibility of the planets' being carried by merely mathematical circles: "the movement of the stars should not be explained by literally tying them to circles each of which moves around its own particular center and carries the attached star with it. After all, how could such bodies be tied to incorporeal circles?" (quoted in Duhem 1969, pp. 13-15) Since epicyclic motions violate the Aristotelian physical principle that all superlunar motions are circular and concentric with the earth, we can see that Theon rejected Aristotelian physics and apparently assumed that the celestial spheres roll and rotate in the same manner as terrestrial ones. This, then, is at least part of the reason why Theon would have been unmoved by the Aristotelian physical arguments on the basis of which Proclus adopted an instrumentalist interpretation of epicycles.

Let us sum up our results on the main types of cognitive attitude adopted towards the Ptolemaic theory in the period between Ptolemy and Copernicus, and the main reasons given in support of these attitudes. One possible attitude, of course, is that the theory should be rejected as even a device for the determination of observed angles. But because of the lack of a superior alternative, Ptolemy's accuracy was generally conceded before Copernicus. Another position is that the theory is acceptable as such a device but should not be accepted on a literal interpretation. Those who took this view generally did so on such grounds as that the theory (literally interpreted) is inconsistent with the physics they accepted, deals with a realm whose true nature is inaccessible to human minds, fits the observations no better than some alternative theories, or has false observational consequences. Persons who thought the theory was a convenient device but was untrue sometimes did, and sometimes did not, think it therefore needed to be replaced by a convenient device that was also a true theory. Finally, some persons thought some Ptolemaic type of theory gave a correct description of the actual physical orbits. Possible supporting grounds were that false theories always have false observational consequences, and that epicyclic orbits are compatible with the laws of a true (non-Aristotelian) physical theory.

3. Harmony

In the period between Copernicus and Kepler, a gradual transition occurred in the dominant view among astronomers of the purpose and content of planetary theory. Eventually certain Copernican theories came to be accepted as true descriptions of the actual orbits of the planets in physical space. Before attempting to determine the reasoning involved in this transition in acceptance-status, I shall explain and criticize what I take to be the most important existing attempts to explain the rationale for the acceptance of Copernicus's theory. Most of these do not distinguish between instrumental and realistic acceptance and so do not attempt to discern separate grounds for each. Still they provide a useful starting point in dealing with my more "refined" question.

There is a prima facie difficulty in accounting for such appeal as the Copernican system—at the time of its initial publication by Rheticus and Copernicus—had to the scientific community, or even to Copernicus and Rheticus themselves. Briefly, the difficulty is that the most obvious factors in appraisal—simplicity and observational accuracy—give no decisive verdict in Copernicus's favor. True, Copernicus did say of his theory and certain of its implications, "I think it is much easier to concede this than to distract the understanding with an almost infinite multiplicity of spheres, as those who have kept the Earth in the middle of the Universe have been compelled to do." (1976, Book I, chapter 10) As Gingerich (1973a) has shown by recomputing the tables used by the leading Ptolemaic astronomers of Copernicus's time, the familiar story that they were forced to use ever-increasing numbers of epicycles on epicycles is a myth. In fact, they used only a single epicycle per planet. Since Copernicus's theory of longitudes used two and sometimes three circles per planet (Kuhn 1957, pp. 169-170), his total number of circles—the most obvious but not the only measure of simplicity—is roughly comparable to Ptolemy's. Gingerich's computations of the best Ptolemaic predictions of Copernicus's time, together with the earliest Copernican predictions and with computations of actual positions based on modern tables, also enabled him to conclude that the latest Ptolemaic and earliest Copernican predictions did not differ much in accuracy (1973a, pp. 87-89). Indeed, Copernicus himself conceded that except in regard to the length of the year, his Ptolemaic opponents "seem to a great extent to have extracted...the apparent motions, with numerical agreement...." (1976, prefatory letter)

Because of the difficulties in using accuracy and simplicity to account for the appeal of the Copernican system to its early supporters, a tradition has grown up among highly respectable historians that the appeal was primarily aesthetic, or at least nonevidentiary. Some of Copernicus's remarks certainly lend themselves to such an interpretation. For example, he wrote: "We find, then, in this arrangement the marvellous symmetry of the universe, and a sure linking together in harmony of the motion and size of the spheres, such as could be perceived in no other way." (1976, Book I, chapter 10) Koyré interprets this passage as reflecting Copernicus's alleged reliance upon "pure intellectual intuition"—as opposed, presumably, to superior evidence. (1973, pp. 53-54) Similarly, Kuhn—who says (1970, p. vi) Koyré was a principal influence on him—remarks that in his main arguments for his theory, including the foregoing quotation, Copernicus appealed to the "aesthetic sense and to that alone." (1957, p. 180) Gingerich falls in with this line when he remarks that Copernicus's defense of his theory in Book I, chapter 10, is "based entirely on aesthetics." (1973a, p. 97)

This interpretation of the Copernican revolution may well have been one of the things that produced the irrationalist tendencies in Kuhn's general theory of scientific revolutions—i.e., his comparisons of such revolutions to conversion experiences, leaps of faith, and gestalt-switches. Certainly the Copernican case is one he frequently cites in such contexts. (1970, pp. 112, 151, 158)

4. Positive Heuristics and Novel Predictions

Clearly the most effective way to counter such a view is to show that Copernicus's aesthetic language really refers to his theory's evidentiary support, the particular terms having perhaps been chosen merely for a scientifically inessential poetic effect. One attempt to show just this is contained in a paper by Lakatos and Zahar (1975). They try to show that acceptance of Copernicus's theory was rational in the light of Lakatos's "methodology of scientific research programs," as modified by Zahar (1973). According to this view, the history of science should be discussed, not it terms of individual theories, but in terms of "research programs," each of which is a sequence of theories possessing a common "hard core" of fundamental assumptions and a "positive heuristic" guiding the construction of variant theories. One research program will supersede another if the new program is, and the old one is not, "theoretically" and "empirically progressive." These terms mean that each new theory in the series exceeds its predecessor in content and also predicts some "novel" fact not predicted by its predecessor, and that such predictions are from time to time confirmed. (Lakatos 1970, pp. 118, 132-134) It is implicit in this claim that the only observational phenomena that have any bearing on the assessment of a research program are those that are "novel." Accordingly Lakatos claimed that a program's success or failure in accounting for non-novel facts has little or no bearing on its assessment. (1970, pp. 120-121, 137)

In their joint paper on Copernicus, Lakatos and Zahar (1975) use Zahar's (1973) criterion of novelty: a fact is novel with respect to a given hypothesis if "it did not belong to the problem-situation which governed the construction of the hypothesis." Elsewhere (1982) I have argued that a different criterion is preferable, but I shall discuss only Lakatos's and Zahar's attempt to use this one, and other Lakatosian principles, to give a rational rather than aesthetic interpretation of the Copernican revolution. They assert that Ptolemy and Copernicus each worked on a research program and that each of these programs "branched off from the Pythagorean-Platonic program." Evidently, then, we have three programs to discuss. Now the positive heuristic of the last-mentioned program, they say, was that celestial phenomena are to be saved with the minimum number of earth-centered motions of uniform linear speed. Ptolemy, however, did not follow this principle, they continue, but only a weaker version of it, which allowed (epicyclic and eccentric) motions to have centers other than the earth, and motions to be uniform in angular speed about a point (the "equant") other than their centers.

It is well known that Copernicus objected to this feature of Ptolemaic astronomy. (1976, prefatory letter) The objection might be taken as aesthetic in character, and/or as symptomatic of Copernicus's intellectual conservatism. (Kuhn 1957, pp. 70, 147) But Lakatos and Zahar think they can show that the objection is reasonable in light of their methodology, which requires that the positive heuristic of a program provide an "outline of how to build" sets of auxiliary assumptions through some "unifying idea" which gives the program "continuity." New theories that violate this requirement are said to be "ad hoc$_3$." (Lakatos 1970, pp. 175-176) In particular, Ptolemy's use of the equant, because it violates the Platonic heuristic, is said to be ad hoc$_3$, or to be an example of "heuristic degeneration." (Lakatos and Zahar 1975, Section 4)

What I do not understand is why, from the standpoint of the Lakatos-Zahar methodology, it is supposed to be an objection to one program that it deviates from the positive heuristic of an *earlier* program. I take it that it would not be at all plausible to say—and Lakatos and Zahar do not seem to be saying—that Ptolemy was working on the Platonic research program, since (despite his use of circular motions that are uniform in some sense) his assumptions and methods were so different from Plato's. If they *do* want to say this, they owe us a criterion of identity for research programs that makes the claim true. As matters stand, their criticism of Ptolemy has as much cogency as a criticism of Einstein for deviating from the Newtonian heuristic by using laws of motion analogous but not identical to Newton's.

They also assert that Copernicus had two further criticisms of the Ptolemaic program: (1) that it was un-Platonic in giving two motions to the stellar sphere, and (2) that it failed to predict any novel facts. But they fail to give any references to support the attribution to Copernicus of (1) and (2), which do not appear in his (1976, prefatory letter) principal summary of his criticisms of Ptolemy et al., or elsewhere in his writings, as far as I know.

In attempting to discern what reasons favored the Copernican program upon publication of *On the Revolutions*, Lakatos and Zahar first claim that Copernicus "happened to improve on the fit between theory and observation." (p. 374) Why they make this claim is puzzling, since relative numbers of anomalies are supposed, as we have seen, to be irrelevant to the rational assessment of a research program. In any case, they then proceed to list a group of facts which they claim to be predictions of the Copernican program that were previously known but novel in Zahar's sense: (a) the "planets have stations and retrogressions"; (b) "the periods of the superior planets, as seen from the Earth are not constant"; (c) each planet's motion relative to the earth is complex, and has the sun's motion relative to the earth as one component; (d) the elongation of the inferior planets from the sun is bounded; (e) "the (calculated) periods of the planets strictly increase with their (calculable) distances from the Sun." (Section 5)

But there are grave difficulties in construing these consequences of Copernicus's assumptions as novel in the sense of not being in the problem-situation, the facts he was trying to account for. Certainly (a), (b), and (d) were long-known facts, which any astronomer of Copernicus's time would have expected any planetary theory to account for. Thus Copernicus certainly was trying to account for them. Perhaps Lakatos and Zahar nonetheless think that Copernicus designed his theory to account for other facts and then discovered to his pleasant surprise that it also accounted for (a), (b), and (d). But they present not a single argument to show that this was the case. Instead they make an entirely irrelevant point: that (a) and (b) follow easily from Copernicus's assumptions. On the other hand, (c) is merely a result of a coordinate-transformation of Copernicus's theory, not a piece of (independently obtainable) evidence for the theory. Point (e) presents a different problem. Since, as Lakatos and Zahar are aware, the calculations it refers to are done by means of the Copernican theory, it is impossible to claim that the correlation (e) describes is a long-known observational result that supports Copernicus's theory. Since Copernicus's theory provided the first and, at the time, the *only* reasonably satisfactory way to determine the planetary distances, there was certainly no question of a prior or subsequent independent determination of the distances, whose agreement with Copernicus's results could support his theory. If his determination of the planetary distances was a crucial consideration in its favor (and we shall see later that it was), Lakatos and Zahar have not shown why—or much (if anything) else about the Copernican Revolution.

5. Simplicity and Probability

I now turn to Roger Rosenkrantz's (1977) attempt to give a Bayesian account of the rationality of preferring Copernicus's to Ptolemy's theory within a few decades after the former was published.

Rosenkrantz is a Bayesian in the sense that he thinks that the probabilities of hypotheses form the basis for comparing them, and that probability $P(H)$ of hypothesis H changes to $P(H/x)$ when evidence x is obtained, where this conditional probability is computed using Bayes's theorem. He is not a Bayesian, however, in the stronger sense that he thinks prior probabilities are subjective. Instead he maintains that the objectively correct distribution is to be computed by assuming that entropy is maximized subject to given constraints. If we have a partition of hypotheses H_i, then the *support* each receives from observation x is defined as the *likelihood* $P(x/H_i)$. By Bayes's rule the posterior probability is

$$P(H_i/x) = P(H_i)P(x/H_i)/P(x). \text{ (1977, pp. vii-ix)}$$

As a Bayesian, Rosenkrantz must hold that all supposed virtues of theories—in particular, simplicity—are such only to the extent that they manifest themselves somehow in high probabilities. His theory of simplicity begins from the observation that we make a theory more complex if we add one or more adjustable parameters. For example,

(1) $(\exists\ a)\ (\exists\ b)\ (y = a + bx)$

is less complex than

(2) $(\exists\ a)\ (\exists\ b)\ (\exists\ c)\ (y = a + bx + cx^2).$

Moreover, if we take a *special case* of a theory—by which Rosenkrantz seems to mean that we set some of its free parameters equal to particular values—then we obtain a simpler theory, as when we set $c = 0$ in our examples. To make sense of such intuitions, Rosenkrantz introduces the concept of *sample coverage* of theory T for experiment X, defined as "the chance probability that the outcome of the experiment will fit the theory, a criterion of fit being presupposed." (1977, pp. 93-94) A criterion of fit will be of the form "$x \in R$," where $P(x \in R/T) \geq \alpha$, so that outcomes outside a certain high-likelihood region are said not to fit T. (1976, p. 169) "A 'chance' probability distribution is one conditional on a suitable null hypothesis of chance (e.g., an assumption of independence, randomness, or the like), constrained perhaps by background information." In the straightforward case in which the chance distribution is uniform, the sample coverage is just the proportion of experimental outcomes which fit the theory. In any case, simplicity (relative to the experiment, criterion of fit, and null hypothesis) is measured by smallness of sample coverage (1977, p. 94). It follows from this that (1) is simpler than (2) relative (e.g.) to an experiment yielding three points $\langle x, y \rangle$, to a reasonable criterion of fit, and to a uniform chance distribution. To deduce this (though Rosenkrantz does not say so), we need to make the number of outcomes finite by assuming that x and y have finite ranges divided into cells corresponding to the precision of measurement. Then the proportion of triples of these cells that fall close to some quadratic curve will obviously be larger than the proportion that fall near the subset of such curves that are straight lines.

To see the bearing of this on the evaluation of theories, consider (for ease of exposition) a theory H with adjustable parameter θ and special cases H_i ($i = 1, 2, \ldots, n$) corresponding to the n possible values of θ. Since $H \Leftrightarrow H_i \vee \ldots \vee H_n$,

$$P(x/H) = \sum_{i=1}^{n} P(x/H_i \wedge H) \, P(H_i/H).$$

Now let $\hat{H}$ be the H_i that maximizes $P(x/H_i)$—i.e., the best-fitting (best-supported) special case of H. Since $P(x/H)$ is a weighted average of $P(x/\hat{H} \wedge H)$ and quantities that are no greater, we can infer that "H is never better supported. . . than its best-fitting special case," and will be less well-supported if any of its special cases fit worse than $\hat{H}$, as happens in practice. The same conclusion holds if θ varies continuously, but the sum must be replaced by $\int P(x/\theta \wedge H) \, dp \, (\theta/H)$. We can therefore infer that the maxim "simplicity is desirable ceteris paribus" has its basis in the greater support which x affords $\hat{H}$ than H, $\hat{H}$ being a special case and hence simpler. (1977, p. 97)

Dissatisfied as we were with the accounts of Kuhn and of Lakatos and Zahar, Rosenkrantz now attempts to use his theory to show the rationality of the Copernican revolution. He represents the Copernican, Tychonic, and Ptolemaic systems for an inferior planet in Figures 1-3 respectively.

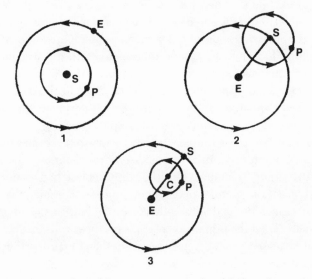

Figures 1-3. Reproduced by permission of publisher from Rosenkrantz 1977.

Since in the *Almagest* Ptolemy specified no values for the planetary distances, Rosenkrantz assumes that the distance EC to the epicyclic center C for planet P is a free parameter, constrained only in that C must lie on the line from the earth E to the sun S. Since he wants to maintain that all special cases of the Ptolemaic system fit the angular variations of P equally well, however, he evidently has in mind that the epicyclic radius extends proportionally as C moves towards S. Figure 2 is a special case of Figure 3, and "can be transformed into an equivalent heliostatic picture by the simple device of fixing S and sending E into orbit around S (Figure 1)." Thus Copernicus's theory is a special case of Ptolemy's. Considering only observations of planetary angles, all the special cases are equally well supported. Hence the Ptolemaic system (with free parameter) has exactly the same support as any of its special cases, including Figure 1. But if we consider Galileo's famous telescopic observation that Venus has a full set of phases, the only special cases of Figure 3 that fit at all well are those that have *P* moving in a circle centered (at least approximately) at *S*, as in Figure 2 or the "equivalent" Figure 1. (See Kuhn 1957, Figure 44.) Thus Copernicus's system is much better supported than Ptolemy's, whose support is obtained by averaging that of Copernicus's theory with many much smaller numbers. (1977, pp. 136-138) Note that this Bayesian view denies that there is any importance in the fact that the nature of Venus's phases followed from a theory proposed *before* they were observed. Cf. the common opposite opinion of Lakatos and Zahar 1975, p. 374 and Kuhn 1957, p. 224.

There are a number of simplifications in Rosenkrantz's version of the history. As he is aware (1977, pp. 156-159), all three astronomical systems were more complex than Figures 1-3 show, and they did not fit the angular data equally well. Let us assume that these simplifications are designed to enable us to concentrate on establishing the significance of the phases of Venus. This aside, one difficulty immediately strikes one. How could Copernicus's theory possibly be considered a special case of Ptolemy's, given that one asserts and the other denies that the earth moves? It would be irrelevant and false to say (and Rosenkrantz does not) that from a relativistic standpoint there is no distinction, since we and Rosenkrantz (1977, p. 140) are trying to articulate the underlying basis of the pro-Copernican arguments given in Copernicus's time. As far as I know, no one in the period covered by this paper—specifically, no one before Descartes—held that all motion is relative. And anyway, both special and general relativity distinguish accelerated from unaccelerated frames of reference, and Copernicus said the earth is accelerated. Nor can we say Copernicus's theory is a "special case" of Ptolemy's, meaning just that it has a smaller sample coverage; for Rosenkrantz's argument for the greater support of some special case requires defining "special case" as above in terms of fixing one or more parameters. But we cannot get a moving earth by fixing the position of *C* in a geostatic theory.

There is an easy way for Rosenkrantz to neutralize this objection. All he has to do is to replace the one-parameter theory of Figure 3 by a new theory obtained by disjoining it with its transformation into a heliostatic system. Then the disjunction of the Copernican and Tychonic systems does become a special case of the new theory—and, indeed, the best-fitting special case.

This change would cohere well with Rosenkrantz' remark that what Galileo's observations did is to narrow "the field to the Copernican and Tychonic alternatives." (1977, p. 138)

We can now face the really serious problems with Rosenkrantz's account. The first is that Copernicus did not in fact take himself to be faced with a Ptolemaic alternative with a free parameter EC. (1976, prefatory letter; Bk. 1, chap. 10) For he was well aware of the long tradition among Ptolemiac astronomers (section 2 above) of determining EC on the basis of the nesting-shell hypothesis. His objection to them on this score was not that EC was unspecified, but that it was specified on the basis of an "inappropriate and wholly irrelevant," "fallacious" assumption that caused their arguments on planetary order and distances to suffer from "weakness and uncertainty." In effect, Rosenkrantz has loaded the dice against Ptolemy and friends by inserting into their theory a nonexistent free parameter that lowers its support. He therefore fails to give an adequate analysis of the reasons why Copernicus attributed such importance to the relative merits of his and Ptolemy's determinations of the planetary distances.

The second serious problem with Rosenkrantz's account is that he fails to consider the bearing of the supposed prior probabilities of the Copernican and Ptolemaic theories on their evaluation. He writes that "simplifications of the Copernican system" such as the supposed one under discussion "are frequently cited reasons for preferring it to the Ptolemaic theory. Yet, writers from the time of Copernicus to our own, have uniformly failed to analyze these simplifications or account adequately for their force. The Bayesian analysis... fills this lacuna in earlier accounts...." (1977, p. 140) But his Bayesian analysis asserts that preference among theories is governed by their probabilities, and support must be combined via Bayes's theorem with the prior probabilities of the theories and data to yield the posterior probabilities. And Rosenkrantz gives no indication of how these priors are to be computed from the maximum entropy rule (what constraints are to be used, etc.). Nor does he show how to compute the numerical values of the conditional probabilities so that the Bayes-theorem calculation can actually be done. Finally, he neglects to say what null hypothesis of chance and what background information define the chance distribution.

The last argument by Rosenkrantz (1977, pp. 143-148) we shall consider relates to a crucial passage by Copernicus (1976, Book I, chapter 10), which I shall discuss later:

We find, then, in this arrangement the marvellous symmetry of the universe, and a sure linking together in harmony of the motion and size of the spheres, such as could be perceived in no other way. For here one may understand, by attentive observation, why Jupiter appears to have a larger progression and retrogression than Saturn, and smaller than Mars, and again why Venus has larger ones than Mercury; why such a doubling back appears more frequently in Saturn than in Jupiter, and still more rarely in Mars and Venus than in Mercury; and furthermore why Saturn, Jupiter and Mars are nearer to the Earth when in opposition than in the region of their occultation by the Sun and re-appearance.... All these phenomena proceed from the same cause, which lies in the motion of the Earth. (Copernicus 1976, Bk. I, chap. 10)

Rosenkrantz attempts to explain the force of the part of this argument which concerns the frequencies of retrogression as follows:

...within the Ptolemaic theory, the period of an outer planet's epicycle can be adjusted at will to produce as many retrogressions in one circuit of the zodiac as desired.... while both theories fit this aspect of the data, the Copernican theory fits *only* the actually observed frequency of retrogression for each planet.... Hence,... qua special case of the geostatic theory, the heliostatic theory is again better supported. (Rosenkrantz 1977, pp. 143-148)

Let us handle the problem about "special case" as above, so that the argument is understood (as Rosenkrantz intends) as being for the disjunction of the Copernican and Tychonic theories. We can then easily see that the argument fails for essentially the same reason as Rosenkrantz's argument concerning the planetary distances. The best Ptolemaic theory of Copernicus's time did not, of course, merely assert that a given planet has one epicycle with *some* radius and *some* period. If this theory had been existentially quantified in this way, it would not have sufficed for the computation of the tables of observed positions, which Copernican astronomers sought to improve upon. Similarly, the Copernican theory could not fit just the observed frequencies of retrogression without specifying the radii and periods of the planets' revolutions. Obviously, then, Rosenkrantz is making an irrelevant and unfair comparison: between the Copernican theory *with* its specifications of radii and speeds, and a Ptolemaic theory weaker than Copernicus's real rival because of existential quantification over its fixed parameters. To obtain a Copernican account of the durations and angular widths of the retrograde motions, we need to specify radii and speeds; and we get a Ptolemaic account of the same appearances if we specify the comparable quantities.

I also cannot see why Rosenkrantz thinks his Bayesian account explains the basis of Copernicus's claim that his theory exhibits a "symmetry of the universe, and a sure linking together in harmony," or in Rosenkrantz's terms an "economy of explanation." Copernicus, as he explicitly said, was evidently referring to the fact that his theory explains a wide variety of phenomena on the basis of the hypothesis that the earth moves. Metaphorically, his theory "symmetrizes" or "harmonizes" these phenomena. But this virtue of his theory has nothing to do with the supposed fact that his theory is compatible only with certain specific observations. It could have this latter property even if it accounted for different sets of these observations on the basis of entirely different hypotheses.

To sum up then: despite the considerable merits that Rosenkrantz's Bayesianism has in connection with other problems in the philosophy of science, I do not think it helps us understand the Copernican revolution.

6. Bootstraps

The last published account of the Copernican revolution I shall discuss is based on Glymour's presentation of what he calls the "bootstrap strategy" of testing, which he says is a common but not universal pattern in the history of science. (1980, chapter 5) Briefly and informally, the strategy is this. One confirms a hypothesis by verifying an instance of it, and one verifies the instance by measuring or calculating the value of each quantity that occurs in it. In calculating values of quantities not themselves measured, one may use *other* hypotheses of a given theory—i.e., the theory can "pull itself up by its bootstraps." However, no hypothesis can be confirmed by values obtained in a manner that guarantees it will be satisfied. We cannot, then, test "$F = ma$" by measuring m and a and using this very law to calculate F; whereas we *can* use this procedure to obtain a value of F to substitute into "$F = GMm/r^2$" along with measured values of M, m, and r, and thereby test the latter law. So, Glymour concludes, Duhem-Quine holists are right in maintaining that hypotheses can be used jointly in a test, but mistaken in concluding that the test is therefore an indiscriminate one of the entire set. The described procedure, e.g., uses two laws jointly, but tests only the latter.

The falsification of an instance of a hypothesis may result solely from the falsity of other hypotheses used in testing it. Again, confirmation may result from compensatory errors in two or more hypotheses. Hence there is a demand for variety of evidence: preferably, a hypothesis is tested in many different ways—i.e., in conjunction with many different sets of additional hypotheses. Preferably also, these additional hypotheses are themselves tested.

As Glymour points out, his strategy of testing makes it possible to understand, e.g., why Thirring's theory of gravitation is not taken seriously by physicists. It contains a quantity

$$\eta_{\mu\nu} - f\psi_{\mu\nu}$$

which can be calculated from measurements using rods, clocks, test particles, etc. But the quantities $\psi_{\mu\nu}$ and $\eta_{\mu\nu}$ can be neither measured nor calculated from quantities that can. Either may be assigned any value whatsoever, as long as compensatory changes are made in the other. We might express this by saying these quantities are *indeterminate:* neither measurable nor computable from measured quantities via well-tested hypotheses. Consequently no hypotheses containing them can be confirmed, and Thirrings's theory is rejected.[1]

Glymour attempts to establish the importance of the bootstrap strategy by showing that it played a role in a variety of episodes in the history of science. He also applies it to the comparison of the Ptolemaic and Copernican planetary theories, but says that in this case (unlike his others) his arguments are merely intended as illustrative of the strategy and not as historically accurate accounts of arguments actually given around Copernicus's time. (Glymour 1980, chapter 6) I shall therefore not discuss here what Glymour does, but I shall consider instead to what extent the bootstrap strategy is helpful in enabling one to see the basis of Copernicus's own arguments against Ptolemaic astronomy and in favor of his own theory. Copernicus gives a helpful summary of his negative arguments in his prefatory letter:

> . . . I was impelled to think out another way of calculating the motions of the spheres of the universe by nothing else than the realisation that the mathematicians themselves are inconsistent in investigating them. For first, the mathematicians are so uncertain of the motion of the Sun and Moon that they cannot represent or even be consistent with the constant length of the seasonal year. Secondly, in establishing the motions both of the Sun and Moon and of the other five wandering stars they do not use the same principles or assumptions, or explanations of their apparent revolutions and motions. For some use only homocentric circles, others eccentric circles and epicycles, from which however the required consequences do not completely follow.

For those who have relied on homocentric circles, although they have shown that diverse motions can be constructed from them, have not from that been able to establish anything certain, which would without doubt correspond with the phenomena. But those who have devised eccentric circles, although they seem to a great extent to have extracted from them the apparent motions, with numerical agreement, nevertheless have in the process admitted much which seems to contravene the first principle of regularity of motion. Also they have not been able to discover or deduce from them the chief thing, that is the form of the universe, and the clear symmetry of its parts. They are just like someone including in a picture hands, feet, head, and other limbs from different places, well painted indeed, but not modelled from the same body, and not in the least matching each other, so that a monster would be produced from them rather than a man. Thus in the process of their demonstrations, which they call their system, they are found either to have missed out something essential, or to have brought in something inappropriate and wholly irrelevant, which would not have happened to them if they had followed proper principles. For if the hypotheses which they assumed had not been fallacious, everything which follows from them could be indisputably verified. (1976, p. 25)

As Copernicus said more explicitly elsewhere, his first point is that his opponents had inaccurate theories of the precession of the equinoxes, which (with the solstices) define the seasons. (1976, Book III, chapters 1, 13) In particular, Ptolemy attempted to define a constant year by reference to the equinoxes, failing to account for the supposed fact that they recur (as Copernicus thought) at unequal intervals, and that only the sidereal year (with respect to the stars) is constant. Copernicus's second complaint is that his opponents disagreed among themselves in regard to the use of homocentric spheres as against epicycles and eccentrics. Moreover, he argued, those who used homocentrics had failed to achieve observational accuracy, because "the planets... appear to us sometimes to mount higher in the heavens, sometimes to descend; and this fact is incompatible with the principle of concentricity." (Rosen 1959, p. 57) Copernicus undoubtedly was referring here not to variations in planetary latitude, whose limits (but not times) Eudoxus *could* account for (Dreyer 1953, p. 103), but to variations in planetary distance ("height") and brightness, a long-standing problem for homocentrics. Although those who used epicycles and eccentrics (realistically interpreted) could handle this problem, they also used the equant, which violates the principle that the basic celestial motions are of uniform linear speed and which makes their system "neither sufficiently absolute nor sufficiently pleasing to the mind." (Rosen 1959, p. 57)

At first glance, it seems difficult to be sympathetic to this last argument—that is, to try (like Lakatos and Zahar) to find some generally acceptable pattern of scientific reasoning into which it fits. Kuhn explains part of what is going on at this point when he says that here Copernicus showed himself to be intellectually conservative—to feel that on this question at least "our ancestors" were right. (Copernicus, in Rosen, 1959, p. 57) Copernicus also evidently felt that his principle of uniformity had a "pleasing" intellectual beauty that lent it plausibility. The principle also derived some of its appeal from the false idea that trajectories that "pass through irregularities...in accordance with a definite law and with fixed returns to their original positions" must necessarily be compounded of uniform circular motions. Finally, he had arguments based on vague, implicit physical principles: irregularity would require "changes in the moving power" or "unevenness in the revolving body," both of which are "unacceptable to reason." (Copernicus 1976, Book I, chapter 4) The principle of uniform circularity does, then, rest after all upon considerations of a sort generally considered scientifically respectable: theoretical conservativism, theoretical beauty, mathematical necessities imposed by the phenomena, and consistency with physical principles and conditions.

In the remainder of the long quotation above, Copernicus made a fourth and a fifth critical point, without distinguishing them clearly. The fourth is that his opponents were not able satisfactorily to compute from their principles (together, presumably, with observational data) the distances of the planets from the earth and therefore the overall arrangement ("form") of the planetary system. They either omitted these "essential" quantities altogether, or else computed them using an "inappropriate" and "irrelevant" assumption—viz., the nesting-shell hypothesis (stated and criticized explicitly in Book I, chapter 10).

It is plain that we can interpret this Copernican criticism (which he regarded as his most important—"the chief thing") in the light of the principle, suggested by Glymour's theory, that it is an objection to the acceptance of a theory that it contains indeterminate quantities—i.e., nonmeasurable quantities that either cannot be computed from observational data at all, or cannot be computed via well-tested hypotheses. I wish to leave it open for the moment, however, whether well-testedness in the Copernican context can be interpreted on the basis of Glymour's approach.

Let us first consider why Copernicus thinks his own theory avoids his fourth objection to Ptolemy's. Copernicus showed how to use his heliostatic system of orbits, together with such data as the maximum angle between an inferior planet and the sun, to compute the relative distances between the sun and each planet. (1976, Book V; see also Kuhn 1957, pp. 174-176) He was referring to this fact about his theory when he said that it "links together the arrangement of all the stars and spheres, and their sizes, and the very heaven, so that nothing can be moved in any part of it without upsetting the other parts and the whole universe." (1976, prefatory letter) Contrary to Kuhn, this argument has little if anything to do with "aesthetic harmony," but concerns the determinateness of quantities. But were the hypotheses used to obtain the radii of these orbits well tested in a way in which the nesting-shell hypothesis was not? And does the bootstrap strategy describe this way? Or does the more familiar hypothetico-deductive (HD) strategy—roundly criticized by Glymour (1980, chapter 2) —provide a better basis for the comparison?

One of the cardinal points of the bootstrap strategy is that a hypothesis cannot be confirmed by values of its quantities that are obtained in a manner that guarantees they will agree with the hypothesis. By this standard the nesting-shell hypothesis was almost entirely untested. Except for the case of the sun (see section 2), Ptolemy had only one way to compute planetary distances—via the nesting-shell hypothesis—and there was therefore no way he could possibly obtain planetary distances that would violate it. If we ignore the difficulties in stating the HD method and follow our intuitions about it, we will say similarly that the nesting-shell hypothesis was untested by HD standards; for it was not used (perhaps in conjunction with other hypotheses) to deduce observational claims which were then verified. Ptolemy adopted it on the entirely theoretical ground that a vacuum between the shells is impossible (section 2 above).

But were the hypotheses Copernicus used in computing planetary distances well tested—and, if so, in what sense(s)? Let us begin with his principal axiom—that the earth moves (in a specified way). In his prefatory letter Copernicus had made a fifth and last criticism of his opponents: that they treated the apparent motion of each planet as an entirely separate problem rather than exhibiting the "symmetry" or "harmony" of the universe by explaining a wide variety of planetary phenomena on the basis of a single assumption used repeatedly. As the long quotation in section 5 (above) makes clear, Copernicus thought that his own theory did have this particular sort of unity. Essentially his claim was that his hypothesis E concerning the actual motion (around the sun) of the earth is used in conjunction with his hypothesis P_k describing the actual motion of each planet k in order to deduce a description O_k of the observed motion of planet k.

$$(*) \qquad \begin{array}{ccc} E & E & E \\ P_1 & P_2 & P_3 \\ \therefore O_1 & \therefore O_2 & \therefore O_3, \text{ etc.} \end{array}$$

From the O_k's one can deduce various relations among the planets' sizes and frequencies of retrograde loops, and the correlation of maximum brightness with opposition in the case of the superior planets. One of Copernicus's main arguments for his theory, then, is that E is very well tested in that it is used (in conjunction with other hypotheses) to deduce the occurrence of all these observed phenomena. As he put it: "All these phenomena proceed from the same cause, which lies in the motion of the Earth." (Copernicus 1976, Book I, Chap. 10) Although Copernicus did not say so explicitly, the phenomena he listed are not only numerous, but also varied. Perhaps the phrase "all these phenomena" was intended to express this. And although he invoked no criterion of variety explicitly, a criterion that his argument satisfies is: the derivations (*) provide *varied* tests for E because different sets of hypotheses are conjoined with E to deduce the O_k's. In any case, the appeal in the argument from "harmony" is not merely to the "aesthetic sense and to that alone" (Kuhn 1957, p. 180), but to the nature of the observational support for the hypothesis E.

Despite the cogency of Copernicus's argument, Ptolemy could have made the reply that each of his determinations of a planetary angle was based on the hypothesis that the earth is stationary at the approximate center of the planet's motion, and that his hypothesis was therefore well tested in the same way E was. But I am just trying to discern Copernicus's testing strategy, and so shall ignore possible replies to it.

Now in some respects the argument from harmony sounds like the bootstrap strategy in operation: for Glymour's principle of variety of evidence also favors a hypothesis that is tested in more ways (other things being equal), where the ways are individuated by the sets of additional hypotheses used in the test. But the bootstrap strategy requires the measurement or computation (using other hypotheses) of every quantity occurring in a hypothesis in order to see if it holds in a given instance. And this does not appear to be what Copernicus had in mind in the argument we are discussing. We are not to observe the apparent position of some planet k at some time and then use this in conjunction with P_k to compute the actual position of the earth at that time to see if E holds then. Rather, the deductions are to proceed in the opposite direction: the "phenomena proceed from the cause" E, as Copernicus put it, and not vice versa. His method here is hypothetico-deductive, and involves no bootstrapping. That Copernicus thought his argument for the earth's motion was HD is also shown clearly by his statement (1976, Book I, chapter 11): "so many and substantial pieces of evidence from the wandering stars agree with the mobility of the Earth...the appearances are explained by it as by a hypothesis."

The set of deductions (*) can be thought of as parts of HD tests not only of E, but also of the descriptions P_k of the motions of the other planets. These, however, are not tested as thoroughly by (*) and the relevant data, each one being used to obtain only one of the O_k. But Copernicus also had other tests aimed more directly at the P_k, and they turn out to involve a complex combination of bootstrapping and hypothetico-deduction. We can illustrate his procedure by examining his treatment of Saturn (Copernicus 1976, Book V, chapter 5-9), which has the same pattern as his treatment of the other two superior planets. (see also Armitage 1962, chapter 6.) He began with three longitudes of Saturn observed at opposition by Ptolemy. He then hypothesized a circular orbit for Saturn eccentric to the earth's. From these data and hypotheses he deduced the length and orientation of the line between the two orbits' centers. He then repeated the procedure using three longitudes at opposition observed by himself, and noted with satisfaction that he obtained about the same length for the line between the centers. (Its orientation, however, seemed to have shifted considerably over the intervening centuries.) This is a bootstrap test of his hypothesis h concerning the length of the line. He used one triple of data and other hypotheses to obtain the quantity (length) in h. To have a test, the length had then to be obtained in a way that could falsify h; hence he used another triple of data and the same auxiliary hypotheses, computed the length anew, and thereby verified h.

This bootstrapping was only a preliminary step. Copernicus's final move was to use the computed length and orientation of the line between the centers to construct what he considered an exact description P_5 of Saturn's orbit (with one epicycle), and showed that P_5 and E yield (very nearly) the triples of data: "Assuming these [computed values] as correct and borrowing them for our hypothesis, we shall show that they agree with the observed appearances." (Book V, chapter 5) The positive results gave Copernicus sufficient confidence in P_5 to use it (with E) to construct extensive tables of observed positions for Saturn.

This procedure is somewhat peculiar, since Copernicus used the very same triples of data to obtain some of the parameters in P_5 and also to test P_5. Still the exercise is nontrivial since (a) one of the parameters could be obtained from either triple of data; and (b) the additional parameters of P_5 might have resulted in inconsistency with the original data.

Peculiarities aside, the final step of the procedure is plainly HD: deducing from P_5 and E values for certain angles, comparing these with the data, and finally concluding that the hypotheses, because they agree "with what has been found visually [are] considered reliable and confirmed." (1976, Book V, chapter 12)

In arguing that Copernicus's method was mainly the HD and not the bootstrap, I have not been arguing against Glymour. For he freely admits (1980, pp. 169-172) that despite the difficulties in giving a precise characterization of the HD strategy, a great many scientists have nonetheless (somehow) managed to use it. What, then, are the morals of my story? The first moral is that the principle suggested by the bootstrap strategy, that indeterminate quantities are objectionable, has a range of application that transcends the applicability of the bootstrap concept of well-testedness. That is, we can see from Copernicus's arguments concerning the planetary distances that the computability of quantities from measured ones via well-tested hypotheses is considered desirable even if the testing in question is done mainly by the HD method. Indeed, Copernicus himself made this connection between the HD method and determinateness when he wrote that if his hypothesis is adopted, "not only do the phenomena agree with the result, but also it links together the arrangement of all the stars and spheres and their sizes...." (1976, prefatory letter) Unfortunately the rationale for this principle within the bootstrap strategy—that its satisfaction makes tests possible or makes them better—is unavailable in an HD framework. What its rationale there might be I do not know.

To say exactly when a hypothesis is well tested in HD fashion would be a task beyond my means. But Copernicus's argument for a moving earth does at least suggest one principle whose applicability in other historical cases is worth examining—namely, that *a hypothesis is better tested (other things equal) if it is used in conjunction with a larger number of sets of other hypotheses to deduce statements confirmed by observation.* This principle of variety of evidence has a rationale similar to the one Glymour gives for his: that its satisfaction reduces the chance that one hypothesis is HD-confirmed in conjunction with a set of others only because of compensatory errors in them both.

7. Copernicus's Realistic Acceptance

Having completed our survey of various ways of accounting for such acceptance as the Copernican theory enjoyed down through the time of Galileo, and having made some preliminary suggestions about why the theory appealed to its author, I should now like to refine the problem by asking whether various astronomers accepted the theory as true or as just a device to determine observables such as angles, and what their reasons were. I shall begin with Copernicus himself.

It is well known that the question of Copernicus's own attitude concerning the status of his theory was initially made difficult by an anonymous preface to *De Revolutionibus*, which turned out to have been written by Osiander. The preface rehearses a number of the traditional arguments discussed earlier against the acceptability of a planetary theory (Copernicus's in this case) on a realistic interpretation: that "the true laws cannot be reached by the use of reason"; that the realistically interpreted theory has "absurd" observational consequences—e.g., that Venus's apparent diameter varies by a factor of four (Osiander ignored the phases' compensatory effect.); and that "different hypotheses are sometimes available to explain one and the same motion." (Copernicus 1976, pp. 22-23) Hence, Osiander concluded, Copernicus (like any other astronomer) did not put forward his hypotheses "with the aim of persuading anyone that they are valid, but only to provide a correct basis for calculation." And for this purpose it is not "necessary that these hypotheses should be true, nor indeed even probable." However, the older hypotheses are "no more probable"; and the newer have the advantage of being "easy." We should note that there is an ambivalence in Osiander's position. At some points he seems to be saying that while Copernicus's theory is at least as probable as Ptolemy's, we are uncertain and should be unconcerned whether it or any astronomical theory is true. But at other points he seems to be saying that the new theory is plainly false.

Despite Osiander's efforts, as he put it in a letter, to "mollify the peripatetics and theologians" (Rosen 1959, p. 23), it is quite clear that Copernicus believed that his hypothesis that the earth moves (leaving aside for the moment some of his other assertions) is a literally true statement and not just a convenient device. For example, in his statement of his seven basic assumptions in *Commentariolus*, he distinguished carefully between the *apparent* motions of the stars and sun, which he said do not in fact move, and the *actual* motions of the earth, which explain these mere appearances. (Rosen, pp. 58-59) Moreover, he felt compelled to answer the ancients' physical objections to the earth's motion, which would be unnecessary on an instrumentalist interpretation of this hypothesis. (1976, Book I, chap. 8)

Unconvinced by such evidence, Gingerich claims that Copernicus's writings are in general ambiguous as to whether his theory, or all astronomical theories, are put forth as true or as only computational "models." However, Gingerich holds that in at least three places, Copernicus was clearly thinking of his geometrical constructions as mere models "with no claim to reality": Copernicus sometimes mentioned alternative schemes (e.g., an epicycle on an eccentric vs. an eccentric on an eccentric) and "could have hardly claimed that one case was more real than the other"; and he did not use the same constructions to account for the planet's latitudes as for their longitudes. (Gingerich 1973b, pp. 169-170)

But the mere fact that someone mentions alternative theories does not show that he thinks that neither is true—i.e., that each is a mere device. He may simply think the available evidence is insufficient to enable one to decide which is true. Contrary to Gingerich, such an attitude towards particular alternatives is no evidence of a tendency in Copernicus to think that all his constructions—much less astronomical theories in general—are mere devices. And in fact, Copernicus appears just to have suspended judgment regarding his alternative systems for Mercury: the second, he wrote, was "no less credible" or "reasonable" than the first and so deserved mention. He did not say or imply that neither is true, or that no astronomical theory is true. (1976, Book V, chap. 32)

Perhaps my disagreement with Gingerich reflects only our differing concepts of what it is to regard a theory as a mere computation-device or (his term) "model"—i.e., to accept it instrumentally. If a person thinks that a theory permits the derivation of a given body of information but is simply unsure whether it is true, I do not say that he regards it as a mere device (accords it instrumental acceptance). I say this only if he actually *denies* that the theory is true—i.e., asserts its falsity or lack of truth value. I adopt this terminological stipulation to retain continuity with the traditional philosophical position of (general) instrumentalism, according to which all theories (referring to unobservables) are mere prediction-devices and are not true—that is, are false or truth-valueless (e.g., Nagel 1961, p. 118). But Gingerich's instances from Copernicus certainly have the merit of indicating that an intermediate attitude towards a particular theory sometimes occurs: that the theory is convenient for deriving certain information but may or may not be true.

In regard to the mechanism for determining a planet's longitudes vs. that for determining its latitudes, it is a mistake to interpret Copernicus either as holding that both are physically unreal or as being uncertain which of the two is physically real. Rather, Copernicus regarded the alternatives as two independent mechanisms that can be described separately for the sake of simplicity, but that operate together to produce the planet's exact trajectory. Thus, in giving his famous nonepicyclic account of retrogression, he wrote: "On account of the latitude it [the orbit of an inferior planet] should be inclined to AB [that of the earth]; but for the sake of a more convenient derivation let them be considered as if they are in the same plane." (1976, Book V, chapter 3) After describing all the longitude constructions as coplanar to a first approximation in Book V, he introduced the assumption that in fact these mechanisms reside in oscillating planes: "their orbits are inclined to the plane of the ecliptic. . . at an inclination which varies but in an orderly way." And he remarked in the case of Mercury:

> However, as in that case [Bk. V] we were considering the longitude without the latitude, but in this case the latitude without the longitude, and one and the same revolution covers them and accounts for them equally, it is clear enough that it is a single motion, and the same oscillation, which could produce both variations, being eccentric and oblique simultaneously. (1976, Book VI, chapter 2)

So in this case we do not have two equally "credible" mechanisms, but one approximate and one supposedly exact total mechanism.

Now the use of approximations or idealizations is connected with the traditional philosophical question of instrumentalism as a general thesis, and is thus at least indirectly related to our successor question of instrumental vs. realistic acceptance of particular theories. For example, one of the arguments for general instrumentalism discussed by Nagel is that "it is common if not normal for a theory to be formulated in terms of ideal concepts such as...perfect vacuum, infinitely slow expansion, perfect elasticity, and the like." (1961, p. 131) Contrary to Nagel, if this practice is only "common" and not universal, it does not provide much reason to say that scientific theories in general are just untrue prediction-devices. But certainly if a particular theory, such as Copernicus's coplanar treatment of longitudes, is presented explicity as an idealization, we must (obviously) conclude that it is being put forward for instrumental acceptance—as untrue but nonetheless convenient for accounting (at least approximately) for a specific body of information. But Copernicus's coplanar theory is not a very interesting example of a theory with instrumental acceptance, since Copernicus had in hand a replacement for it which he appears to have accepted realistically—in contrast with Proclus and Osiander, who asserted that realistic acceptance is always unreasonable.

I have claimed—what is not these days disputed—that Copernicus accorded realistic acceptance to the hypothesis that the earth moves. But it is quite another matter whether he accorded realistic acceptance to all the other features of his geometrical constitutions—leaving aside Gingerich's cases in which he was undecided or presented first approximations. In particular, did Copernicus really believe that the planets performed the remarkable acrobatic feat of moving in physical space on epicycles? Or were the epicycles mere computation-devices?

Let us consider this question briefly from the standpoint of generalized instrumentalism. Carnap held that a theory's "theoretical terms," unlike its "observational terms," receive no rules of designation. Hence sentences containing such terms can be "accepted" only in the sense that they can play a role in "deriving predictions about future observable events." They cannot also be accepted as true (of certain "theoretical entities") in some further sense than this. (Carnap 1956, pp. 45, 47) From such a philosophical perspective, it is perfectly clear which sentences of someone's theory are to be regarded as mere prediction-devices—viz., those which contain theoretical terms. This is clear, anyway, if we can somehow get around the well-known difficulties in distinguishing this class of terms.

But by anyone's standards it is a theoretical and not an observational question whether the earth moves at all or whether a planet moves on an epicycle (assuming our observations are done on the earth). Contrary to Carnap, however, this does not settle the question of whether Copernicus's assertions on these motions were put forward as true or as mere devices. Instead we must look at the specific scientific grounds (explicit or implicit) Copernicus had for adopting one view or the other in regard to each assertion. (Here I am indebted to Shapere 1969, p. 140)

Unfortunately I do not know of any place where Copernicus explicitly raised the issue of whether his epicycles were real physical orbits or convenient fictions, and then came down on one side or the other. Still, such evidence as there is favors the view that Copernicus considered his epicycles physically real. First, there is no place, as far as I know, where Copernicus (like Osiander) pointed out that his theory would have some false observational consequences if interpreted realistically. Second, recall that another sort of argument we have seen for instrumental acceptance is that the realistically interpreted theory is physically impossible. But Copernicus's physical theory (which was not fully articulated) allowed the compounding of uniform circular motions in epicyclic fashion. I call this principle of uniformity "physical" because part of the argument for it asserts the impossibility of "changes in the moving power." (Copernicus 1976, Book I, chapter 4) The third and most persuasive piece of evidence that Copernicus thought the actual physical orbits of some planets are epicyclic occurs in a passage of the sort Gingerich referred to, in which Copernicus discussed alternative constructions for the earth: "the same irregularity of appearances will always be produced and . . . whether it is by an epicycle on a homocentric deferent or by an eccentric circle equal to the homocentric circle will make no difference. . . . It is therefore not easy to decide which of them exists in the heaven." (1976, Book III, chapter 15) The implication here is that the epicyclic and the eccentric theories are logically incompatible despite the fact that they yield the same longitudes, that the two are about equally credible, and that exactly one of them is true. All this entails that the epicyclic theory has asserted content beyond its implications regarding longitudes—that its purpose is to describe the actual physical orbit of the earth.

Thus far I have said nothing about Copernicus's reasons for regarding his theory as true (except for some idealizations and uncertainties). Since he did not explicitly raise the issue of realistic vs. instrumental acceptance, it is not easy to determine what factors influenced his choice between the two. It will be best to defer consideration of this question, then, until we have seen what factors were operative in the thinking of those in whose writings the issue did arise explicitly. This will provide some basis, at least, for conjectures regarding Copernicus's own reasons.

8. The Instrumentalist Reception of Copernicanism

In the half-century or so after the publication of *De Revolutionibus*, Copernicus's theory was widely perceived as an improvement upon Ptolemy's in regard to observational accuracy and theoretical adequacy, and yet was almost universally regarded as untrue or at best highly uncertain. (Kuhn 1957, pp. 185-188; Westman 1972a, pp. 234-236) A concise expression of instrumental acceptance occurs in an astronomical textbook of 1594 by Thomas Blundeville: "Copernicus. . . affirmeth that the earth turneth about and that the sun standeth still in the midst of the heavens, by help of which false supposition he hath made truer demonstrations of the motions and revolutions of the celestial spheres, than ever were made before." (Quoted in Johnson 1937, p. 207)

Crucial in promoting this point of view, especially in the leading German universities, was a group of astronomers led by Phillipp Melanchthon (1497-1560) of the University of Wittenberg. Generally speaking, their opinion was that Copernicus's theory was credible primarily just in regard to its determinations of observed angles; that it was preferable to Ptolemy's in that it eschewed the abhorrent equant; but that the new devices needed to be transformed into a geostatic frame of reference, since the earth does not really move (Westman 1975a, pp. 166-167). For example, Melanchthon praised parts of Copernicus's theory in 1549 for being "so beautifully put together" and used some of his data, but held that the theory must be rejected on a realistic interpretation because it conflicts with Scripture and with the Aristotelian doctrine of motion. (Westman 1975a, p. 173) Similarly, Melanchthon's distinguished disciple Erasmus Reinhold was plainly more impressed by the fact that "we are liberated from an equant by the assumption of this [Copernican] theory" (as Rheticus had put it in Rosen 1959, p. 135) than by the theory's revolutionary cosmology. On the title page of his own copy of *De Revolutionibus*, Reinhold wrote out Copernicus's principle of uniform motion in red letters. And in his annotations he consistently singled out for summary and comment Copernicus's accomplishments in eliminating the equant, because of which (he said) "the science of the celestial motions was almost in ruins; the studies and works of this author have restored it." Thus Reinhold saw Copernicus entirely as the reactionary thinker he in some respects was, returning astronomy to its true foundations on uniform circular motions. In contrast, the paucity of Reinhold's annotations on the cosmological arguments of Book I indicates little interest, and in an unpublished commentary on Copernicus's work he maintained a neutral stance on the question of whether the earth really moves. But there was no doubt in his mind that Copernicus's geometric constructions provided a superior basis for computing planetary positions, and the many users of Reinhold's *Prutenic Tables* (1551) found out he was right, whatever their own cosmological views. (Westman 1975a, pp. 174-178)

An especially influential advocate of instrumental acceptance in our sense—i.e., with an explicit denial of truth—of the Copernican theory was Caspar Peucer, Melanchthon's successor as rector at Wittenberg. Like his predecessor and mentor, Peucer used Copernican values for various parameters, but denied the theory's truth on Scriptural and Aristotelian physical grounds in his popular textbook of 1553. He also suggested in 1568 that if certain parts of Copernicus's theory were reformulated in a geostatic frame, "then I believe that the same [effects] would be achieved without having to change the ancient hypotheses." (Quoted in Westman 1975a, pp. 178-181)

We have already seen one reason why the Wittenberg school refused to grant realistic acceptance to the Copernican theory: namely, that on a realistic interpretation the theory conflicts with Aristotelian physics and with Holy Scripture. (In section 2 above we saw that a parallel argument from Aristotelian physics had often been given against a realistically interpreted Ptolemaic theory.) But let us consider whether something else was involved as well. Westman makes some intriguing suggestions about this:

> . . . what the Wittenberg Interpretation *ignored* was as important as that which it either asserted or denied. In the writings both public and private of nearly every author of the generation which first received the work of Copernicus, the new analysis of the relative *linear* distances of the planets is simply passed over in silence. . . . questions about the Copernican ordering of the planets were not seen as important topics of investigation. In annotated copies of *De Revolutionibus* which are datable from the period *circa* 1543-1570, passages in Book I extolling the newly discovered harmony of the planets and the eulogy to the sun, with its Hermetic implications, were usually passed over in silence. (Westman 1975a, pp. 167, 181)

The reference is to the long quotation in section 5 above, and to adjacent passages.

Although Westman deserves our thanks for pointing this out, he makes no effort to explain why a lack of interest in Copernicus's planetary harmony and distances was associated with a witholding of realistic acceptance. Can we get any deeper? Consider first the question of distances, deferring that of harmony. In another paper (1979) I argued that a principle *P* operative in the nineteenth-century debates about the reality of atoms was that *it is an objection to the acceptance of a theory on a realistic interpretation that it contains or implies the existence of indeterminate quantities*. We have already seen that this principle played at least some role in astronomy: Ptolemy's theory had indeterminate planetary distances and tended to be refused realistic acceptance. One might well wonder, however, how this principle could explain refusal to accept the literal truth of Copernicus's theory, since he did make the planetary distances determinate. The answer appeals to a variant of *P* called *P'*: *persons who either reject or ignore a theory's determinations of magnitudes from measurements via its hypotheses will tend to refuse it realistic acceptance. P'*, though not precisely a corollary of *P*, is plausible given *P*: someone who rejects a theory's magnitude-determinations is likely to do so because he regards the hypotheses used as not well tested and hence regards the magnitudes as indeterminate; and someone who ignores the determinations is unaware of some of the support for realistic acceptance. Principle *P'* certainly fits the behavior as described by Westman of the first-generation response to Copernicus, especially among the Wittenberg group.

It also fits Johannes Praetorius (1537-1616), who studied astronomy at Wittenberg and later taught there and elsewhere. He expressed his instrumental acceptance of the Copernican theory as follows in a manuscript begun in 1592:

> Now, just as everyone approves the calculations of Copernicus (which are available to all through Erasmus Reinhold under the title *Prutenic Tables*), so everyone clearly abhors his hypotheses on account of the multiple motion of the earth. . . . we follow Ptolemy, in part, and Copernicus, in part. That is, if one retains the suppositions of Ptolemy, one achieves the same goal that Copernicus attained with his new constructions. (Westman 1975b, p. 293)

Like others associated with Wittenberg, Praetorius was most impressed by the improvements in observational accuracy over Ptolemy and even over Copernicus himself, that were achieved by Reinhold on Copernican assumptions, and by Copernicus's elimination of the "absurd" equant. But unlike the first generation at his school, he paid careful attention to

Copernicus's determinations of the planetary distances and to his evoca-
tions of the planetary system's "harmony" or "symmetry" (i.e., unified
overall structure) that the new theory makes evident. Thus in lectures
written in 1594 he listed Copernicus's values for the planetary distances
and remarked about them: "this symmetry of all the orbs appears to fit
together with the greatest consonance so that nothing can be inserted
between them and no space remains to be filled. Thus, the distance from
the convex orb of Venus to the concave orb of Mars takes up 730 earth
semidiameters, in which space the great orb contains the moon and earth
and moving epicycles." (quotations etc. in Westman 1975b, pp. 298-299)

Another point emerges from this quotation: Praetorius evidently followed the tradition of thinking of the planets as moving on solid spheres.
This assumption created difficulties for him when he attempted to
transform Copernicus's system into a geostatic one. For he found that using
Copernicus's distances, "there would occur a great confusion of orbs
(especially with Mars). . . . because it would then occupy not only the Sun's
orb but also the great part of Venus'. . . ." Since intersections of the spheres
are impossible, he argued, Copernicus's distances "simply cannot be
allowed." He therefore roughly doubled the distance to Saturn, on the
ground that there will still be plenty of distance to the stars, and claimed
that with that done, "nothing prohibits us. . . from making Mars' orb
greater so that it will not invade the territory of the Sun." (Westman 1975b,
p. 298) Plainly, on Praetorius's version of the Copernican theory, the
planetary distances are indeterminate: they are set through entirely
theoretical considerations regarding the relative sizes of the spheres,
instead of being computed from observational data via well-tested hypotheses. Because he rejected Copernicus's determinations of the distances, it
is in accord with principle P' that he rejected Copernicus's theory on a
realistic interpretation. And since the distances are indeterminate (even
though specified) on his own theory, it is in accordance with our principle P
that he granted instrumental acceptance to his own astronomical theory,
since he refused realistic acceptance to any:

> . . . the astronomer is free to devise or imagine circles, epicycles and
> similar devices although they might not exist in nature. . . . The
> astronomer who endeavors to discuss the truth of the positions of these
> or those bodies acts as a Physicist and not as an astronomer—and, in
> my opinion, he arrives at nothing with certainty. (Westman 1975b, p.
> 303)

This quotation reveals that Praetorius was influenced by two additional factors we have observed earlier as counting against realistic acceptance of an astronomical theory. One is that *the theory is independent of physics:* that it is either outside the domain of physics, or if literally interpreted is inconsistent with the true principles of physics. (We have noted above, section 2, that Ptolemy, who seems sometimes to have been thinking of parts of his theory as mere devices, held to the first kind of independence; and that Proclus used the second as an argument against realistic acceptance of a Ptolemaic theory.) The second factor influencing Praetorius was the argument, also found in Proclus and others, that no realistically interpreted astronomical theory can be known to be true.

Another argument we found in the pre-Copernican period (e.g., in Proclus) against realistic acceptance of any planetary theory was that alternative systems of orbits may be compatible with the appearances; and that no particular system, therefore, can be asserted as literally true. A strengthened version of principle *P*—no theory *T* containing indeterminate quantities should receive realistic acceptance—is closely related to Proclus's principle, since various settings of the indeterminate parameters in *T* would in some cases produce various alternative theories equally compatible with the data. But although the notion that indeterminate quantities count against realistic acceptance continued to play a role in post-Copernican astronomy, Proclus's principle came under attack and seems not to have played much (if any) role in the thinking of instrumentally Copernican astronomers. For example, the influential Jesuit astronomer Christopher Clavius wrote in 1581 that it is not enough merely to speculate that there *may* be some other method than ours of accounting for the celestial appearances. For the argument to have any force, our opponents must actually produce the alternative. And if it turns out to be a "more convenient way [specifically, of dealing with the appearances] . . . we shall be content and will give them very hearty thanks." But failing such a showing, we are justified in believing that the best theory we actually have (Ptolemy's, he thought) is "highly probable"; for the use of Proclus's principle would destroy not just realistically interpreted astronomy, but all of natural philosophy: "If they cannot show us some better way, they certainly ought to accept this way, inferred as it is from so wide a variety of phenomena: unless in fact they wish to destroy. . . Natural Philosophy. . . . For as often as anyone inferred a certain cause from its observable effects, I might say to him precisely what they say to us—that forsooth it may be

possible to explain those effects by some cause as yet unknown to us."
(Blake 1960, pp. 31-37) Although this gets us ahead of and even beyond our
story, eventually a principle very much like Clavius's appeared as Newton's fourth rule of reasoning in philosophy:

> *In experimental philosophy we are to look upon propositions inferred
> by general induction from phenomena as accurately or very nearly
> true, notwithstanding any contrary hypotheses that may be imagined,
> till such time as other phenomena occur by which they may either be
> made more accurate, or liable to exceptions.*
> This rule we must follow, that the argument of induction may not be
> evaded by hypotheses. (Newton 1934, p. 400)

Newton's rule is a stronger critical tool than Clavius's, since even if the
alternative "hypothesis" is actually produced (but, by definition of "hypothesis," not by deduction from phenomena), Newton refused it consideration, whereas Clavius might even have preferred it if it proved to be
more convenient, or better in accord with physics and Scripture.

Clavius thought that although Copernicus's theory was approximately as
accurate as Ptolemy's, it was false because it conflicted with physics and
Scripture. So he accorded the Copernican theory instrumental acceptance,
and the Ptolemaic theory realistic acceptance as an approximation. In
addition to consistency with physics and Scripture, he used one other
consideration in favor of realistic acceptance which, I have argued elsewhere (1979), was also operative in the nineteenth-century atomic debates.
It is progressiveness—i.e., the power of a theory to inform us of "novel"
facts, of facts the theory's inventor did not know at the time of the
invention. (See my 1982 paper on this definition of "novel.") Thus Clavius
argued in favor of a realistic acceptance of Ptolemy's theory:

> But by the assumption of Eccentric and Epicyclic spheres not only are
> all the appearances already known accounted for, but also future
> phenomena are predicted, the time of which is altogether unknown:
> thus, if I am in doubt whether, for example, the full moon will be
> eclipsed in September, 1583, I shall be assured by calculation from the
> motions of the Eccentric and Epicyclic spheres, that the eclipse will
> occur, so that I shall doubt no further. . . . it is incredible that we force
> the heavens (but we seem to force them, if the Eccentrics and
> Epicycles are figments, as our adversaries will have it) to obey the
> figments of our minds and to move as we will or in accordance with our
> principles. (Blake 1960, p. 34)

Here we have as clear an example as could be desired of an explicit distinction being made between realistic and instrumental acceptance, and of progressiveness being used to decide between them.

The instrumentally Copernican astronomers discussed so far—i.e., astronomers who preferred Copernican angle-determinations but thought the theory needed a geostatic transformation—either ignored or rejected Copernicus's determinations of the planets' distances. But this is not true of the most famous of their group, Tycho Brahe. Like the members of the Wittenberg circle, with whom he had extensive contact, Tycho wrote in 1574 that although Copernicus "considered the course of the heavenly bodies more accurately than anyone else before him" and deserved further credit for eliminating the "absurd" equant, still "he holds certain [theses] contrary to physical principles, for example, . . . that the earth . . . move(s) around the Sun. . . ." He therefore invented his own system, which was essentially Copernicus's subjected to a transformation that left the earth stationary, the sun in orbit around it, and the other planets on moving orbits centered at the sun. Having become convinced that there are no solid spheres carrying the planets, since the comets he had observed would have to penetrate them, he did not share Praetorius's motivation for altering the Copernican distances, and therefore retained them. (Westman 1975b, pp. 305-313, 329; see Kuhn 1957, pp. 201-204)

Now the case of Tycho may seem to be anomalous from the standpoint of principles *P* and *P'*: for Tycho accorded Copernicus's theory only instrumental acceptance, and yet neither ignored nor rejected Copernicus's determinations of the planetary distances. He would have conceded that these quantities were determinate, since he knew they could be computed from observations and certain of Copernicus's hypotheses on the *relative* positions of the planets, hypotheses that Tycho accepted and regarded as well tested. But this objection to *P* ignores its implicit ceteris paribus clause: *P* requires only that the determinateness of a theory's quantities should count in favor of realistic acceptance, and that indeterminateness should count against; there may nonetheless be countervailing considerations. Tycho's reasoning is entirely in accord with this notion. In one of his own copies of *De Revolutionibus*, Tycho underlined the passage (quoted in section 6 above) in which Copernicus stated that his theory links together the planetary distances, and commented on the passage: "The reason for the revival and establishment of the Earth's motion." And next to the passage (quoted in section 5 above) in which Copernicus spoke of the "symmetry of the universe" made evident by explaining so many varied phenomena in terms of the earth's motion, Tycho wrote: "The testimonies of the planets, in particular, agree precisely with the Earth's motion and thereupon the hypotheses assumed by Copernicus are strengthened." (Westman 1975b, p. 317) Despite these favorable remarks, Tycho rejected the Copernican theory on the sorts of grounds with which we are now familiar: that it conflicts with Scripture, physical theory, and certain observational data—specifically, Tycho's failure to detect the annual stellar parallax entailed by the earth's motion, and the relatively large apparent sizes of the stars given the great distances entailed by the undetectability of parallax. (Dreyer 1953, pp. 360-361) Although these considerations prevailed in his mind, it is still plain from the marginal notes just quoted that in accordance with principle *P* he counted it in favor of realistic acceptance of the hypothesis of the earth's motion that it made the planetary distances determinate and also (a point not yet discussed in this context) that *it satisfies Copernicus's principle of variety of evidence*, which in section 6 above I tentatively proposed explicating in terms of number of sets of auxiliary hypotheses.

That variety of evidence (metaphorically, "symmetry" or "harmony") counted in favor of realistic acceptance is also indicated by Westman's remark that in the period of instrumental acceptance of Copernicus's theory his remarks on harmony tended to be ignored.

We can sum up our discussion of the instrumental acceptance of the Copernican theory by listing the factors that, in the immediately post-Copernican period, were counted in favor of, or whose absence was counted against, acceptance on a realistic interpretation:

On such an interpretation, the theory

(1) satisfies the laws of physics,
(2) is consistent with other putative knowledge (e.g., the Scriptures),
(3) is consistent with all observational data,
(4) contains only determinate quantities,
(5) is able to predict novel facts,
(6) has a central hypothesis supported by a large variety of evidence,
(7) is within the realm of possible human knowledge.

Failing any of (1) - (3), a theory (if we assume it is still a convenient prediction-device in a certain domain) will tend to be accorded instrumental acceptance (with denial of truth). Supposed failure of (4) - (7) leads only to scepticism regarding truth.

We can now turn to those who accepted the Copernican theory on a realistic interpretation. If the foregoing is correct and complete, we should not find anything new.

9. Realistic Acceptance of Copernicanism

We saw in section 7 that the first person to accept the Copernican theory as literally true was Copernicus himself. We also saw that he argued explicitly that his theory satisfies the laws of physics, makes the planetary distances determinate, and has a central hypothesis supported by a wide variety of evidence. He also mentioned no observational data inconsistent with his theory, and implied in his prefatory letter to the Pope that Scripture conflicts with his theory only if "wrongly twisted." Since these considerations were all "in the air" in Copernicus's period as counting in favor of realistic acceptance, I think it is plausible to regard them as his reasons, although he did not make this more obvious by citing them in the context of an explicit distinction between realistic vs. instrumental acceptance.

The second astronomer to give realistic acceptance to Copernicanism was undoubtedly Georg Joachim Rheticus. He left Wittenberg to live and study with Copernicus (from 1539 to 1541), during which time he became familiar with the still unpublished Copernican theory. In 1540 he published *Narratio Prima*, the first printed account of the new theory of "my teacher," as he called Copernicus. In this work he nowhere indicated that he thought the theory to be just a convenient device. Moreover, he claimed (falsely) for unspecified reasons that at least some aspects of the Copernican theory could not be subjected to the sort of geostatic transformation (permitting instrumental acceptance) favored by others associated with Wittenberg: "I do not see how the explanation of precession is to be transferred to the sphere of stars." (Rosen 1959, pp. 4-5, 10, 164) Finally, in two copies of *De Revolutionibus* he crossed out Osiander's preface with red pencil or crayon. (Gingerich 1973c, p. 514) So it is obvious enough that his acceptance was realistic.

But why? First, he thought the theory is consistent with the most important relevant law of physics—uniform circularity of celestial motion—and all observational data:

> . . . you see that here in the case of the moon we are liberated from an equant by the assumption of this theory, which, moreover, corresponds to experience and all the observations. My teacher dispenses with equants for the other planets as well. . . . (Rosen 1959, p. 135)

> . . . my teacher decided that he must assume such hypotheses as would contain causes capable of confirming the truth of the observations of

previous centuries, and such as would themselves cause, we may hope, all future astronomical predictions of the phenomena to be found true. (Rosen 1959, pp. 142-143)

It is somewhat puzzling that Copernicus's repudiation of the equant was a basis for both realistic and instrumental acceptance. The explanation is perhaps that his principle of uniformity can be thought of as an aesthetic virtue of a calculation-device—it is "pleasing to the mind" (Copernicus, in Rosen 1959, p. 57)—or as a physical principle (Copernicus 1976, Book I, chapter 4).

In the last quotation from Rheticus, he invoked the criterion of progressiveness (5), since he implied that the predictions were not known to be correct on some other ground (such as simple induction). He also contrasted the Copernican and Ptolemaic hypotheses in regard to the determinateness of planetary distances:

> ...what dispute, what strife there has been until now over the position of the spheres of Venus and Mercury, and their relation to the sun.... Is there anyone who does not see that it is very difficult and even impossible ever to settle this question while the common hypotheses are accepted? For what would prevent anyone from locating even Saturn below the sun, provided that at the same time he preserved the mutual proportions of the spheres and epicycle, since in these same hypotheses there has not yet been established the common measure of the spheres of the planets....
>
> However, in the hypotheses of my teacher,...(t)heir common measure is the great circle which carries the earth.... (Rosen 1959, pp. 146-147)

It will be noted that Rheticus ignored the nesting-shell hypothesis and took the Ptolemaic distances as entirely unspecified. Finally, Rheticus argued that Copernicus's central hypothesis that the earth moves was supported by a wide variety of evidence—specifically, the apparent motions of the five visible planets. "For all these phenomena appear to be linked most nobly together, as by golden chain; and each of the planets, by its position and order and every inequality of its motion, bears witness that the earth moves...." (Rosen 1959, p. 165) God arranged the universe thus "lest any of the motions attributed to the earth should seem to be supported by insufficient evidence." (Rosen 1959, p. 161) Rheticus, then, appealed to criteria (1), (3), (4), (5), and (6) for realistic acceptance—criteria, I have argued, that were widely accepted in his period.

Westman is dissatisfied with "rational" explanations (such as the above) of Rheticus's behavior—i.e., explanations in terms of generally applicable and generally applied criteria regarding the theory and the evidence for it: "If today we might defend the rationality of this argument [from unity or harmony] on grounds of its empirical adequacy, its simplicity, and hence its considerable promise for future success, Rheticus went much further: he took it as evidence of the *absolute* truth of the entire theory of Copernicus." (Westman 1975a, pp. 184-186) Rheticus's "excessive zeal," then, requires not a rational but a "psychodynamic" explanation. We should note that Westman appears at this point to be following the psychohistorical analog of the "arationality assumption" widely accepted by sociologists of knowledge: *"the sociology of knowledge may step in to explain beliefs if and only if those beliefs cannot be explained in terms of their rational merits"* (This formulation of the assumption and references to writers espousing it are in Laudan 1977, p. 202).

Westman finds himself at a disadvantage by comparison with professional psychoanalysts in that he cannot put Rheticus on the couch and get him to free-associate. Still, the psychohistorian has considerable information with which to work, including the fact that when Rheticus was fourteen his father was convicted of sorcery and beheaded. Westman thinks that part of the reason why Rheticus was attracted to Copernicus and thus to his theory was that "in Copernicus, Rheticus had found a kind and strong father with a streak of youthful rebellion in him: a man who was different, as Rheticus' father had been . . . ." But just as important was the fact that the Copernican system had the sort of *unity* that Rheticus's father so notably lacked after he had been beheaded. Rheticus had made "determined efforts—in the search for wholeness, strength, and harmony—to unconsciously repair the damage earlier wrought on his father." And Copernicus was a substitute father "who, like *the system he created*, had a head and a heart which were connected to the same body." (Westman 1975a, pp. 187-189)

Despite my considerable reliance upon and admiration for Professor Westman's brilliant contributions to our understanding of the Copernican revolution, I cannot follow him here. Leaving aside the question of the scientific merit of the general psychological theory on which his explanation of Rheticus's behavior is premised, the main problem with the explanation is that it simply is not needed. Rheticus argued explicitly that Copernicus's theory met certain criteria, and these were criteria used by many writers of his time (and other times) for realistic acceptance. Rheticus himself said that the importance of the "golden chain" unifying the planetary appearances was that it assures that there is sufficient evidence for the earth's motion. There is simply no need (and certainly no direct evidence) for an appeal to symbolic posthumous surgery on his father.

To make this criticism of Westman is not to accept the arationality assumption. In fact, I consider it absurd to deny that there can be *both* scientific reasons and psychosocial causes for a given cognitive attitude of a scientist. Still,.when a psychosocial explanation lacks support for its initial conditions (here, the symbolic import of "unity"), so that the most that can be said for it is that we cannot think of any other explanation of the explanandum, then the existence of a documented rational account undermines the psychosocial one.

Laudan points out that a difficulty in using the arationality assumption is that one cannot apply it correctly unless one has an adequate theory of the rational merits of scientific theories; and, he says, use of an overly simple theory of rationality has been the cause of much confusion in the sociology (and presumably, the psychohistory) of ideas. (1977, pp. 205; 242, n.l) This, I believe, is what has gone wrong in Westman's discussion of Rheticus. "Empirical adequacy. . . simplicity. . . and. . . promise" is just not an adequate description of the scientific merits of Copernicus's arguments. Moreover, it is beside the point that today we think Copernicus's theory has only approximate and not "absolute" truth. Since we have much more astronomical evidence and know much more physics than Rheticus did, the rationality of his beliefs (in the light of his knowledge) cannot be assessed by reference to our knowledge.

Perhaps Westman would concede in response to my criticism that although the unity of the Copernican theory may have provided Rheticus with *reason* to believe it, still we need a psychodynamic explanation of why Rheticus was impressed by this unity in a way others of his generation generally were not. To this I would reply that a perfectly straightforward and plausible explanation of Rheticus's attitude is that he learned the theory from Copernicus's own lips; and since Copernicus obviously thought the "harmony" of his system was one of its main virtues, he no doubt forcefully called it to Rheticus's attention. This explanation is at any rate considerably less speculative than Westman's.

We find realistic acceptance and reasoning similar to that of Copernicus and Rheticus in another early Copernican, Michael Mästlin (1550-1631), whose support was crucial because of the pro-Copernican influence Mästlin exerted on his student Kepler. In his annotations of a copy of *De Revolutionibus*, which are consistently approving of Copernicus, he complained that Osiander's preface had made the mistake of "shattering [astronomy's] foundations" and suffered from "much weakness in his meaning and reasoning." This indicates that his acceptance was realistic. Commenting on Copernicus's attempts in Book I of *De Revolutionibus* to answer Ptolemy's physical arguments against the earth's motion, he wrote: "He resolves the objections which Ptolemy raises in the Almagest, Book I, chapter 7." So he evidently thought the Copernican theory satisfied the laws of physics. Finally, referring to Copernicus's arguments from determinateness of distances and variety of evidence, he wrote: "Certainly this is the great argument, *viz.* that all the phenomena as well as the order and magnitude of the orbs are bound together in the motion of the earth. . . . moved by this argument I approve of the opinions and hypotheses of Copernicus." (Quoted in Westman 1975b, pp. 329-334) Mästlin's realistic acceptance, then, was based at least on criteria (1), (4), and (6).

10. Kepler

The trend toward realistic acceptance of a heliostatic theory culminated in the work of Kepler. As Duhem wrote, "the most resolute and illustrious representative [of the realistic tradition of Copernicus and Rheticus] is, unquestionably, Kepler." (1960, p. 100) Kepler was quite explicit in making the distinction between realistic and instrumental acceptance of astronomical theories in general and of Copernicus's in particular, and explicit in saying where he stood on these issues. Not only Osiander, but Professor Petrus Ramus of the University of Paris, had asserted that Copernicus's theory used hypotheses that were false. Ramus wrote: "The fiction of hypotheses is absurd. . . would that Copernicus had rather put his mind to the establishing of an astronomy without hypotheses." And he offered his chair at Paris "as the prize for an astronomy constructed without hypotheses." Kepler wrote in his *Astronomia Nova* of 1609 that had Ramus not died in the meantime, "I would of good right claim [his chair] myself, or for Copernicus." He also indignantly revealed that Osiander was the author of the anonymous preface to *De Revolutionibus* and asserted: "I confess it a most absurd play to explain the processes of nature by false causes, but there is no such play in Copernicus, who indeed himself did believe his hypotheses true. . . nor did he only believe, but also proved them true." (Blake 1960, p. 43) Kepler intended to proceed in the same realistic spirit: "I began this work declaring that I would found astronomy not upon fictive hypotheses, but upon physical causes." (Quoted in Westman 1971, p. 128) Astronomy "can easily do without the useless furniture of fictitious circles and spheres." (Kepler 1952, p. 964) In taking such a view Kepler was consciously aware of contributing to a revolution not only in the prevailing theories in astronomy but also in the prevailing view of the field's purposes. He described his work as involving "the unexpected transfer of the whole of astronomy from fictitious circles to natural causes, which were the most profound to investigate, difficult to explain, and difficult to calculate, since mine was the first attempt." (Quoted in Gingerich 1973d, p. 304)

According to Kuhn, disagreement over the problems, aims, and methods of a field is one of the things that makes competing schools "incommensurable"—i.e., makes arguments for either of them circular, the choice a matter of "faith" or a "conversion experience" based, sometimes, on "personal and inarticulate aesthetic considerations" rather than on "rational" argument. (1970, pp. 41, 94, 151, 158) But in fact Kepler did present rational (if not always decisive) arguments for the shift in astronomical aims that he advocated, grounds for asserting that these aims could and should be achieved. Most of these grounds were neither inarticulate nor merely personal, but were explicitly stated attempts to show that a heliostatic theory could meet the widely used criteria for realistic acceptance.

We saw (section 8 above) that one such criterion, in this historical period and others, is progressiveness—prediction of novel facts. Thus we are not surprised to find Kepler arguing as follows in his *Mysterium Cosmographicum* (1596) for an essentially Copernican theory—i.e., one that has the planets somehow orbiting a stationary sun:

> My confidence was upheld in the first place by the admirable agreement between his conceptions and all [the objects] which are visible in the sky; an agreement which not only enabled him to establish earlier motions going back to remote antiquity, but also to predict future [phenomena], certainly not with absolute accuracy, but in any case much more exactly than Ptolemy, Alfonso, and other astronomers. (Quoted in Koyré, 1973, p. 129)

Kepler (*ibid*, p. 133) also echoed Copernicus's and Rheticus's appeals to criterion (6), well-testedness: "Nature likes simplicity and unity. Nothing trifling or superfluous has ever existed: and very often, one single cause is destined by itself to [produce] several effects. Now, with the traditional hypotheses there is no end to the invention [of circles]; with Copernicus, on the other hand, a large number of motions is derived from a small number of circles." (1973, p. 133) As Koyré pointed out, this claim about numbers of circles is an overstatement which had become traditional among Copernicus and his followers. But we can accept the part of this argument alluding to the variety of evidence for the earth's motion.

We also saw (sections 8-9 above) that realistic acceptance tended to be associated with belief and interest in the determinate values of the planetary distances that Copernicus's theory and data provided. There can be no doubt that Kepler had both belief and the most intense interest. In 1578 Kepler's teacher Michael Mästlin had published a theory of the motion of the comet of 1577 that asserted that it moved within the heliocentric shell of Venus, a theory that presupposed the Copernican arrangement of the inferior planets and his values for their distances: "I noticed that the phenomena could be saved in no other way than if. . . [the comet's radii] were assumed to be 8420 parts when. . . the semidiameter of the [earth's orbit] is 10,000; and likewise, when the semidiameter of Venus' eccentric is 7193." Kepler wrote in 1596 that Mästlin's theory that the comet "completed its orbit in the same orb as the Copernican Venus" provided "the most important argument for the arrangement of the Copernican orbs." (Quotations in Westman, 1972b, pp. 8,22) A second argument for Kepler's acceptance of (at least as approximations) and interest in Copernican determinations of the distances is that the main purpose of his *Mysterium Cosmographicum* was to explain the distances (as well as the number) of the planets on the basis of the assumption that their spheres are inscribed within a nest formed by the five regular solids. This work was published with an appendix by Mästlin containing improved Copernican calculations of the distances based upon the Prutenic Tables (Westman 1972b, pp. 9-10; Koyré, 1973, pp. 146-147). The third way in which approximately Copernican values for the sun-to-planet distances were important for Kepler was that he held that a planet's orbital period is proportional to its mean distance raised to the 3/2 power, and that its linear speed is inversely proportional to its distance from the sun. And from either premise he deduced his crucial conclusion that the planets are made to revolve in their orbits neither by their supposed souls (Plato) nor their intrinsic nature (Copernicus), but by a physical force orginating in the sun. (Kepler 1952, p. 895)

A corollary of our principle (4) governing realistic acceptance (section 8 above) is that a theory's failure to provide any means at all—let alone well-tested hypotheses—to determine some of its parameters counts against realistic acceptance. Kepler appealed to this corollary when he tried to show that Ptolemy believed his theory was more than a prediction-device: "to predict the motions of the planets Ptolemy did not have to consider the order of the planetary spheres, and yet he did so diligently."[2] (Kepler, forthcoming) Since the hypothesis from which Ptolemy obtained the order (that it corresponds to increasing orbital period) was entirely untested, I do not say that the order was determinate, but only specified.

We have seen earlier that such writers as Proclus and Praetorius argued against realistic acceptance of any astronomical theory on the grounds that knowledge of the full truth of such a theory exceeds merely human capacities and is attainable only by God. Aware of this traditional instrumentalist argument, Kepler felt obligated to argue on theological grounds that there is no *hubris* in claiming to know the true geometry of the cosmos:

> Those laws [governing the whole material creation] are within the grasp of the human mind; God wanted us to recognize them by creating us after his own image so that we could share in his own thoughts. For what is there in the human mind besides figures and magnitudes? It is only these which we can apprehend in the right way, and... our understanding is in this respect of the same kind as the divine.... (Letter of 1599, in Baumgardt 1951, p. 50.)

To supplement his theological arguments, Kepler also attempted to undermine such support as astronomical scepticism received from the unobservability of the planets' orbits in physical space: "But Osiander... (i)f [you say that] this art knows absolutely nothing of the causes of the heavenly motions, because you believe only what you see, what is to become of medicine, in which no doctor ever perceived the inwardly hidden cause of a disease, except by inference from the external bodily signs and symptoms which impinge on the senses, just as from the visible positions of the stars the astronomer infers the form of their motion." (Kepler, forthcoming)

Another epistemological argument we saw was popular among astronomical instrumentalists asserts that since there are, or may be, alternative systems of orbits equally compatible with all observational data, no one system can be asserted as physically correct. About twenty years after the Ptolemaic astronomer Clavius attacked the argument from observational equivalence, Kepler mounted his own attack from a Copernican viewpoint. First, he argued, sets of hypotheses—some true, some false—which have exactly the same observational consequences are found (if ever) far less frequently than those who use the argument from equivalence suppose:

> In astronomy, it can scarcely ever happen, and no example occurs to
> me, that starting out from a posited false hypothesis there should
> follow what is altogether sound and fitting to the motions of the
> heavens, or such as one wants demonstrated. For the same result is
> not always in fact obtained from different hypotheses, whenever
> someone relatively inexperienced thinks it is. (Kepler, forthcoming)

For example, Kepler argued, Magini attempted (1589) to produce a
Ptolemaic theory agreeing with the *Prutenic Tables*, but failed to obtain
Copernicus's prediction that Mars has a greater parallax than the sun.
Kepler believed that whenever two conflicting hypotheses give the same
results for a given range of phenomena, at least one of them can be refuted
by deriving observational predictions from it in conjunction with new
auxiliary hypotheses:

> "And just as in the proverb liars are cautioned to remember what they
> have said, so here false hypotheses which together produce the truth
> by chance, do not, in the course of a demonstration in which they have
> been applied to many different matters, retain this habit of yielding
> the truth, but betray themselves." (Kepler, forthcoming)

Thus a false hypothesis may occasionally yield a true prediction, but only
when it chances to be combined with an auxiliary hypothesis containing a
compensatory error, as when Copernicus proposed a lunar latitude and a
stellar latitude, both too small by the same amount, and thus obtained a
correct prediction for an occultation of the star by the moon. (Kepler,
forthcoming) Conjoined with different auxiliary hypotheses, this one on
the moon would certainly "betray itself." Kepler was arguing—exactly in
accordance with (6) of section 8 above—that we have good reason to accept
a hypothesis realistically when it is supported by a variety of phenomena,
where this means "in conjunction with many sets of auxiliary hypotheses."
Moreover, he appealed to the rationale for this idea, which in section 6
above we took from Glymour and adapted to the HD context—namely,
that it reduces the chance of compensatory errors.

Kepler also considered a somewhat different way of stating the instrumentalist argument from observational equivalence. The Copernican and
Ptolemaic hypotheses are concededly both compatible with, and both
imply, e.g., the daily motion of the whole heaven. Ptolemy apparently
inferred, partly because of this property of his theory, that it is true. But,
the objection runs, the Copernicans think Ptolemy was thus led into error.
"So, by the same token, it could be said to Copernicus that although he
accounts excellently for the appearances, nevertheless he is in error in his

hypotheses." (Trans. Jardine 1979, p. 157)

Kepler's rejoinder to this objection is as follows:

> For it can happen that the same [conclusion] results from two suppositions which differ in species, because the two are in the same genus and it is in virtue of the genus primarily that the result in question is produced. Thus Ptolemy did not demonstrate the risings and settings of the stars from this as a proximate and commensurate middle term: 'The earth is at rest in the centre'. Nor did Copernicus demonstrate the same things from this as a middle term: 'The earth revolves at a distance from the centre'. It sufficed for each of them to say (as indeed each did say) that these things happen as they do because there occurs a certain separation of motions between the earth and the heaven, and because the distance of the earth from the centre is not perceptible amongst the fixed stars [*i.e.*, there is no detectable parallactic effect]. So Ptolemy did not demonstrate the phenomena from a false or accidental middle term. He merely sinned against the law of essential truth *(kath'auto)*, because he thought that these things occur as they do because of the species when they occur because of the genus. Whence it is clear that from the fact that Ptolemy demonstrated from a false disposition of the universe things that are nonetheless true and consonant with the heavens and with our observations—from this fact I repeat—we get no reason for suspecting something similar of the Copernican hypotheses. (Trans. Jardine 1979, p. 158)

Kepler's point is this. Let H = "The heaven moves (in certain way) about a stationary earth." E = "The earth moves (in a certain way) within a stationary heaven," R = "The earth and heavens have a (certain) relative motion," and O = "The stars rise and set (in certain ways)." Since H implies O, it may appear that the verification of O supports H. But Aristotle required that in a premise of a "demonstration," or "syllogism productive of scientific knowledge," the predicate must belong to the subject "commensurately and universally," which he took to entail both that it belongs "essentially" and also that the subject is the "primary subject of this attribute." Aristotle's example of a violation of this latter requirement is the statement that any isosceles triangle has angles equal to 180°. (*Post. Anal.* I, 2,4; in McKeon 1941). Evidently, then, it would be a "sin" against this requirement to use a specific predicate when a more general one would yield the conclusion. (Thus far I follow Jardine 1979, p. 172). Kepler's application of this idea to Ptolemy is as follows: the syllogism that leads validly from H to O is not a proper demonstration, since the weaker premise R would suffice; hence the verification of O gives HD-support only to R, and the falsity of H casts no doubt on HD reasoning.

The supposedly erroneous reasoning that Kepler here attributed to Ptolemy is the same as what occurs in the "tacking paradox," an objection to the HD account of scientific reasoning discussed by Glymour (1980, chapter 2). The paradox is that if hypothesis h entails and is hence HD-supported by evidence e, then the same holds for "h & g," where g is an arbitrary, "tacked on" sentence. Ptolemy obtained support for R (i.e., "H or E"), but then tacked on $\sim E$ and claimed he had support for the conjunction and hence for H.

The difficulty with Kepler's argument (and with the HD account generally) is that it is quite unclear when tacking on is allowed and when it is not. We do not want to say, as Kepler's argument suggests, that no premise is ever HD-confirmed when a weaker one would have sufficed for the deduction. This would entail that whenever one applies, say, Newton's law of gravitation to objects within a certain region, one obtains no support for the law, but only for the hypothesis that the law holds within that region. On the contrary, we can tack on the law's restriction to the region's complement. But why? Thus Kepler's reply to the objection to HD reasoning answers it by revealing another, more damaging one.

When two hypotheses seem to be observationally indistinguishable, one way Kepler thought they could be distinguished is by relating them to what he called "physical considerations": "And though some disparate astronomical hypotheses may yield exactly the same results in astronomy, as Rothmann insisted...of his own mutation [geostatic transform] of the Corpernican system, nevertheless a difference arises because of some physical consideration." (Kepler, forthcoming). Physics, in this context, includes dynamics and cosmology—theories of the causes of motion and of the large-scale structure of the universe. His example here is cosmological: Copernicus's system and its geostatic transform differ in that the latter can "avoid postulating the immensity of the fixed stars" required by the former to explain the absence of detectable stellar parallax. Since Kepler had no way to show (without appeal to Copernicus's theory) that more sensitive instruments would detect stellar parallax due to the earth's revolution, this consideration provided no reason to prefer Copernicus's theory to a geostatic transform of it such as Tycho's. But this neutrality was appropriate in a piece titled "A Defence of Tycho Against Ursus."

In other works, however, Kepler used physical considerations—specifically, about the causes of planetary motions—to argue in favor of his own theory, which was Copernican in using a moving earth, non-Copernican in using elliptical orbits. In ancient and medieval astronomy the problem of why the planets move had either not arisen, or had been solved in a very simple way. To the extent that hypothesized motions were viewed as mere computation-devices, the problem of explaining them dynamically did not arise. Motions that were considered physically real, such as those of the spheres carrying the stars or planets, were usually explained as due to the spheres' "nature" or to spiritual intelligences attached to them. But Kepler, in part because be thought that God created nothing haphazardly but followed a rational plan knowable by man, sought to understand why the planets move as they do. (Koyré 1973, pp. 120-122)

In accordance with principle (1)—consistency with the laws of physics counts in favor of realistic acceptance—the ideas of physical explanation and realistic interpretation of planetary orbits were intimately connected in Kepler's mind:

> Consider whether I have made a step toward establishing a physical astronomy without hypotheses, or rather, fictions, ... The force is fixed in the sun, and the ascent and descent of the planets are likewise fixed according to the greater or lesser apparent emanation from the Sun. These, therefore, are not hypotheses (or as Ramus calls them) figments, but the very truth, as the stars themselves; indeed, I assume nothing but this. (Quoted in Gingerich 1975, p. 271)

> . . . Astronomers should not be granted excessive license to conceive anything they please without reason: on the contrary, it is also necessary for you to establish the probable cause of your Hypotheses which you recommend as the true causes of Appearances. Hence, you must first establish the principles of your astronomy in a higher science, namely Physics or Metaphysics. . . . (quoted in Westman, 1972a, p. 261)

In addition to the foregoing arguments based on criteria (1) and (4)-(7) of section 8 above for realistic acceptance—criteria that had been used by earlier writers—Kepler formulated two criteria of his own. One of these might be called "explanatory depth": it counts in favor of realistic acceptance of a theory that

(8) it explains facts that competing theories merely postulate.

After asserting the superior accuracy of Copernicus's retrodictions and predictions, Kepler remarked: "Furthermore, and this is much more important, things which arouse our astonishment in the case of other(s) [astronomers] are given a reasonable explanation by Copernicus. . . ." In particular, Ptolemy merely postulated that the deferent motions of the sun, Mercury, and Venus have the same period (one year). Copernicus, in contrast, could explain this equality on the basis of his theory that the planets revolve around the sun. Transformed into an earth-centered system, this theory yields components of the sun's and each planet's motions that are, as Kepler put it, "projections of the earth's proper motion on to the firmament." (Quoted in Koyré 1973, pp. 129, 136-137)

Second, Kepler argued that readers of Ptolemy should be astonished that the five planets, but not the sun and moon, show retrograde motion; whereas Copernicus can explain these facts, specifically by saying the five planets have epicycles of such speeds and sizes as to produce retrograde motion but the sun and moon do not. Kepler presumably meant that Copernicus's theory, when transformed geostatically, yields these statements about epicycles. Using Copernicus's figures for the radii and periods of all the planets' heliostatic orbits, one can show that their geostatic transforms will contain combinations of circles producing retrogression. In contrast, the moon, since it shares the earth's heliocentric motion, does not have a component of its geostatic motion that mirrors the earth's orbit and thereby yields retrogression, as does the epicycle of a superior planet. Finally, Kepler argued, since the earth's heliostatic orbit is circular with constant speed, it follows that the sun's geostatic orbit shows no retrogression. (Koyré 1973, pp. 136-137)

Kepler's third and fourth points are similar to ones made by Copernicus himself (section 5 above). The third is that Ptolemy postulates but cannot explain the relative sizes of the planets' epicycles, whereas their ratios can be obtained by transforming Copernicus's system into a geostatic one. The fourth is that Ptolemy postulates that, but does not explain why, the superior planets are at the closest point on their epicycles (and hence brightest) when at opposition with the sun, whereas this fact too results from a geostatic transformation of Copernicus's system.

Kepler sometimes said these four arguments are designed to show Copernicus' theory is preferable to Ptolemy's (Koyré 1973, p. 136), and sometimes to unnamed "other" astronomers' (p. 129). It is worth noting, however, that arguments from explanatory depth do not establish the superiority of Copernicus's system to Tycho's. Either of these can explain anything (regarding relative motions within the solar system) asserted by the other, by means of the appropriate transformation. (This symmetric relation does not hold between Copernicus's theory and Ptolemy's, since a heliostatic transformation of the latter would yield few if any features of the former. For example, the transform would have the planets orbiting a moving earth.) Of all Kepler's arguments, it is only the dynamical ones considered above—that there is a plausible physical explanation why the planets should move around the sun, but not why the sun should move around the earth carrying the other planets' orbits with it—that favor Copernicus's theory over Tycho's.

The last argument by Kepler that I shall consider has a remarkably contemporary ring. One of Ursus's arguments against Tycho had been an induction on the falsity of all previous astronomical theories. (Jardine 1979, p. 168) Kepler's reply was that despite the continuing imperfection of astronomical theories, cumulative progress had nonetheless been made at least since Ptolemy. Erroneous in other respects, Ptolemy's theory at least taught us that "the sphere of the fixed stars is furthest away, Saturn, Jupiter, Mars, follow in order, the Sun is nearer than them, and the Moon nearest of all. These things are certainly true. . . ." Tycho taught us at least that "the Sun is the centre of motion of the five planets," Copernicus that the earth-moon distance varies less than Ptolemy said, and unspecified astronomers established the "ratios of the diameters of the Earth, Sun and Moon. . . . Given that so many things have already been established in the realm of physical knowledge with the help of astronomy, things which deserve our trust from now on and which are truly so, Ursus' despair is groundless." (Kepler, forthcoming) Similarly, Kepler argued in favor of Copernicus that he "denied none of the things in the [ancient] hypotheses which give the cause of the appearances and which agree with observations, but rather includes and explains all of them." (Quoted in Jardine 1979, pp. 157-158). This last statement is part of Kepler's reply to the objection that Copernicus's theory might be false even though it saves the phenomena.

Clearly, then, Kepler was appealing to a principle we have not previously come across: it counts in favor of realistic acceptance of a theory that

> (9) it agrees with some of the nonobservational claims of some previous theories purporting to explain the same observations.

I say "nonobservational" because the agreed-upon claims Kepler mentioned here were not observable phenomena such as brightnesses and angular positions, but unobservables such as the planets' orbits. The tacit assumption behind (9) seems to be that if the astronomical theories produced through history were merely a series of devices for predicting observations, there would be relatively little reason to expect them to contain any common non-observational parts: whereas this is what we would expect if the sequence of principal theories contains a growing set of true descriptions of astronomical reality.

11. Contemporary Realisms

I said this argument sounds contemporary because it is echoed with little change in an argument Hilary Putnam has recently discussed, attributing it to R. Boyd.[3] (Forthcoming) According to this argument, a new and better theory in a given field of science usually implies "the *approximate truth of the theoretical laws of the earlier theories in certain circumstances.*" Further, scientists usually require this feature of new theories in part because they believe that (a) the "laws of a theory belonging to a mature science are typically approximately *true*"; and meeting this requirement is fruitful in part because this belief is true. (Putnam 1978, pp. 20-21) Like Kepler, then, Boyd thinks that there is a degree of agreement in the nonobservational hypotheses of successive theories in a given field, and that this tends to show that these hypotheses are (partially or approximately) true and are not just prediction-devices. And like Kepler, Putnam (1978, p. 25) thinks that this consideration helps rescue contemporary science from the charge, based on induction from past theories, that it is probably false. (1978, p. 25)

Boyd labels as "realism" the conjunction of (a) above with (b): "terms in a mature science typically *refer.*" (Putnam 1978, p. 20) This essay is mainly about a quite different thesis also called "realism"—the thesis that *scientific theories in general are put forward as true, and accepted because they are believed to be true*. But we have just seen that our astronomical case study has at least some connection with Boyd's kind of realism as well. I should like to conclude with some remarks about the more general question of the relevance of this case-study to various versions of realism.

Let us call the version of realism I have mainly been discussing "purpose-realism," since it is based on a thesis about the purpose of any scientific theory, and identifies acceptance with belief that that purpose is fulfilled. This is the kind of realism stated (and criticized) by van Fraassen: "*Science aims to give us, in its theories, a literally true story of what the world is like; and acceptance of a scientific theory involves the belief that it*

is true." (1980, p. 8) Van Fraassen's own opposing position, which he calls
"constructive empiricism," is: "*Science aims to give us theories which are
empirically adequate* [true regarding observables]; *and acceptance of a
theory involves as belief only that it is empirically adequate.*" Now I claim
to have shown above and in my 1979 paper that purpose-realism and
constructive empiricism are both over-generalizations, and that each holds
for some theories but not others. Neither of the theses stated by van
Fraassen can accommodate the extensive historical evidence that scientists
sometimes believe that a theory is true and sometimes only that it is
empirically adequate, and that there are different sets of grounds for these
two beliefs.

Consider now the question whether this case study has any further
relevance for Boyd's thesis, which we might call "approximation-realism"
because it appeals to the concept of approximate truth. We should note first
that approximation-realism is very different from purpose-realism. It is one
thing to talk about the purpose of scientific theories and what those who
accept them therefore believe, and quite another to say when this aim has
to some degree actually been accomplished. That this distinction is
nonetheless insufficiently appreciated is clear from the fact that Boyd
(1976) puts approximation-realism forward as a rival to an earlier version of
van Fraassen's (1976) thesis, which refers to purpose. Another difference
between the two theses is that although purpose-realism is sufficiently
clear to be refuted, the obscurity of approximation-realism makes its
assessment difficult and perhaps impossible. I leave aside the question of
what might be meant by a "mature science," and say nothing of the
vagueness of "typically." The more difficult question is whether it makes
any sense to speak of a law or theory (Boyd 1976) as "approximately true." It
certainly makes sense, although the statement is somewhat vague, to say
that a particular *value* for a given magnitude is approximately correct
—e.g., that "$\theta = 3.28$" is approximately true, because in fact $\theta = 3.29$. But
speaking of a *law* or *theory* as "approximately true" raises serious problems. Usually a law or theory refers to several different magnitudes—
their values and relations at various times and places. Given two theories,
one may be more accurate with respect to some magnitudes, and the other
theory more accurate with respect to some others. If this happens, it is
quite unclear what it would mean to say that one theory is *on the whole*
more accurate than another. Some weights would have to be assigned to
the various quantities, and no general way to do this springs readily to

mind. It might seem that this problem could be obviated at least in the special case in which, for theories T_1 and T_2 and *every* common magnitude, T_1's predicted value is never further from (and is sometimes closer to) the true value than T_2's. But Miller (1975) has shown that uniformly greater accuracy in this sense is impossible—at least where the two predictions never lie on different sides of the true value.

To be fair, we should note that Boyd himself concedes that (approximation-) realists still have considerable work to do in developing a concept of approximate truth suitable for stating their thesis. Boyd does not mention that some work relevant to this task has been done by Popper and various of his critics. (See Popper 1976, for references.) This body of work provides little help to Boyd, however. For Popper's critics have found fatal weaknesses in a number of his explications of "verisimilitude." Worse, Popper's original intuitive concept of verisimilitude was a confusing amalgamation of two quite distinct desiderata of theories—namely, *accuracy* and *strength*. As he put it, "we combine here the ideas of truth and content into. . . the idea of *verisimilitude*" (Popper 1968, pp. 232-233) Thus he supposed it to be evidence for greater verisimilitude that one theory has passed tests that another has failed (greater accuracy), and also that it explains more facts than the other does (greater strength). But combining these two properties into one has at least three disadvantages: (1) it makes the term "verisimilitude" (whose etymology suggests only accuracy) highly misleading; (2) it obscures the question of the relative importance of accuracy vis-à-vis strength as desiderata of theories; and (3) it makes verisimilitude irrelevant to Boyd's problem of explicating the concept of overall accuracy of a theory. If we had a theory which was perfectly accurate (over its domain), we would not say its *accuracy* could be increased by adding to it an arbitrary additional true statement.

I conclude, then, that approximation-realism is too obscure to be assessed and is likely to remain so. (This stricture does not apply to Kepler's somewhat similar view, since he says only that there is a cumulatively growing set of truths upon which the principal astronomers up to any given time have agreed—and this does not presuppose a concept of approximate truth.)

I shall end this essay by considering two final "realistic" theses. Boyd remarks, "What realists really should maintain is that *evidence for a scientific theory* is evidence that both its theoretical claims and its empirical claims are. . . approximately true with respect to certain issues."[4] (Boyd 1976, pp. 633-634). Similarly, Glymour (1976) defines "realism" as "the thesis that to have good reason to believe that a theory is empirically adequate is to have good reason to believe that the entities it postulates are real and, furthermore, that we can and do have such good reasons for some of our theories." (Glymour has in mind van Faassen's (1976) concept of empirical adequacy: roughly, that all measurement-reports satisfy the theory.) To avoid the difficulties just discussed, I shall ignore Boyd's use of "approximately" and define "empirical realism" as the thesis that *evidence for a theory's empirical adequacy is evidence for its truth*. This claim is logically independent of claims as to which theories are in fact approximately true, or as to what the purpose of science is, although it might have affinities to such claims. As we saw in section 6 above, Glymour rejects empirical realism, since he thinks two theories (e.g., Einstein's and Thirring's) might both be compatible with given evidence that nonetheless fails to test one of them—the one containing nonmeasurable, noncomputable quantities. And I also claim to have shown above and in my 1979 paper that a given body of data may be regarded as good evidence for the empirical adequacy but not for the truth of a theory—as when the theory conflicts with physics, contains indeterminate quantities, lacks proven predictive capability, etc.

Perhaps Boyd would not be much bothered by this argument and would say that the main concern of a "realistically"-minded philosopher is to assert the less specific thesis, which I will call "evidential realism," that *we sometimes have evidence that a theory is true (and not just empirically adequate)*. I claim to have shown that evidential realism is correct, and moreover to have spelled out at least some of the reasons—criteria (1)-(9) above, and acceptability of explanatory basis (in my 1979 paper)—that scientists have counted in favor of a theory's truth over very long historical periods. If someone wants to say that what scientists have considered to be reasons are not really reasons, or are not good enough, I can only reply that such a claim clashes with what I take to be one of the purposes of the philosophy of science—to state explicitly, clearly, and systematically the principles of reasoning that have been and are used in actual scientific practice.

Notes

1. Most of the last three paragraphs are taken from my (1979).
2. I am grateful to Nicholas Jardine for allowing me to see and quote his unpublished draft translation. His final, published version may be different.

3. I shall assume for the sake of argument that Putnam gives an accurate account of Boyd's thinking, or at least of some stage thereof.

4. From the context, and to avoid triviality, the phrase I have italicized has to be interpreted to mean "instances of (i.e., evidence for) a scientific theory's empirical adequacy."

References

Armitage, Angus. 1962. *Copernicus, The Founder of Modern Astronomy*. New York: A.S. Barnes.

Baumgardt, Carola. 1951. *Johannes Kepler: Life and Letters*. New York: Philosophical Library.

Bernardus de Virduno. 1961. *Tractatus super total Astrologiam*. Werl/Westf: Dietrich-Coelde-Verlag.

Blake, Ralph M. 1960. Theory of Hypothesis among Renaissance Astronomers in R. Blake (ed.), *Theories of Scientific Method*. Seattle: University of Washington Press, pp. 22-49.

Bluck, R.S. 1955. *Plato's Phaedo*. London: Routledge & Kegan Paul.

Boyd, Richard. 1976. Approximate Truth and Natural Necessity. *Journal of Philosophy*. 73:633-635.

Boyd, Richard. Forthcoming. *Realism and Scientific Epistemology*. Cambridge: Cambridge University Press.

Carnap, Rudolf. 1956. The Methodological Character of Theoretical Concepts. *The Foundations of Science and the Concepts of Psychology and Psychoanalysis*, in Herbert Feigl and Michael Scriven (eds.), *Minnesota Studies in the Philosophy of Science*, vol. I. Minneapolis: University of Minnesota Press, pp. 33-76.

Copernicus, Nicolaus. 1976. *On the Revolutions of the Heavenly Spheres*. Trans. Alistair M. Duncan. New York: Barnes & Noble. Orig. ed. 1543.

Cornford, Francis M. 1957. *Plato's Cosmology*. Indianapolis: Bobbs-Merrill.

Dreyer, J.L.E. 1953. *A History of Astronomy from Thales to Kepler*. New York: Dover.

Duhem, Pierre. 1969. *To Save the Phenomena*. Trans. Stanley L. Jaki. Chicago: University of Chicago Press. Orig. ed. 1908.

Gardner, Michael R. 1979. Realism and Instrumentalism in 19th-Century Atomism. *Philosophy of Science* 46:1-34.

Gardner, Michael R. 1982. Predicting Novel Facts. *British Journal for the Philosophy of Science*. 33:1-15.

Gingerich, Owen. 1973a. Copernicus and Tycho. *Scientific American* 229:87-101.

Gingerich, Owen. 1973b. A Fresh Look at Copernicus. In *The Great Ideas Today 1973*. Chicago: Encyclopaedia Brittanica, pp. 154-178.

Gingerich, Owen. 1973c. From Copernicus to Kepler: Heliocentrism as Model and as Reality. *Proceedings of the American Philosophical Society* 117:513-522.

Gingerich, Owen. 1973d. Kepler. In *Dictionary of Scientific Biography*, ed. C.C. Gillespie, vol. 7. New York: Scribners, pp. 289-312.

Gingerich, Owen. 1975. Kepler's Place in Astronomy. In *Vistas in Astronomy*, ed. A. & P. Beer, vol. 18. Oxford: Pergamon, pp. 261-278.

Glymour, Clark. 1976. To Save the Noumena. *Journal of Philosophy* 73:635-637.

Glymour, Clark. 1980. *Theory and Evidence*. Princeton: Princeton University Press.

Goldstein, Bernard. 1967. The Arabic Version of Ptolemy's *Planetary Hypotheses*. *Transactions of the American Philosophical Society*, n.s. vol. 57, part 4:3-12.

Hamilton, Edith and Cairns, Huntington. 1963. *The Collected Dialogues of Plato*. New York: Bollinger Foundation.

Hanson, Norwood R. 1973. *Constellations and Conjectures*. Dordrecht: Reidel.

Hutchins, Robert M. 1952. *Great Books of the Western World*, vol. 16. Chicago: Encyclopaedia Brittanica.

Jardine, Nicholas. 1979. The Forging of Modern Realism: Clavius and Kepler against the Sceptics. *Studies in History and Philosophy of Science* 10:141-173.

Johnson, Francis R. 1937. *Astronomical Thought in Renaissance England*. Baltimore: Johns Hopkins Press.

Kepler, Johannes. 1952. *Epitome of Copernican Astronomy*. Trans. C.G. Wallis. Orig. ed. 1618-1621. In Hutchins, 1952, pp. 841-1004.

Kepler, Johannes. Forthcoming. "A Defense of Tycho against Ursus." Trans. Nicholas Jardine. Written about 1601.

Koyré, Alexander. 1973. *The Astronomical Revolution*. Trans. R.E.W. Maddison. Ithaca: Cornell University Press. Orig. ed. 1961.

Kuhn, Thomas S. 1957. *The Copernican Revolution*. Cambridge, Mass: Harvard University Press.

Kuhn, Thomas S. 1970. *The Structure of Scientific Revolutions*. Chicago: University of Chicago Press.

Lakatos, Imre. 1970. Falsification and the Methodology of Scentific Research Programmes. In *Criticism and the Growth of Knowledge*, ed. Imre Lakatos and Alan Musgrave, pp. 91-195. London: Cambridge University Press.

Lakatos, Imre and Zahar, Elie. 1975. Why Did Copernicus' Research Program Supersede Ptolemy's? In Westman (1975c), pp. 354-383.

Laudan, Larry. 1977. *Progress and its Problems*. Berkeley: University of California Press.

Lloyd, G.E.R. 1978. Saving the Appearances. *Classical Quarterly* 28:202-222.

McKeon, Richard. 1941. *The Basic Works of Aristotle*. New York: Random House.

Miller, David. 1975. The Accuracy of Predictions. *Synthèse* 30:159-191.

Nagel, Ernest. 1961. *The Structure of Science*. New York: Harcourt, Brace and World.

Neugebauer, Otto. 1952. *The Exact Sciences in Antiquity*. Princeton: Princeton University Press.

Newton, Isaac. 1934. *Mathematical Principles of Natural Philosophy*. Trans. Andrew Motte and Florian Cajori. Berkeley: University of California Press. Orig. ed. 1687.

Pedersen, Olaf. 1974. *A Survey of the Almagest*. Odense: Odense University Press.

Popper, Karl. 1968. *Conjectures and Refutations*. New York: Harper.

Popper, Karl. 1976. A Note on Verisimilitude. *British Journal for the Philosophy of Science* 27:147-164.

Proclus. 1903. *In Platonis Timaeum Commentaria*. Ed. E. Diehl. Leipzig.

Proclus. 1909. *Hypotyposis Astronomicarum Positionum*. Ed. C. Manitius. Leipzig.

Ptolemy, Claudius. 1952. *The Almagest*. Trans. R.C. Taliaferro. In Hutchins 1952, pp. 1-478.

Ptolemy, Claudius. 1907. *Planetary Hypotheses*, In *Claudii Ptolemae: opera quae extant omnia, volumen II, opera astronomica minora*, ed. J.L. Heiberg, pp. 69-145. Leipzig: B.G. Teubneri.

Putnam, Hilary. 1978. *Meaning and the Moral Sciences*. London: Routledge & Kegan Paul.

Rosen, Edward, trans. 1959. *Three Copernican Treatises*. New York: Dover.

Rosenkrantz, Roger. 1976. Simplicity. In *Foundations of Probability Theory, Statistical Inference, and Statistical Theories of Science*, ed. William A. Harper and C. Hooker, vol. I, pp. 167-203. Dordrecht: Reidel.

Rosenkrantz, Roger. 1977. *Inference, Method, and Decision*, Dordrecht: Reidel.

Sambursky, Samuel. 1962. *The Physical World of Late Antiquity*. London: Routledge & Kegan Paul.

Shapere, Dudley. 1969. Notes Towards a Post-Positivistic Interpretation of Science. In *The Legacy of Logical Positivism*, ed. Stephen Barker and Peter Achinstein, pp. 115-160. Baltimore: Johns Hopkins Press.

van Fraassen, Bas. 1976. To Save the Phenomena. *Journal of Philosophy* 73:623-632.

van Fraassen, Bas. 1980. *The Scientific Image*. Oxford: Clarendon.

Vlastos, Gregory. 1975. *Plato's Universe*. Seattle: University of Washington Press.

Westman, Robert. 1971. *Johannes Kepler's Adoption of the Copernican Hypothesis*. Unpublished doctoral dissertation. Ann Arbor: University of Michigan.

Westman, Robert. 1972a. Kepler's Theory of Hypothesis and the 'Realist Dilemma'. *Studies in History and Philosophy of Science* 3:233-264.

Westman, Robert. 1972b. The Comet and the Cosmos: Kepler, Mästlin and the Copernican
    Hypothesis. In *The Reception of Copernicus' Heliocentric Theory*, ed. J. Dobrzycki,
    pp. 7-30. Dordrecht: Reidel.
Westman, Robert. 1975a. The Melanchthon Circle, Rheticus, and the Wittenberg Interpre-
    tation of the Copernican Theory. *Isis* 66:165-193.
Westman, Robert. 1975b. Three Responses to the Copernican Theory: Johannes Praetorius,
    Tycho Brahe, and Michael Maestlin. In Westman (1975c), pp. 285-345.
Westman, Robert. 1975c. *The Copernican Achievement*. Berkeley: University of California
    Press.
Zahar, Elie. 1973. Why Did Einstein's Research Programme Supersede Lorentz's? *British
    Journal for the Philosophy of Science* 24:95-123, 223-262.

V. SOME ALTERNATIVE VIEWS ON TESTING THEORIES

Testing Theoretical Hypotheses

1. Introduction

Philosophers of science concerned with theories and the nature of evidence tend currently to fall into several only partially overlapping groups. One group follows its logical empiricist ancestors at least to the extent of believing that there is a "logic" in the relation between theories and evidence. This logic is now most often embedded in the theory of a rational (scientific) agent. Bayesian agents are currently most popular, but there are notable dissenters from The Bayesian Way such as Henry Kyburg and Isaac Levi. Another group derives its inspiration from the historical criticisms of logical empiricism begun a generation ago by such writers as Gerd Buchdahl, Paul Feyerabend, N. R. Hanson, Thomas Kuhn, and Stephen Toulmin. Partly because their roots tend to be in intellectual history, and partly in reaction to logical empiricism, this group emphasizes the evolution of scientific ideas and downplays the role of empirical data in the development of science. For these thinkers, the rationality of science is to be found in the historical process of science rather than in the (idealized) minds of scientists. If there is something that can rightfully be called a middle group, it consists mainly of the followers of the late Imre Lakatos, who skillfully blended Popper's version of empiricism with elements of Kuhn's account of scientific development. Yet Lakatos's "methodology of scientific research programmes" also locates the ultimate rationality of science in a larger historical process rather than in relations between particular hypotheses and particular bits of data.

I shall be arguing for a theory of science in which the driving rational force of the scientific process is located in the testing of highly specific theoretical models against empirical data. This is not to deny that there are elements of rationality throughout the scientific enterprise. Indeed, it is only as part of an overall theory of science that one can fully comprehend what goes on in tests of individual hypotheses. Yet there is a "logic" in the

The author's research has been supported in part by a grant from the National Science Foundation.

parts as well as in the whole. Thus I agree with contemporary students of probability, induction, and the foundations of statistics that the individual hypothesis is a useful unit of analysis. On the other hand, I reject completely the idea that one can reduce the rationality of the scientific process to the rationality of individual agents. The rationality of science is to be found not so much in the heads of scientists as in objective features of its methods and institutions.

In this paper I shall not attempt even to outline an overall theory of science. Rather, I shall concentrate on clarifying the nature of tests of individual hypotheses, bringing in further elements of a broader theory of science only when necessary to advance this narrower objective. My account of how individual hypotheses are tested is not entirely new. Indeed, it is a version of the most ancient of scientific methods, the method of hypothesis, or, the hypothetico-deductive (H-D) method. But some elements of the account are new, and some have been borrowed from other contexts.

2. Models, Hypotheses, and Theories

Views about the nature of evidence and its role in science depend crucially on views about the nature of hypotheses and theories. The major divergences of current opinion in the philosophy of science are correlated with strong differences as to just what the highly honorific title "theory" should apply. For the moment I shall avoid the term "theory" and speak of "models" and "hypotheses" instead.

My use of the term "model" (or "theoretical model") is intended to capture current scientific usage—at least insofar as that usage is itself consistent. To this end, I would adopt a form of the "semantic" or definitional view of theories (hereafter, models). On this view, one creates a model by defining a type of system. For most purposes one can simply identify the model with the definition. But to avoid the consequence that rendering the definition in another language would create a different model, it is convenient to invent an abstract entity, the system defined, and call it the model. This move also preserves consistency with the logician's and mathematician's notion of a model as a set of objects that satisfies a given linguistic structure. For present purposes it will make no difference whether we focus on the definition itself or its nonlinguistic counterpart, so long as there is no presumption that in referring to "the model" one is thereby committed to there being any such thing in the empirical world.

Philosophers differ as to the appropriate form of these definitions. I much prefer the state-space approach of van Fraassen or Suppe to the set-theoretical approach of Suppes, Sneed, and Stegmüller, partly because the former seems better to correspond to scientific practice.[1] Here a system is defined by a set of state variables and system laws that specify the physically possible states of the system and perhaps also its possible evolutions. Thus, for example, classical thermodynamics may be understood as defining an ideal gas in terms of three variables, pressure, volume and temperature, and as specifying that these are related by the law PV = kT. Similarly, classical mechanics defines a Newtonian particle system in terms of a 6n-dimensional space (three components each of position and momentum for each particle) and Newton's laws of motion. A wide variety of models in population genetics (with states given by gene frequencies and development governed by Mendel's laws) are also easily expressed in this framework. So also are learning models in psychology and models of inventory and queuing in economics.

Viewed as definitions, theoretical models have by themselves no empirical content—they make no claims about the world. But they may be *used* to make claims about the world. This is done by identifying elements of the model with elements of real systems and then claiming that the real system exhibits the structure of the model. Such a claim I shall call a *theoretical hypothesis*. These are either true or false. From a logical point of view, the definition of a model amounts to the definition of a predicate. A theoretical hypothesis, then, has the logical form of predication: This is an X, where the corresponding definition tells what it is to be an X.

Our primary concern here is with the testing of theoretical hypotheses and the role of such tests in the overall practice of science. For such purposes, the *logical* differences between statements and definitions are not very important. More important are the implications of this difference for what we take to be the form of the major claims of science.

Since Aristotle it has been assumed that the overall goal of science is the discovery of *true universal generalizations* of the form: All A's are B. Moreover, it has often been supposed that the wider the scope of the antecedent the better. Thus Newton's Law of Universal Gravitation, interpreted as a generalization covering "all bodies," is seen as the epitome of a scientific conclusion. Philosophers, beginning with Hume, have reduced the concept of physical necessity to that of universality, and scientific explanation has been analyzed in terms of derivation from a generalization. Within this framework it is easy to regard a theory as simply a conjunction of universal generalizations. This would mean that testing a theory is just testing universal generalizations.

The distinction between models and hypotheses permits a view of the goals of science that is more particularized, or at least more restricted—and therefore, I think, more applicable to the contemporary practice of science. The simplest form of a theoretical hypothesis is the claim that a particular, identifiable real system fits a given model. Though extremely limited in scope, such claims may be very complex in detail and wide-ranging in space and time. The claim that the solar system is a Newtonian particle system (together with a suitable set of initial conditions) contains the whole mechanical history of this system—so long as it has been or will be a system of the designated type. Moreover (although this is more controversial), the same hypothesis contains all the different *possible histories* of this system that could result from different, but physically possible, initial conditions. Thus even a very particular theoretical hypothesis may contain a tremendous amount of empirical content.

Contrary to what some philosophers have claimed, one can have a science that studies but a single real system. Current geological models of the earth are not less than scientific, or scientifically uninteresting, simply because the only hypotheses employing these models refer to a single entity limited in time and space. Nor would models of natural selection be in any way scientifically suspect if there were no life anywhere else in the universe. Geology, however, is atypical. The models of a typical science are intended to apply to one or more *kinds* of systems, of which there are numerous, if only finitely many, instances.

What then is a "theory"? It is tempting to identify a theory with a generalized model; for example, the theory of particle mechanics with a generalized Newtonian model (i.e., one in which the number of particles is left unspecified). But most physicists would immediately reject the suggestion that "Newton's theory" is just a definition. And most scientists would react similarly concerning the theories in their fields. They think that "theories" have empirical content. This is a good reason to use the term "theory" to refer to a more or less generalized theoretical hypothesis asserting that one or more specified kinds of systems fit a given type of model.[2] This seems broad enough to encompass all the sciences, including geology and physics.

Testing a theory, then, means testing a theoretical hypothesis of more or less restricted scope. This is an important qualification because the scope of a hypothesis is crucial in any judgment of the bearing of given evidence on that hypothesis. Knowing what kind of thing we are testing, we can now turn to an analysis of empirical tests. Here I shall not be challenging, but defending, a time-honored tradition.

3. The Hypothetico-Deductive Tradition

To put things in proper perspective, it helps to recall that the hypothetico-deductive method had its origins in Greek science and philosophy. Its most successful employment, of course, was in astronomy. Recast in the above terminology, the goal of astronomy was to construct a model of the heavens that one could use to deduce the motions of the various heavenly bodies as they appear from the earth. "Saving the phenomena" was thus a *necessary* requirement for an acceptable hypothesis. The methodological issue, then as well as now, was whether it is also *sufficient*.

Greek astronomers were well aware that the phenomena could be equally well saved by more than one hypothesis. This methodological fact was exemplified by the construction of both heliocentric and geocentric models. But it was also evident on general logical grounds. Every student of Aristotle's logic knew that it is possible to construct more than one valid syllogism yielding the same true conclusion, and that this could be done as easily with false premises as with true. Truth of the conclusion provides no logical ground for truth of the premises. This obvious logical principle generated a methodological controversy that continues to this day. If two different hypotheses both saved the phenomena, there could be no *logical* reason to prefer one to the other. Some thinkers seemed content to regard any empirically adequate hypothesis acceptable and did not attempt to argue that one was fundamentally better. Others, however, wished to regard one model as representing the *actual* structure of the heavens, and this requires some way of picking out the correct hypothesis from among those that merely save the phenomena.

In the ensuing centuries of debate, the antirealists clearly had logic on their side. The realists, however, did offer several suggestions as to what, in addition to saving the phenomena, justified regarding a hypothesis as uniquely correct. Some appealed to the internal simplicity, or harmony, of the model itself. But this suggestion met the same objections it meets today. There is no objective criterion of simplicity. And there is no way to justify thinking that the simpler of two models, by whatever criterion, is more likely to provide a true picture of reality. Of course one may prefer a model regarded as simpler for *other* reasons having nothing to do with truth, but we shall not be concerned with such reasons here.

Another suggestion, and one I shall explore further in this paper, is that true hypotheses are revealed by their ability to *predict* phenomena before they are known. This suggestion appears explicitly in the late sixteenth century in the writings of Christopher Clavius, although it must certainly have been advanced earlier.[3] In any case, it became a standard part of the methodology of continental philosophers in the seventeenth century. Thus Descartes, in *Principles of Philosophy* (that is, *natural philosophy*) writes: "We shall know that we have determined these causes correctly only when we see that we can explain in terms of them, not merely the effects we had originally in mind, but also all other phenomena of which we did not previously think."[4]

Leibniz agrees as follows: "Those hypotheses deserve the highest praise (next to truth) . . . by whose aid predictions can be made, even about phenomena or observations which have not been tested before . . . ."[5] The best statement I know, however, occurs in the preface to Huygens's *Treatise on Light* (1690). Having carefully distinguished the deduction of theorems from "certain and incontestable principles" (as in geometry) from the testing of principles by verifying their consequences (the method of science), he continues:

> It is possible in this way to attain a degree of probability which very often is scarcely less than complete certainty. This happens when the things which have been demonstrated by means of the assumed principles agree perfectly with the phenomena which experiment brings to light; especially when there are a great number of them, and, furthermore, principally, when one conceives of and foresees new phenomena which must follow from the hypotheses one employs, and which are found to agree with our expectations."[6]

Huygens refers to the three conditions (agreement, number, and anticipation of new phenomena) as "proofs of probability"—meaning that their satisfaction confers probability on the assumed hypotheses. This is noteworthy because Huygens was one of the first students of probability in its modern form. Yet it would be at least a century, and arguably two, before there were any serious attempts to develop and justify the hypothetical method using ideas from the theory of probability.

In the eighteenth century, the success of Newton's physics sanctified Newton's methodology, including his professed abhorrence of "hypotheses." The method of hypothesis was apparently thought to be almost as discredited as Cartesian physics and Ptolemaic astronomy. Inference to general laws "by induction" from the phenomena was the methodological rule of the day. Interest in the hypothetical method did not revive until the triumphs of wave theories of optics in the nineteenth century—association with scientific success being the apparent standard against which methodological principles are in fact judged. Thus by the third quarter of the nineteenth century we find such eminent methodologists as Whewell and Jevons expounding the virtues of hypotheses with explicit reference to the remarkable predictions that had been based on the wave theory of light. Whewell, for example, writes: "If we can predict new facts which we have not seen, as well as explain those which we have seen, it must be because our explanation is not a mere formula of observed facts, but a truth of a deeper kind."[7] This passage is typical of many of Whewell's writings.

Whewell's homage to the methodological virtues of successful prediction did not go unchallenged. Mill, in particular, denigrated the celebrated predictions of the wave theory as "well calculated to impress the uninformed," but found it "strange that any considerable stress should be laid upon such coincidences by persons of scientific attainments." Moreover, Mill goes on to explain why "coincidences" between "prophecies" and "what comes to pass" should not count for a hypothesis any more than simple agreement with the predicted occurrence. I shall pass over the details of his argument here.[8] Of more interest for this brief survey is that the essentials of the exchange between Whewell and Mill were repeated more than a half-century later in a similar exchange between Peirce and John Maynard Keynes.

In many of his scattered writings, Peirce advocated versions of the following "rule of prediction": "A hypothesis can only be received upon the ground of its having been *verified* by successful *prediction*."[9] Unlike his many predecessors who either lacked the necessary concepts, did not think to apply them, or did not know how, Peirce attempted to justify his rule by explicit appeal to considerations of *probability*. But even this appeal was not decisive. Keynes, whose own view of scientific reasoning incorporated a theory of probability, examined Peirce's arguments for the rule of prediction and concluded that "the peculiar virtue of prediction" was "altogether imaginary." Addressing the details of Keynes's argument would again take us too far afield.[10] I shall only pause to suggest that there must be methodological principles beyond a commitment to concepts of probability that separate the tradition of Huygens, Whewell, and Peirce from that of Bacon, Mill, and Keynes.

Among contemporary methodolgists, the main defenders of the hypothetico-deductive method seem to be Popper and his intellectual descendents. Elie Zahar and Alan Musgrave have even advocated a special role for successful "novel predictions" in a Lakatosian research programme.[11] Yet these writers seem to me not to be the legitimate heirs of Huygens, Whewell, or Peirce. For the main stream of the hypothetico-deductive tradition, confirmation of a hypothesis through the verification of its consequences, particularly its *predicted* consequences, provides a reason to believe or accept that the hypothesis is *true*. Popper explicitly denies that there can be any such reasons. No matter how "severely tested" and "well-corroborated" a hypothesis might be, it remains a "conjecture" whose truth we have no more reason to believe than we did on the day it was first proposed. Similarly, for Lakatos or Zahar the success of a novel prediction is merely one sign of a "progressive" research program—not a sign of the truth of any particular theory or hypothesis. Only if one accepts the "problem shift" that replaces "reasons to regard as true" with the very different notions of "corroboration" or "progress" can one place these methodological suggestions firmly within the hypothetico-deductive tradition.

Similar remarks apply to those who take their methodological cues from Quine. Insofar as Quine belongs in the hypothetico-deductive tradition, it is that of the antirealists among the classical astronomers. Saving the phenomena is the main thing. Simplicity in one's hypotheses is desirable, but not because of any supposed link between simplicity and truth. Simplicity is desirable in itself or because it contributes to some pragmatic end such as economy of thought. Similarly with prediction. Hypotheses that are useful in making reliable predictions are desirable, but not because this makes them any more likely to be true. Rather, there is pragmatic value in being able to foresee the future, and we value hypotheses with this virtue without thereby ascribing to them any "truth of a deeper kind."

In championing the hypothetico-deductive method of testing scientific hypotheses, I am adopting only the "realist" tradition of Huygens, Whewell, Peirce, and, in part, Popper. I am not defending the more pragmatic or conventionalist versions represented by Quine.[12] Nevertheless, most of the following account is compatible with the subtle antirealism of van Fraassen's *The Scientific Image*.[13] Just how I would differ from van Fraassen will be explained later.

4. Tests of Theoretical Hypotheses

The secret to understanding the rationale of the H-D method is to focus not on the meager logical relations between hypothesis and data but on the notion of a *test* of the hypothesis. What kind of a thing is such a test? What is the purpose of testing hypotheses? What are the possible results of a test? How one answers these fundamental questions will to a large extent determine one's view of the legitimacy of the H-D method.

The ancient idea of a test as an *ordeal* is suggestive because it implies that a test is a process, a procedure to which a hypothesis is subjected. Furthermore, the idea that the purpose of the ordeal is to determine a person's guilt or innocence suggests that the purpose of testing a hypothesis is to determine its truth or falsity. Finally, to complete the analogy, those subjected to ordeals are pronounced guilty or innocent depending on whether they pass or fail the test. This suggests that hypotheses may be pronounced true or false depending on whether they pass or fail the test. Analogies, of course, are not arguments. But by developing the analogy we may be led to a better understanding of tests of theoretical hypotheses.

One way in which we scientific philosophers of the twentieth century have advanced beyond our predecessors is that we now accept the idea that no empirical data could possibly determine *with certainty* that any theoretical hypothesis is *true*. Opinions still differ over whether *falsity* may be so determined. I shall take the liberal position that neither truth nor falsity can be determined with certainty. In the language of testing, no test of a theoretical hypothesis can be completely reliable.

Almost all contemporary students of scientific method would agree that the relevant notion of reliability is to be understood and explicated using concepts of *probability*. But just what role probability plays is a matter of deep disagreement. Many philosophers assume that probability is to be introduced as a measure applied to hypotheses themselves. Thus any test would result in the assignment of a probability to the hypothesis in question. That no test is completely reliable means that this probability is always less than one. I am convinced that this is not the way to go. No model of science that places the relations between evidence and hypotheses *within* the probability calculus will prove adequate. Since I cannot argue so general a thesis here, however, I shall proceed with the constructive task of developing an alternative account. This account *uses* probability without attempting to make scientific inference itself a species of probability calculation.[14]

The way probability enters our account is through the characterization of what constitutes an "appropriate" test of a theoretical hypothesis. So far we have concluded only that a test of a hypothesis is a process whose result provides the basis for our "verdict" either that the hypothesis is true or that it is false. This general characterization, however, is satisfied by the procedure of flipping a coin and calling the hypothesis true if heads comes up and false if tails. This procedure has the virtue that our chances of reaching the *correct* conclusion are fifty-fifty regardless of the truth or falsity of the hypothesis. But no one would regard this as a satisfactory way of "testing" hypotheses. It does, however, suggest that an "appropriate" test would be one that has *higher* probabilities for leading us to the correct conclusion. We shall follow this suggestion.

Thinking about tests in this way throws new light on the classical objections to the method of hypothesis. Let us assume for the moment that our powers of deduction and observation are perfect. This will allow us to concentrate on the nature of tests themselves. At least some realists among the classical astronomers may be viewed as advocating a testing procedure that recommended calling a hypothesis true if and only if it saves the phenomena. Following this procedure, the chances of calling a hypothesis false if it is in fact true are (ideally) zero. A true hypotheses cannot have false consequences. The defect of the procedure is that the chances of calling a false hypothesis true are at best simply not known. One might even argue that this probability is high on the ground that there are many, perhaps even infinitely many, false hypotheses that would also save the phenomena. The odds seem overwhelming that the hypothesis in question is one of these. What is needed to improve the procedure, therefore, is some way of increasing the chances that a false hypothesis will be rejected. This must be done in such a way, however, that the probability of rejecting a true hypothesis is not increased. It would be trivially easy to design a procedure guaranteed to reject false hypotheses: simply reject *any* proposed hypothesis, regardless of the evidence. Unfortunately this procedure is also guaranteed to reject any true hypothesis as well.

The above considerations suggest characterizing an *appropriate test* as a procedure that has *both* an appropriately high probability of leading us to accept true hypotheses as true and to reject false hypotheses as false. Alternatively, an appropriate test of a hypothesis is a procedure that is reasonably *unlikely* to lead us either to accept a false hypothesis or to reject a true one. This characterization still requires considerable elaboration and refinement, but it makes clear the kind of account we seek.

One immediate task is to clarify the interpretation of probability assumed in the above characterization of a good test. By a "procedure" I mean an actual process in the real world. If such a procedure is to have probabilities for leading to different results, these must be *physical* probabilities. Our account thus presupposes an acceptable physical interpretation of probability, something many philosophers regard as impossible. Here I would agree with those who reject attempts to reduce physical probability to relative frequency, and opt for some form of "propensity" interpretation. But since this is again too big an issue to be debated here, I shall proceed under the assumption that there is *some* acceptable physical interpretation of probability.[15] Moreover, we must assume that we can at least sometimes have good empirical grounds for judging the relevant physical probabilities to be high.

5. Example: Fresnel's Model of Diffraction

An example may help to flesh out the relatively abstract outline presented so far. This example is appropriate in many ways, one being its historical association with the re-emergence of the H-D method of testing in the early eighteen-hundreds after a century in the shadows of Newtonian methodological orthodoxy.

Wave models of optical phenomena had been developed by Hooke and Huygens at the end of the seventeenth century. Particle models were favored by Newton and the later Newtonians. At that time, the evidence for either type of model was genuinely ambiguous. Each type of model explained some phenomena better than the others. The then recently discovered phenomenon of polarization, for example, was an embarrassment to both, though perhaps more so to wave theorists. In general, particle models dominated eighteenth-century theorizing, perhaps partly because of greater empirical success but also, I think, because of the general triumph of Newtonianism. In any case, for most of the eighteenth century there was little serious work on wave models until Thomas Young took up the cause around 1800. The scientific establishment, including the French Academy of Sciences, was then dominated by particle theorists. Laplace, for example, published a particle model of double refraction in 1808. But interest in optics was obviously high, since the Academy prizes in 1810 and 1818, for example, were for treatments of double refraction and diffraction respectively.

The diffraction prize eventually went to Augustin Fresnel for a *wave* model. In Fresnel's models, diffraction patterns are produced by the interference of secondary wave fronts originating at the edges of an object placed between a point light source and screen. Fresnel developed special cases of this general model for a single straight edge, a narrow body with parallel edges, and a narrow slit. The calculated patterns agreed well with known experimental results.

Fresnel's memoir was referred to a Commission in which well-known advocates of particle models, Laplace, Poisson, and Biot, held a majority. The commission was apparently not fully convinced by the evidence Fresnel had presented, and Poisson devised a further test. He applied Fresnel's model to the case of a shadow produced by a circular disk and deduced that the resulting diffraction pattern would have a bright spot at the center of the shadow. Even from superficial accounts of this incident it seems clear that no one involved had ever seen such a spot. Moreover, it seems that Poisson and his fellow Commissioners did not expect the spot to appear. It certainly was not a consequence of any current particle models that such a spot should exist. The experiment was performed by François Arago, and apparently also by Fresnel. The spot appeared as predicted, and the Commissioners yielded.[16]

Now let us consider this episode in light of the framework outlined earlier. Assuming sufficient familiarity with the kind of *model* Fresnel proposed, the next question is just what *hypotheses* were at issue. There are a number of possibilities. (i) The specific set-up in Arago's laboratory fits the model. (ii) Any similar set-up with a circular disk, etc., would fit the model. (iii) All types of diffraction phenomena fit this sort of model. (iv) All optical phenomena fit a similar wave model. Which of these hypotheses were tested by Arago's experiments, and which did the Commission accept in awarding Fresnel the prize?

The episode, I suggest, is best understood taking the hypothesis of most direct concern to be the third of the above four: Fresnel's models capture diffraction phenomena in general. Following the terminology suggested earlier, this hypothesis could be designated "Fresnel's theory of diffraction." Of course everyone was also concerned with whether Arago's set-up fit the model, and this is a consequence of Fresnel's theory. The second hypothesis is also a consequence of Fresnel's theory, but once people were convinced that the model applied to Arago's apparatus, this generalization was not problematic. Enough was known of the general stability of optical phenomena by that time that this simple generalization could legitimately be taken for granted once it was firmly established for a single case. The emphasis placed by empiricist philosophers on such generalizations is quite misleading.

One reason for focusing on Fresnel's theory of diffraction rather than on a broader wave theory of light is that Arago's experiments provide an appropriate test of Fresnel's theory, but not of the broader theory—in spite of the fact that the former is a logical consequence of the latter. This follows from our characterization of an appropriate test of a hypothesis, as we shall now see.

At the time of Arago's experiments, techniques for dealing with optical phenomena were sufficiently well developed that it was very probable that the spot would be observed—given that the Fresnel-Poisson model does fit this situation. So, given that Fresnel's theory is true, it was very unlikely that it should mistakenly have been rejected. This aspect of the test was entirely appropriate to the circumstances. But what if Fresnel's theory had been false? How probable was it that the testing process should have yielded the predicted spot even if Fresnel's models did not really capture diffraction phenomena? To answer this question we must first decide just how much of the episode to include within the "testing process."

According to common interpretations of the discovery/justification distinction, the decisive testing process began when Poisson constructed a Fresnel-style model for the circular disk and deduced that the spot should appear. Nothing that happened earlier is relevant to the confirmation of any of the hypotheses we have considered. I expect that many who reject a discovery/justification distinction would nevertheless agree with this conclusion. And indeed, this view of the matter follows naturally from the doctrine that there is a "direct" evidential relationship, analogous to deduction, between hypothesis and evidence. But this is not our view. On our account, the relationship between hypothesis and evidence is medi- ated by the testing process, and there is no a priori reason why incidents that occurred before the actual formulation of the hypothesis should not be relevant to the character of this process. In particular, the process by which a hypothesis is selected for consideration might very well influence its content and thus the likelihood of discovering a further consequence to be true.

The Commissioners apparently did not regard Fresnel's success in explaining the diffraction pattern of straight edges as decisive. Why? Was this just prejudice? Or did they have good reasons for not regarding these familiar patterns as being part of a good test of Fresnel's models? I think the latter is the case. From Fresnel's own account it is clear that the straight-edge pattern acted as a constraint on his theorizing. He was unwilling to consider any model that did not yield the right pattern for straight edges. Thus we know that the probability of *any* model he put forward yielding the correct pattern for straight edges was near unity, independently of the general correctness of that model. Since the straight-edge pattern thus had no probability of leading to rejection of any subsequently proposed hypothesis that was in fact false, this pattern could not be part of a good test of any such hypothesis.

We could regard agreement with the straight-edge pattern as a test of a hypothesis if we knew the probability that Fresnel should pick out a satisfactory model using this, together with similar data, as a constraint. At best this probability is simply unknown. And given the frequency with which even experienced scientists come up with unsatisfactory models, there is reason to judge such probabilities to be fairly low. In either case we fail to have a good test of the hypothesis.

The case with the spot is quite different. We know that this result did *not* act as a constraint on Fresnel's choice of models. Suppose, then, that Fresnel had come up with a model that applied satisfactorily to straight edges and the like but was *not* correct for diffraction phenomena in general. The corresponding theory would therefore be false. What is the probability that any model selected in this way should nevertheless yield the correct answer for the disk? In answering this question we must also take into account the fact that the disk experiment was specifically chosen because it seemed to Poisson and others that no such phenomenon existed. So the consequence selected for the test was one that knowledgeable people thought unlikely to be true. Given all these facts, it seems clear that the test was quite likely to lead to a rejection of any false theory that Fresnel might have proposed. And this judgment about the test was one that could easily be made by everyone involved. My view is that they did make this judgment, implicitly if not explicitly, concluded that Poisson's proposed test was quite adequate, and, when the result came in, acted accordingly.

In thinking about this and similar examples it is crucial to remember that the probabilities involved are physical probabilities inherent in the actual scientific process itself. If one slips into thinking in terms of probability relations among hypotheses, or between evidence and hypotheses, one will necessarily misunderstand this account of the nature of empirical testing. In particular, one must not imagine that to estimate the probability of not finding the spot one must be able to calculate the probability of this result as a weighted average of its probabilities relative to all possible alternative theories. No such probabilities are involved. Rather, one needs only to estimate the chances that *any* model generated and tested in this way should fail to cover the general class of diffraction phenomena and nevertheless give the right result for an experiment devised as the disk experiment was in fact devised. My contention is that the participants' knowledge of the whole situation justified concluding that this chance was low.

There remains the question of why Arago's experiment does not provide an appropriate test of a more general wave theory of all optical phenomena. Let us regard this more general theory as a conjunction of theories restricted to various types of optical phenomena: reflection, refraction, polarization, etc. A Fresnel theory of light would say that Fresnel's models are adequate for this whole range of phenomena. Now since Fresnel's theory of diffraction is one conjunct in this larger theory, and since the test is a diffraction experiment, the probabilities of mistaken acceptance or rejection of a more general theory are identical to those for the more restricted theory. Is the experiment not then an equally good test of the broader theory? A positive answer would be a strong objection to this account of empirical testing since it seems intuitively clear that it would not have been correct to accept the broader theory on the basis of these experiments.

The objection fails, however, because the *appropriateness* of a test need not be solely a function of the absolute *magnitude* of the relevant probabilities. It may depend also on what other tests of the theory might be possible, and in this case there were much better tests available. Since Fresnel's model was selected using mainly diffraction phenomena as constraints, the falsity of a general Fresnel theory of light would be much *more likely* to be demonstrated by experiments on phenomena *other* than diffraction. In particular, one would want a phenomenon to which such wave models had not yet been applied. Many such phenomena were familiar at the time. In fact, no such experiments were necessary because it was almost immediately apparent that there were many phenomena, e.g., polarization, for which Fresnel's model gave no account whatsoever.

Many recent philosophers have objected to the H-D method because it satisfies a "converse consequence condition." That is, if T is confirmed by the truth of some consequence, O, then, if T′ implies T, T′ is equally confirmed. In particular, T and H, for any H, is confirmed. And, granted that a logical consequence of any hypothesis is at least as well confirmed as the hypothesis itself, by confirming T we can equally well confirm any H whatsoever. The above discussion shows that such objections are based on an oversimplified view of the H-D method—indeed, a version to which few if any serious defenders of the H-D method ever subscribed.[17]

6. The Role of Novel Predictions

As we have seen, many champions of the H-D method have suggested that successful predictions are sufficient for the confirmation of hypotheses; some, such as Peirce, have taken them to be necessary as well. Critics argued that successful predictions were neither necessary nor sufficient. From our present perspective we can see why the defenders were on the right track even though their critics were technically correct. First let us give the critics their due.

That successful predictions are *not sufficient* is easily seen by imagining other possible sequences of events in the Fresnel example. Suppose that Biot had repeated Fresnel's calculations for at straight edge and persuaded Arago to repeat these measurements. Of course Biot's prediction would have been verified, but no one would have regarded this replication as providing a decisive test of Fresnel's hypothesis regarding this experiment or of the general adequacy of his approach to diffraction phenomena. Why? Because the imagined process would not have been a good test of the hypothesis. The process had a high probability of supporting Fresnel's hypothesis if it were true. But it also had a high probability of supporting the hypothesis even if it were false. The many previous experiments with straight edges had provided ample evidence for the empirical generalization that this type of experiment yielded the indicated diffraction pattern. So regardless of the truth or falsity of Fresnel's hypothesis, it was highly probable that the hypothesis would be supported. This violates our conditions for a good test of a hypothesis. Both Mill and Keynes used this sort of example in their analyses of the prediction criterion, though each within a quite different framework.

This same counterexample also shows why many H-D theorists have insisted on *novel* predictions. If a predicted result is not novel, there will be a more or less well-justified low-level empirical hypothesis linking the type of experiment and the type of result. This makes it likely that the test will justify the hypothesis no matter whether it is true or false. Thus it is difficult to have a good test unless the prediction is novel. This point seems to have been missed by empiricist critics of the H-D method such as Mill and Keynes, although perhaps the true value of novelty was also not sufficiently understood by either Whewell or Peirce.

That successful predictions are *not necessary* is also easily demonstrated by an imaginary variation on the same example. It has been claimed that the bright spot in the center of a circular shadow was observed in the early part of the eighteenth century by J. N. Delisle.[18] It seems pretty clear that none of the principles in the case had ever heard of these supposed observations. But suppose they did occur and were published. Imagine, then, that Laplace, but not Fresnel, knew of these results, and upon reading Fresnel's memoir recalled Delisle's unexplained observations. It would not have taken him long to apply Fresnel's method to the case and conclude that Fresnel's model explained Delisle's results. I think the Commission would have been equally convinced that Fresnel's theory was correct. But whether or not they would have been, they should have been. It was about as improbable that Fresnel, ignorant of Delisle's results, should have developed an inadequate model that nevertheless explained Delisle's results, as it was that an inadequate model should have happened to predict correctly the result of Arago's later experiment. In either case the conditions for a good test are satisfied.

Returning to the champions of the prediction rule, it is clear that they overstated their case. But they were fundamentally correct in thinking that the fact that a result was predicted successfully may be *relevant* to the confirmation of a hypothesis. The conditions that define an appropriate test of a hypothesis are themselves contingent empirical hypotheses about the actual process by which a particular hypothesis is tested. This is due to the fact that the relevant probabilities are *physical* probabilities embodied in the testing process. Judging a process to constitute a good test of a hypothesis thus requires judging that the relevant physical probabilities are high. All sorts of empirical facts about the case may be relevant to these judgments—including facts about when a specified consequence of a hypothesis became known, and to whom.

Let us imagine yet another variant on the Fresnel example. Suppose that someone named Delisle really did observe the spot and that Fresnel, but not other principles in the case, knew of Delisle's results right from the beginning of his work on diffraction. So in addition to explaining the standard diffraction patterns for straight edges, etc., Fresnel was all along seeking a model that would also explain the existence of the spot—and he succeeded. But suppose further that Fresnel suppressed all mention of disks and spots in his memoir, and the rest of the story proceeds as in real life. On our analysis this would be a case of scientific dishonesty, even fraud. The Commission would have had every reason to think that a good test had been performed. But they would have been deceived, for in fact it would have not been a good test. It is not in fact unlikely that a model designed to accommodate a given result should in fact do so. And this is true no matter whether the corresponding hypothesis is true or false. It is possible, therefore, to be justifiably mistaken about whether a given experiment constitutes a good test or not.

We can go further. One might object that it does not matter whether Fresnel *knew* of Delisle's result, but only whether he *used* this knowledge in selecting a model of diffraction. This is in principle correct. The probability of fit between model and observation is not influenced if the observations play no role in the selection of the model. But it is exceedingly difficult to be confident that no such selection pressure existed when the result was known to the developer of the model. One can always be more confident of the goodness of a test if the result was not previously known. So it is a sound methodological rule to prefer results that were genuinely predicted even though this is not strictly necessary for a good test. The methodological good sense of the champions of prediction was better than their justifications.

Finally, our account of testing theoretical hypotheses reveals the methodological insight in Popper's claims that to test a hypothesis severely, one must attempt sincerely to refute it. Far from introducing irrelevant psychological considerations into a supposed logical relationship between evidence and hypothesis, Popper is highlighting (though in an exaggerated and misleading fashion) one aspect of good tests. Not only is the process by which the *hypothesis* was selected relevant to our judgment of the quality of a test; the process of selecting the *prediction* to be investigated is also relevant. This process can be one that makes it more or less likely that the test will reveal the falsity of the hypothesis—if it is indeed false. In particular, if a knowledgeable scientist such as Poisson investigates a model with the express intent of discovering an inadequacy, and finds a consequence he thinks likely to be false, that is good reason to

think that the test has a high probability of discovering a false consequence—if there are any to be discovered. Of course a scientist need not
be attempting to refute the hypothesis in question; he may just be trying to
devise the best possible test. But the knowledge that a given consequence
was selected for investigation in a well-informed attempt to refute the
hypothesis is relevant to the judgment as to how good the test might be.

7. The Logic of Tests

A good test of a hypothesis is a physical process with specified stochastic
properties, namely, a high probability of one outcome if H is true and
another if H is false. That this process has one outcome rather than the
other, however, also has *epistemic* consequences. If a good test has a
favorable outcome, we are to "conclude" that H is true, "accept" H as being
true, or some such thing. One must provide some account of the rationale,
or "logic," of this step from the physical outcome to the epistemic
conclusion. Here I shall follow those who regard the epistemic step as a
kind of *decision*. This opens the way to a decision-theoretic analysis of
scientific inference. But since we have renounced probabilities of hypotheses, our decision theory must be "classical," or "non-Bayesian," decision
theory.

Casting the problem in decision-theoretic terms, we realize immediately that what really needs to be justified is not so much the decision to
accept (or reject) any particular hypothesis, but the general *decision rule*
that tells us, for each possible outcome of the experiment, whether to
accept or reject the hypothesis. It is so obvious which of the four possible
decision rules is correct that most traditional accounts of the H-D method
do not even note the epistemic step from physical outcome to accepted
conclusion. To understand the logic of the step, however, it is useful to
consider the full range of possibilities.

In any test of a theoretical hypothesis there are four possible epistemic
results, two correct and two incorrect. The correct ones are accepting H if it
is true and rejecting H if it is false. The incorrect ones are rejecting H if it is
true and accepting H if it is false. Now to conceptualize the problem in
decision-theoretic terms we must assume that it is possible to assign some
kind of "value" to these possible results. For the moment we shall not
worry what kind of value this is or whether it has the formal properties of
utility. And just for convenience (it makes no difference to the argument),
we shall suppose that both correct results have the same value (which we

may set arbitrarily at 1) and that both incorrect results also have the same value (which we may set arbitrarily at 0). Finally, let α be the probability that the prediction is false even though H is true and β be the probability that the prediction is true even though H is false. For the moment it will not matter much what these probabilities are so long as they are strictly between zero and one half. With these assumptions we can represent the "decision" to accept or reject H in a two by two matrix (Figure 1).

	Hypothesis true	Hypothesis false
Accept Hypothesis	$Pr = 1 - \alpha$ $V = 1$	$Pr = \beta$ $V = 0$
Reject Hypothesis	$Pr = \alpha$ $V = 0$	$Pr = 1 - \beta$ $V = 1$

Figure 1.

Each of the four possible outcomes is labeled with its respective value and probability (conditional on the hypothesis being true or false).

The meta-decision problem of choosing a decision rule for making the object-level decision is represented as a four-by-two matrix (Figure 2). The obvious decision rule to accept H if and only if the prediction is true is represented as (A, R), and the others are represented accordingly. The outcomes are labeled with the appropriate *expected* values of applying the rule conditional on the truth or falsity of the hypothesis.

	Hypothesis true	Hypothesis false
(A, A)	1	0
(A, R)	$1 - \alpha$	$1 - \beta$
(R, A)	α	β
(R, R)	0	1

Figure 2.

We are now in a position to consider in a systematic way why the obvious rule is correct.

We have arrived in "meta-meta land." What principle do we use to justify a decision rule to guide our actual decisions? The least controversial principle that might possibly apply—namely, dominance—fails to be applicable. Since (A, A) is best if H is true and (R, R) is best if H is false, no one choice of decision rule dominates all others. But the next least controversial principle does apply—namely, *maximin* (it would be "minimax" if we measured our values as losses rather than gains). Maximin recommends the intuitively obvious decision rule (A, R). The justification for this recommendation is that following (A, R) guarantees the greatest "security level," that is, the highest minimum expected gain whether the hypothesis is true or not. In either case, our expected gain can be no lower than the smaller of $(1 - \alpha)$ or $(1 - \beta)$. Since α and β are both less than one-half, no other decision rule can guarantee a greater security level.

The trouble with an appeal to the maximin principle is that it justifies too much. It sanctions the (A, R) rule as long as α and β are less than one-half, which means that any test of a hypothesis is acceptable as long as the probabilities of accepting true hypotheses and rejecting false ones are strictly greater than one-half. This is certainly not in general acceptable for tests of scientific hypotheses. We need a somewhat stronger principle that can force the probabilities of correct acceptance and rejection to be appropriately greater than one-half.

One general decision strategy that has some theoretical backing and seems well suited to the present context is *satisficing*.[19] This strategy may be understood as a strengthening of maximin. To apply this strategy we must assume that there is some way to set a value that is the minimum regarded as a "satisfactory" outcome of the decision process. This is the *satisfaction level* for the particular decision problem at hand. This level is not a function of the given decision matrix but is imposed "from the outside" either by a decision maker or by the decision context. The decision problem will have a *satisfactory* choice only if the security level is at least as great as the imposed satisfaction level. Otherwise there simply will not be any choice sanctioned by this strategy.

Satisficing has not been much studied by decision theorists and philosophers because it does not conform to the standard conditions for a solution to the fundamental meta-decision problem. That problem is: For any decision matrix, devise a general rule that selects a uniquely rational choice from among all the given possible options. Any rule that does not guarantee some choice of options cannot be a solution to this problem. By invoking satisficing when it does not reduce to maximin, we are rejecting the standard formulation of the fundamental problem of decision theory. And in making satisficing the basis for our acceptance or rejection of hypotheses we would similarly be rejecting a standard, if often implicit, formulation of the basic theoretical problem of scientific inference: For any hypothesis and any evidence, define a uniquely rational function that gives the degree to which the evidence "supports" the hypothesis. On our account, the decision matrix corresponding to an appropriate test must have satisfactory payoffs for the evidence to count either for or against the hypothesis. If the evidence is not part of an appropriate test, the hypothesis is simply neither accepted nor rejected. Neither conclusion is justified.

The demand for a unique solution to any decision problem may have some merit in the context of *practical* decision making. Here one often cannot refuse to make some choice. But science is different. It is not required that scientists be able to say at any moment, given whatever happens to be known at that time, whether some specified hypothesis should be accepted or rejected. One may simply say that sufficiently good tests are lacking. The other side of the coin is that scientists are not helpless in the face of inadequate tests. They may devise and carry out tests they have explicitly designed to be adequate according to the standards of their field. The scientific enterprise, after all, is not simply a matter of evaluating hypotheses in light of available information, but an active seeking out of information to answer definite (and often theoretical) questions.

My analysis of the Academy Commission follows the above outline. The evidence presented in Fresnel's memoir did not constitute a sufficient test of Fresnel's theory, and the Commission, rightly, could not decide whether it was correct or not. So they devised a new experiment, one that was sufficient. What made the new experiment sufficient was primarily that it had an adequately high probability of rejecting a mistaken model.

Probabilities, however, are not the only component in a decision-theoretic analysis of the logic of testing. We must also assume the existence of "epistemic" or "scientific" values. I say "values" rather than "utilities" because a satisficing strategy, like maximin, can be applied with as weak a value structure as a mere ordering—a full-fledged utility function is not necessary. This is an additional virtue of satisficing as a general decision strategy. Still, those who question this approach may ask whether there are any such things as scientific values and, if so, whether they can play the role assigned to them by this analysis. Here I can only sketch a reply.

In principle we need appeal to only one scientific value, truth. That is, correctly accepting a true hypothesis or rejecting a false one is valued over the reverse. This seems very difficult to deny, except as part of a radical theory of science in which truth plays little or no role.[20] Otherwise the decision-theoretic analysis requires only that the satisfaction level of expected value be strictly less than the value assigned to accepting truth (or rejecting falsehood). Because the probability of a mistake is strictly greater than zero, our expected payoff is necessarily less than maximal no matter whether H is true or not. If our satisfaction level were equal to the value of a correct decision, no amount of data or care in experimental design would be good enough for us ever to make either choice. Thus we really do not need to delve deeply into questions about the value of truth, e.g., *how much* is truth valued over error. So long as truth is *more* valued, we can assume an arbitrary scale. The interesting question is what determines the satisfaction level. And given a fixed value scale, this question reduces to asking what determines an acceptable level of risk. So we are back to probabilities.

My view, which I can only affirm here, is that the satisfaction level, or level of acceptable risk, is not a function of individuals but of the *institution* of science. The institution decrees that tolerating some risk of error is better than having no theoretical conclusions at all. Yet something more should be forthcoming. It seems that different fields, or the same field at different stages of maturity, have different tolerances for risk of error. Again, it seems not just that scientists in some fields are more risk adverse. Rather, some *fields* are more or less risk adverse. I think there are objective reasons why this is so and why it is proper, but that would be a digression here. The main point is to recognize this as an important question for a theory of science.[21]

One important consequence of introducing values into an account of scientific inference is that it automatically relativizes the acceptance of a theoretical hypothesis to the scientific context characterized by those values. This relativization eliminates what has often been regarded as a fatal objection to the idea that hypotheses are ever "accepted" as being true. The objection is that it would be irrational to regard any hypothesis as true for all purposes or in any possible context. Thus Laplace may have been convinced of the truth of Fresnel's hypothesis, but it would have been irrational to stake the fortunes of the whole French nation on this conclusion. On our analysis, however, Laplace's commitment to the truth of the hypothesis would be restricted to the scientific context, leaving open the question whether the hypothesis was sufficiently well-tested to presume its truth if other values were at stake. This makes the relationship between "pure" and "applied" science more complicated than is often supposed, but that, I would argue, is an added virtue of this approach.[22]

8. The Weight of Evidence

It is a commonplace that the evidence for a theory may be better or worse, and that evidence may accumulate. In short, evidence seems to come in degrees. Our account, however, leads to hypotheses being either "accepted" or "rejected." Does not this account, therefore, run counter to well-established methodological intuitions? I think not, but some explanation is in order.

In the first place, it is not strictly true that our account leaves no room for degrees of support. A test can be more or less stringent (or severe) depending on the probabilities of mistaken acceptance or rejection. The probability of correctly rejecting a mistaken hypothesis, what statisticians call the "power" of a test, is a particularly good measure of the severity of a test. For example, later tests of wave models tended to be better than the early experiments of Fresnel and Arago because they employed fairly precise *quantitative*, rather than merely qualitative, predictions. The famous 1850 experiment of Foucault and Fizeau on the relative velocity of light in air and water was a much better test because wave models gave a quite precise value for this ratio based on the known relative indices of refraction. The chances of a badly mistaken model yielding the right value to within the known experimental error had to have been very small indeed.

Second, the idea that a theory is a generalization over regions of a domain provides ample room for the idea of accumulating evidence for a theory. As more and more regions of the domain are brought under a given type of model, the evidence in favor of the generalization increases. This is how I would understand the history of the wave theory between 1820 and 1850. As wave models were refined, more and more optical phenomena were shown to fit: diffraction, polarization, double refraction, and so on. By 1850 it was reasonable to conclude that all optical phenomena could be accounted for by suitably specialized wave models.[23]

9. Approximation and Scientific Realism

Peirce said that every scientific conclusion should be doubly qualified by the phrase "probably and approximately." Thus far we have been primarily concerned with the role of probability in the testing of theoretical hypotheses, and, by implication, the way probability "qualifies" our conclusions. It is time to turn to the second of Peirce's qualifications. The need for some such qualification is once again well-illustrated by the Fresnel episode. At the time of his original memoir on diffraction, Fresnel's wave models employed *longitudinal* rather than transverse waves. It so happens that for diffraction phenomena the two types of models give at least qualitatively similar results. This is not true in general, however, and it was the switch to transverse wave models several years later that provided the key to a wave explanation of polarization and other previously difficult cases. But what are we to say of the Commission's conclusion after Arago's experiments with the opaque disk? Were they justified in concluding simply that Fresnel's theory is correct, even though this conclusion had later to be modified? Or should the conclusion have been softened by a qualifying "approximately"?

One reason for softening conclusions is the general realization that (with the possible exception of models of microphysical phenomena) no model is likely to capture the full complexity of the phenomena under investigation. Thus no hypothesis asserting a "perfect fit" between a model and a phenomenon is likely to be exactly true. We can be fairly confident on general grounds that there are likely to be at least minor discrepancies, even if these do not show up in existing experiments. Thus, if one desires a conclusion that one does not know in advance to be strictly false, it seems advisable to say that what one accepts as true is not ever H itself, but the more complicated conclusion that "H is approximately true."

This position is reinforced by our account of the testing of theoretical hypotheses. It is required that the testing process be such that a false hypothesis is very likely to lead to a failed prediction and thus to rejection of the hypothesis. But we could never be in the position of knowing that even a very slight difference between the model and the real system would very probably lead to a failed prediction. The most we could ever reasonably claim about a testing process is that it is very likely to detect a type and degree of deviation from the model. The *type* is given by the category of phenomena being studied. Arago's experiment could not possibly detect mistaken features of the model that would not show up in simple diffraction experiments. Nor could it detect *degrees* of deviation beyond the resolving power of his experimental apparatus. If the spot had been of very low intensity, for example, he would have missed it even though the wave model was basically correct. Thus, if the experiment yields a positive conclusion, that conclusion must be correspondingly qualified. What we can say is that the real system exhibits the structure of the model in those respects relevant to the domain in question and to a degree that the experiments performed are capable of detecting. This is an elaboration of the simple phrase "H is approximately true."

It is sometimes supposed that in moving from "H is true" to "H is approximately true," one is trading a probable falsehood for a vacuous truth. That this is not so is well illustrated by the Fresnel example. Models of the type Fresnel used for diffraction simply do not work for polarization experiments. And the reason is clear: longitudinal waves do not polarize. The correct conclusion, as was clear at the time, was that any supposed Fresnel theory of polarization was not even approximately true. Of course much more could and needs to be said about approximation, but the general charge of vacuity can be safely dismissed.

Even doubly qualified, our conclusions are still quite solidly realistic. With all qualifications understood, to accept H as approximately correct is to assert that there is a real structure underlying the phenomena, and that this underlying structure, as far as we can tell, reflects the structure of our model. Bas van Frassen has recently raised general objections that apply even to so qualified a version of scientific realism. Since the view of theoretical models embedded in my account is essentially van Fraassen's, it is important to see how his objections may be avoided.[24]

Reformulated in my terminology, van Fraassen's view is that theoretical hypotheses may *refer* to underlying entities and processes, but that no evidence can *justify* accepting claims about such theoretical goings on. Rather, the most we can ever justifiably claim is that our models are "empirically adequate," i.e., that they "save the phenomena." His main direct argument for so restricting our conclusions is that, for any given evidence, the conclusion of empirical adequacy is necessarily better supported than any realistic conclusion simply because it is logically weaker. The realistic conclusion implies empirical adequacy, but the reverse implication, of course, fails. Van Fraassen tends to state his argument by saying that the empiricist hypothesis of empirical adequacy is *more probable*, for any given evidence, than a realistic hypothesis. The general point, however, does not depend on invoking (logical or subjective) probabilities of hypotheses.

That the empiricists' hypothesis is always *better* supported does not, however, imply that the realists' hypothesis cannot be *adequately* supported. Reformulating the same point, just because empirical adequacy is a *more acceptable* hypothesis, it does not follow that a realistic hypothesis is not *acceptable*. Van Fraassen assumes that only the more probable empiricists' hypothesis could be acceptable. That this assumption is not necessarily justified is reinforced by the logic of satisficing. If we assign *equal* value to the truth of both empiricist and realist hypotheses, the empiricist hypothesis, being more probable, would have greater expected value. But the realist hypothesis might still have "satisfactory" expected value, making it acceptable to a satisficer. But of course a realist would assign *greater* scientific value to true realistic hypotheses, which could give them greater expected value. So a realistic satisficer need not even be in the position of settling for second best.

These considerations do not, of course, settle the issues between realists and empiricists. My objective has been simply to show that one can adopt van Fraassen's formal account of theoretical models without committing oneself to his anti-realist arguments. Van Fraassen himself insists that this separation be possible, but it is useful to see how to make the separation for a particular realistic alternative.

10. Final Considerations

In conclusion, I would like briefly to mention two issues that would have to be treated in any generally adequate account of scientific inference but that cannot be examined in any detail here. One involves a technical elaboration; the other is philosophically much broader.

The above account treats only deterministic models. This leaves out stochastic models, which yield only statistical predictions. But the account also fails to be adequate to the actual testing of deterministic models. Experiments testing theoretical hypotheses typically require multiple measurements on complex apparatus that, when pushed to their limits of accuracy, yield a spread of values. Thus even tests of deterministic models typically involve testing statistical hypotheses. Testing theoretical hypotheses is therefore generally a two-stage affair. One begins by testing a statistical hypothesis or estimating a statistical parameter in order to determine whether the theoretical prediction has been fulfilled or not.

Only after deciding on the truth or falsity of the prediction can one reach a conclusion on the theoretical hypothesis. It turns out, however, that the above account of testing theoretical hypotheses can be extended in a completely natural way to incorporate the statistical level into the whole testing process. No fundamentally new principles need be involved.[25]

The broader issue concerns *justification* of the type involved in traditional philosophical discussions of "the justification of induction." On the above account, whether a proposed test of a theoretical hypothesis is a satisfactory test is itself an empirical question. The judgment that fulfillment of the prediction would be unlikely if the hypothesis tested is not (even approximately) true is an empirical judgment. This opens the way for a typical Humean regress.

Philosophers of science today are much less concerned with Hume's problem than in the recent past. Foundationist justifications, in particular, have largely gone out of fashion. In their place are discussions of the "rationality" of the scientific process. I regard this development as a "progressive problem shift." The new program, however, requires a sound theory of the scientific process before one can fruitfully inquire after the rationality of that process. We do not yet have such a theory. Indeed, one of the main defects of current theories is that they lack good accounts the empirical testing of theoretical hypotheses. We may hope this is just a temporary situation, part of the general over-reaction to positivism. In any case, only when we achieve a more balanced picture of the scientific enterprise itself shall we be in a position to develop better ideas about the nature of scientific justification and rationality.[26]

Notes

1. Bas C. van Fraassen, On the Extension of Beth's Semantics of Physical Theories. *Philosophy of Science* 37 (1970): 325-339; Frederick Suppe, Theories, Their Formulations and the Operational Imperative. *Synthèse* 25 (1973); 129-164; Patrick Suppes, What is a Scientific Theory?, in *Philosophy of Science Today* S. Morgenbesser, ed., (New York: Basic Books, 1967), pp. 55-67; Joseph Sneed, *The Logical Structure of Mathematical Physics* (Dordrecht, Holland: D. Reidel, 1971); Wolfgang Stegmüller, *The Structure and Dynamics of Theories* (New York: Springer, 1976).

2. Here I depart from the view I have taken in previous papers and in my elementary text, *Understanding Scientific Reasoning* (New York: Holt, Reinhart and Winston, 1979). I have generally used the term "theory" to refer to a generalized definition or a model—which has no empirical content. But this usage has met sufficient resistance from scientists and science students that I have decided to compromise in the interests of communication. The underlying view of the scientific enterprise, however, is the same. My view here parallels that of Sneed and Stegmüller, although the parallel is difficult to see through the forest of set theory in which they hide their views.

3. R. M. Blake, Theory of Hypothesis among Renaissance Astronomers, in *Theories of Scientific Method*, ed. E. H. Madden (Seattle: University of Washington Press, 1960).

4. René Descartes, *Principles of Philosophy*, Part III, Sec. 42, in *Descartes: Philosophical Writings*, trans. E. Anscombe and P. Geach (Edinburgh: Nelson, 1954), p. 223.

5. L. E. Loemker, *Leibniz: Philosophical Papers and Letters*, 2 vols. (Chicago: University of Chicago Press, 1956), Vol. 1, p. 288. See also the paper On the Method of Distinguishing Real from Imaginary Phenomena. In Loemker, Vol. 2, p. 604.

6. Christiaan Huygens, *Treatise on Light*, trans. S. P. Thompson (London: Macmillan, 1912).

7. William Whewell, *Philosophy of Discovery* (London: J. W. Parker, 1860), p. 273.

8. John Stuart Mill, *Logic* (8th ed.) (London, 1881), Book III, Ch. XIV, Sec. 6.

9. C. S. Peirce, *Collected Papers of Charles Sanders Peirce*, 8 vols. (Cambridge, Mass.: Harvard University Press, 1931-1958), 2.739.

10. John Maynard Keynes, *A Treatise on Probability* (London: Macmillan, 1921), pp. 304-306.

11. E. G. Zahar, Why Did Einstein's Programme Supersede Lorentz's? *The British Journal for the Philosophy of Science* 24 (1973): 95-123 and 223-262. Alan E. Musgrave, Logical Versus Historical Theories of Confirmation. *The British Journal for the Philosophy of Science* 25 (1974): 1-23.

12. Still less am I defending the logical shadow of the hypothetico-deductive tradition recently criticized by Clark Glymour in *Theory and Evidence* (Princeton: Princeton University Press, 1980), pp. 29-48.

13. Bas C. van Fraassen, *The Scientific Image* (Oxford: Oxford University Press, 1980).

14. I have elaborated on the deep philosophical differences between these two approaches to scientific inference in Testing vs. Information Models of Statistical Inference. In *Logic, Laws and Life*, ed., R. G. Colodny (Pittsburgh: University of Pittsburgh Press, 1977), pp. 19-70. For further references see this article and the review essay Foundations of Probability and Statistical Inference. In *Current Research in Philosophy of Science*, ed. P. D. Asquith and Henry Kyburg, Jr. (East Lansing: Philosophy of Science Association, 1978).

15. I have myself developed a propensity interpretation in several papers beginning with Objective Single-Case Probabilities and the Foundations of Statistics, in P. Suppes, L. Henkin, A. Joja, and Cr. C. Moisil, eds. Logic, Methodology and Philosophy of Science, IV (Amsterdam: North-Holland, 1973), pp. 467-83. For later references see also A Laplacean Formal Semantics for Single Case Propensities, *Journal of Philosophical Logic* 5 (1976): 321-353.

16. The above account is based entirely on secondary sources such as E. T. Whittaker's *A History of the Theories of Aether and Electricity* (London: Thomas Nelson & Sons, 1910; rev. ed. 1951; Torchbook edition, 1960).

17. This objection was first advanced by C. G. Hempel in his classic Studies in the Logic of Confirmation, *Mind* 54 (1945): 1-26 and 97-121. It forms much of the basis of Glymour's discussion in *Theory and Evidence*.

18. The reference to Delisle's observations appears in a footnote in Whittaker's *History* (Torchbook edition, p. 108). However, Whittaker gives no references and there are no other entries under "Delisle" in the index. Here is a case where *historical* research might alter one's *methodological* appraisal of a scientific episode.

19. For an authoritative introduction to satisficing see H. A. Simon, *Models of Man* (New York: Wiley, 1957), pp. 196-206 and also chapters 14 and 15.

20. Such a view is developed in Larry Laudan's influential book *Progress and its Problems* (Berkeley: University of California Press, 1977).

21. In several previous papers I have attempted to distinguish two rough types or stages in scientific inquiry, "exploratory" and "confirmatory," and to argue that the satisfaction level, as reflected in the acceptable risk for a mistaken conclusion, is justifiably higher in confirmatory inquiry. See Empirical Probability, Objective Statistical Methods and Scientific Inquiry, in *Foundations of Probability Theory, Statistical Inference and Statistical Theories of Science*, W. L. Harper and C. A. Hooker, Vol. 2 (Dordrecht, Holland: D Reidel, 1976), pp. 63-101;

and Testing vs. Information Models of Statistical Inference, in *Logic, Laws and Life*, ed. R. G. Colodny, (Pittsburgh: University of Pittsburgh Press, 1977), pp. 19-70.

22. This point is developed is somewhat greater detail in both papers mentioned in n. 21.

23. The useful notion of a "domain" has been developed in a number of papers by Dudley Shapere. See, for example, Scientific Theories and their Domains, in *The Structure of Scientific Theories*, ed: Fred Suppe, (Urbana: University of Illinois Press, 2nd ed., 1977), pp. 518-555.

24. See n. 13.

25. This point is further developed in the two papers mentioned in n. 21 above.

26. My own version of a "nonfoundationist" justification of induction is developed in The Epistemological Roots of Scientific Knowledge, *Minnesota Studies in the Philosophy of Science,* Vol. VI, *Induction, Probability and Confirmation,* G. Maxwell and R. M. Anderson, eds., (Minneapolis: University of Minnesota Press, 1975), pp. 212-261.

The Deductive Model:
Does It Have Instances?

1. By the deductive model, I mean the deductive model of scientific inquiry. There are plenty of deductive systems around, of course: arithmetic, number theory, set theory, probability theory, even axiomatizations of scientific theories, for example in physics, in biology, and even in psychology. What I have in mind in its simplest form is the good old hypothetico-deductive model: formulate your hypothesis clearly; draw testable implications from it; put those implications to the test; if they fail the test, reject the hypothesis; and if they pass, the hypothesis is to that extent confirmed.

Everybody knows that things are a little more complicated than this, and numerous suggestions of alternatives have been made. Karl Popper (1959) suggests that we make sincere efforts to refute hypotheses having great content; but the mechanism of falsification was still deductive. Lakatos (1970) applied the deductive model only within a research program, and offers other criteria for evaluating research programs. Even Clark Glymour (1980), who rejects some aspects of the hypothetico-deductive method in his bootstrap technique, still holds to the general deductive model in many respects. What counts in the straightforward deductive model are deductive relations, given background assumptions, or given a theory T, or given a *really sincere* attempt to look for contrary evidence, or given a paradigm, or given a research program.

What is more curious is that the great modern inductivist Carnap, as well as others who have followed him in focussing on confirmation, also accepts the deductive model. Much of Carnap's work is concerned precisely with the attempt to elucidate the force of conforming instances: refuting on nonconforming instances never seemed problematic.

It is more curious yet that many philosophers and statisticians who have been concerned with statistical inference, and indeed the vast majority of those scientists who employ statistical inference as an everyday scientific tool, accept the deductive model. Significance testing, Neyman's rejection rules, fall right into the classical pattern. Loosely speaking, the idea is that you specify in advance some data such that *if* they are observed, you will reject the hypothesis being tested.

With respect to the issues I propose to focus on, the ordinary Bayesian view of scientific inference is just as wedded to the deductive model as those I have already mentioned. If a hypothesis H entails a statement E, and $\sim E$ is observed or otherwise comes to have probability 1, H must come to have probability 0. Furthermore, in the nonextreme cases the deductive rules of the probability calculus provide the connection between prior and posterior degrees of belief.

There are three aspects of the deductive model to which all the authors I have mentioned seem firmly committed.

First: There is a deductively closed and strictly consistent set of statements accepted by the scientist. This set may be empty (for a superextreme Bayesian); or it may include any statements that in a given context are regarded as "unproblematic"; or it may even include the very general sorts of statements that characterize a paradigm or a research program.

Second: There is a set of statements (axioms of theories, hypotheses) that are considered or contemplated. These are the statements that are up for test. They may be quite general, or they may be quite specific, but in any case they have empirical content, and they are to be confirmed, in some sense or other, by their instances or consequences, or they are to be refuted by counterinstances.

Third: There is a set of statements that is directly testable, statements that express the results of experimentation or observation directly and incorrigibly.

According to this general view, scientific procedure consists in putting statements of this third class to experimental verification or refutation; the result may be the deductive refutation of a statement belonging to the second class (Popper); or the increase in degree of confirmation of the hypothesis in the second class (Carnap); or the general rearrangement of degrees of belief assigned to items in the second class (subjectivist Bayesian); or the verification of a statement in the second class, in virtue of general assumptions embodied in the first class (Lakatos, Sneed); or the basis for a computation providing confirming instances of statements in the second class (Glymour); or it might be the grounds for the rejection of a null hypothesis (for the practicing statistician).

I shall argue that science doesn't work this way at all, not even an idealized version of science, and thus that there are no instances of the deductive model, and should be none. In particular, I shall deny each of the three assertions underlying the deductive model; the result will be one possible *inductive* model:

A: I shall assume that there is a set of statements accepted by the scientist, but I shall deny that the complete set of accepted statements is consistent or deductively closed.

B: I shall deny that scientific laws and hypotheses can be up for test, since I shall deny that they have empirical content at all. Since they have no empirical content, they neither can be "confirmed" by their instances or their consequences, nor can they even be refuted deductively by counterinstances.

C: I shall assume that there is no set of statements that unproblematically express the results of experimentation or observation. I shall indeed suppose that some such statements may be accepted on the basis of observation; but I shall not suppose that they are incorrigible, once accepted.

Nonetheless, I shall assert, what it has recently become fashionable to deny in some quarters, that scientific argument *is* argument, and that it conforms to canons of rationality. The bearing of evidence on the acceptability of laws, on the choice between theories, etc., can be reflected by formal logical relations.

2. First, some preliminaries. Like most proponents of the deductive model, I shall suppose we can represent scientific knowledge in a first-order language; I suppose that the language contains some axioms for set theory, and thus all the mathematics we need, including the measure theory needed for statistics.[1] In the course of scientific inquiry, this representation changes. There is change in the vocabulary; the term "phlogiston" is dropped; the term "quark" is introduced. There is also change in the meaning of terms that persevere through scientific change: the quantity m (interpreted "mass of") changes its meaning in the transition

from Newtonian to relativistic mechanics. This requires some explanation, a bit of which I'll provide later.

Note that there is nothing in the language itself that represents the distinction between observational and theoretical vocabulary.

A rational corpus will simply be a set of statements in the language, subject to a single constraint: that it contain all the consequences of any statement it contains. I include among the consequences those arrived at with the help of axioms of the language. We assume that these axioms are consistent; if we learned that they were inconsistent, we would change them to eliminate the inconsistency. It is already clear that no one's body of actual or dispositional beliefs can be a rational corpus—no one believes all the truths of set theory or is even inclined to believe them explicitly. Nevertheless it is a useful notion, for many people accept (say) Zermelo Frankel axioms for set theory, and thereby *commit* themselves to the consequences of those axioms. A rational corpus represents a set of statements to which one is committed regardless of one's doxastic state.

Note that neither deductive closure nor even consistency is built into the notion of a rational corpus. It is only the set of consequences of the axioms alone that is deductively closed. There is only one rational corpus that is explicitly inconsistent (the corpus containing all statements), but there are any number whose deductive closure is inconsistent. (For example, a corpus could contain the premises for the lottery paradox, but need not contain the conclusion: no ticket will win.)

Probability will be a function from pairs consisting of a rational corpus and a statement to subintervals of $[0, 1]$. It is objective and logical: i.e., given a rational corpus K and a statement S, the probability of S is completely determined to be the interval (p, q) regardless of what anyone knows or believes. Underlying any such probability is a statement of frequency (or, rarely, chance) that is a member of the corpus K. This statement of frequency may be empirical, or it may be a consequence of the axioms of our language.[2] (For example, the axioms of set theory, or the axioms characterizing "fair die.") The definition of probability involves no more than direct inference.

Rational corpora of individuals (or groups, or disciplines) come with indices reflecting criteria of acceptance. A sentence belongs to a corpus of level .9, for example, if its probability relative to a (specified) corpus of level higher that .9 is a subinterval of $[.9, 1]$.[3]

There are two ways in which statements get into rational corpora, other than by being consequences of the axioms of the language: by inference and by observation. Both involve probability.

Acceptance by inference has already been characterized: S belongs to the corpus K of level r, if its probability relative to the specified corpus of level higher than r is greater than r. It is a theorem of probability (in my sense) that if p entails q, then the minimum probability of q is at least as great as that of p.

Acceptance by observation is more complicated. Borrowing some terminology from Mary Hesse (1974), let us take an observation *report* to be a sentence of our language we are motivated to accept by what happens to us. Not all such sentences can be true simultaneously; we know that some must be false, for the set of reports may well be inconsistent with the axioms of our language. From this we can infer that each kind of observation report is characterized by a certain long-run frequency of error. We accept an observation *report* (i.e., we construe it as an *observation sentence*) in the corpus of level r, just in case its probability of *not* being in error, relative to the specified metacorpus of level higher than r, is higher than r.[4]

Note that the *set* of observation sentences in the corpus of level r may still be inconsistent with our axioms. For example, we may have a large collection of observation reports of the form Rx, of which we know that less than 1 percent are erroneous. Still, if we have no way of picking out any particular ones that are in error, they may all be characterized by a probability of error of less than .01, and they may all thus end up in the rational corpus of level .99.

One more piece of machinery, and we'll turn to some examples. The predictive observational content of a rational corpus is the set of sentences in that corpus that get there by inference and not by observation, but that are of the sort that *could* get there by observation. Thus if Rx is a highly dependable form of observation report, and the sentence "Ra" appears in a rational corpus due to inference and not due to observation, then it is a member of the predictive observational content of that corpus. I assume that there is some natural way of constraining the predictive observational content of a rational corpus to be finite, so that we can compare the contents of two rational corpora expressed in different languages.[5]

3. Let us begin with measurement. We might suppose that our judgments of relative length were infallible and incorrigible. We would

have to suppose, then, that few of the conventional properties of relative length actually held. In particular, the transitivity of *being the same length as* would have to be regarded as false. But we would never have to reject an observation statement; and we could even include some predictive observation statements in our rational corpus: it might well be true of *almost* all triples of objects that if the first is longer than the second, and the second longer than the third, then the first is longer than the third.

But we might supposed instead that "longer than" satisfied the ordinary axioms, and that our judgments of relative length were sometimes erroneous. For example, it will follow from these axioms that *being the same length as* is transitive. Once we accept that as axiomatic, then there is no test that will refute the generalization. And that is important, because we have all observed prima facie refutations of it: the most obvious occur when we measure something a number of times and get a number of different values. In fact, under these circumstances (since we have a rather complete theory of error for this case) we do not accept any of the observation *reports* as observation *statements*, but rather use them, together with that theory of error, to infer a "theoretical" sentence (theoretical in scare quates) that we do add to our rational corpus: the length of *x* lies in the interval *L* plus or minus *d*. This is no doubt what an ordinary scientist means by an observation statement.

The result of this change from regarding the predicate *longer than* as strictly observational but without structure, to regarding it as having structure but being subject to observational error, is an enormous increase in the predictive observational content of our rational corpus. I cannot demonstrate this now, but it is at least intuitively plausible. That it is advantageous in this way does depend on facts about the world: on the fact that the errors of observation we introduce to allow ourselves to stick to the handy structure we attribute a priori to "longer than" are ordinarily rather small relative to our concerns with relative length. It also depends on the fact that there are ways of reducing those errors, but this is a complex fact and depends on far more in the way of scientific knowledge than the mere existence of relatively rigid bodies. And it depends importantly on the contribution to communication among people that standard scales of length make possible.[6]

Let us turn to something more empirical than the theory of lengths of rigid bodies. How about the law of thermal expansion: that the change in length of a object is proportional to its change in temperature. Well, of course that isn't really a law: it applies only to homogeneous substances of certain sorts, not undergoing changes of state, and then only within certain limits. So it is hedged around with provisos. And of course if we are to test it with a mercury thermometer we are presupposing that the law holds for mercury. What is going on here? Well, to begin with, we have a way of measuring length. We also have an indirect way of measuring temperature (possibly several ways) that may depend on the uniform expansion (or apparently uniform expansion) of mercury. That in itself might give us the idea of thermal expansion, as applied to metals in general, for example. This is reminiscent of Clark Glymour's idea of bootstrapping. There is no reason why a special case of the law shouldn't enter into the confirmation of the law in its generality. But even taking this for granted, do we now hold this law in a kind of epistemological limbo, and look for refuting instances? De we subject it to thorough test before we accept it?

How would we do it? We might take an iron bar and heat it up, carefully measuring it and its temperature before and after. This would allow us to compute a coefficient of thermal expansion of iron, but it is no *test* of the theory. Well then, we perform the experiment again, using a different bar and heating it a different amount. Does it expand the predicted amount? No. Does that refute the law? No. It gives us a different value for the coefficient of thermal expansion. Wait, though. We've forgotten the errors of measurement of length and temperature. These will be reflected in an error for the coefficient of thermal expansion, and we might demand that the second trial yield a number that is the same as the first, "within experimental error." Let us unpack this. From the distribution of errors of measurement of length and the distribution of errors of measurement of temperature, together with a bit of knowledge concerning the independence of these errors, we can derive a distribution of errors of (indirect) measurement of the coefficient of thermal expansion, *given* the truth of the law we are testing. The trouble is that these distributions are generally taken to be approximately normal, and thus to be unbounded: no conceivable set of measurements can conflict with the law. No set of numbers can fail to be an "instance" of the law "within experimental error."

Let us try again. We know a lot about the distribution of errors of measurement of length; rather less about the distribution of errors of measurement of temperature. Suppose they are both approximately normal. Given an observation report of a measurement, we can accept in a corpus of level r, the assertion that the measured length lies within certain limits. This is an *inductive* inference: it is a direct inference from the distribution of errors in general to a conclusion about the magnitude of an error in a particular case. The same is true for each of our four measurements of length and four measurements of temperature. This is still no help (although it does correspond to the way people sometimes talk), because what we require to achieve a possible counterinstance to the proposed law is the *conjunction* of these eight statements, and it is obvious that each of them can be acceptable at the level r while their conjunction is not. Very well, we can consider the joint distribution of the eight errors, and choose our bounds for each of the quantities in such a way that an appropriate conjunction appears in our corpus of level r; this conjunction might or might not refute the proposed law. But there are a lot of ways of choosing those bounds (we could take some quite narrowly and others quite broadly, or vice versa), some of which might lead to conflict with the proposed law, and some of which might not, and this all at the same level of practical certainty.

Let us try once more. This time we shall assume that the law is true, and that all of our measurements are to be understood as a computation of the coefficient C. From the truth of the law and the error distributions for the measurements of length and temperature, we can derive an error distribution for C. Our first set of measurements yields the result that the assertion that C lies in such and such an interval (I_1) is in our corpus of level r. We might suppose that our second set of measurements could yield the result that the assertion that C lies in I_2 is in our corpus of level r. Suppose that I_1 and I_2 do not overlap at all (or overlap very little); would this refute our law? or test it? This can't happen! Why not? Because in order for "$C \varepsilon I_2$" to have a probability of at least r, the (indirect) measurement of C must be a random member of the set of such measurements with respect to having a given error, and the previous measurement of C precludes this. Just as in sampling from a binomial population to estimate the binomial parameter p, the conditions of randomness require that we not make two inferences based on two parts of the sample, but consider the *whole* sample as the data for our inference, so in this case, the requirements of randomness demand that we base our inference concerning C on the *total* evidence available. Thus in fact what we have in our corpus of level r is neither "$C \varepsilon I_1$" nor "$C \varepsilon I_2$," but "$C \varepsilon I_c$," where I_c is the interval computed from the total evidence—i.e., all sets of observations.

Is there no way that the evidence can refute the law? Let us make one last try. Given that we know the distribution of errors in the measurements of lengths and temperatures, and given that the law is true, the distribution of errors of the indirect measurement of C will have a known distribution. Could a sample of measurements of C lead to the rejection of this distribution of errors? Not if the distribution is *known*. If the distribution is known, an awkward collection of sample values can only be evidence of bad luck. At a certain point, though, doesn't one reject the notion of bad luck, and thus the assumed distribution of errors of measurement of C? Sure. But notice that this assumed distribution depended on three things: the distribution of errors of measurement of length, and distribution of errors of measurement of temperature, and the law of thermal expansion. Let us simplify things by supposing that the evidence on which our knowledge of the distribution of errors of measurement of length is based is so extensive that we can take that statistical generalization for granted.

There are then two courses open to us. We may take our body of evidence as throwing new light on the distribution of errors of measurement of temperature, or we may suppose that it puts our law of thermal expansion in a bad light. How do we decide which? I propose that we decide in the same way as we decide whether or not to accept the transitivity of *being the same length as* or *being the same temperature as*. Thus it seems to me that the law should be regarded as a proposed *linguistic* reform: if we accept it, the data regarding the distribution of C are to be construed as data bearing on the distribution of errors in the measurement of temperature. And surely, if the distribution of observations of C is close to that expected, that is just what we would do. We would regard those observations as a bit more evidence regarding errors of measurement of temperature.

This means that in considering the law and its evidence, we are implicitly choosing between two languages: one that contains the law, and one that does not. To this choice we apply the criterion mentioned earlier: which language yields us a corpus of practical certainties with the greatest predictive observational content? If adopting the language containing the law yields the result that our measurements of temperature are subject to much more error than we thought, then the corpus of practical certainties using the language will have to forego observational predictions concerning temperature that the corpus based on the other language could retain. Note that it would have to forego observational predictions concerning temperature that had nothing to do with the law of thermal expansion, so that there is a holistic element in the criterion. But if adopting the language contianing the law led to the result that our measurements of temperature were only a little, or not at all, more inaccurate than we thought, the predictive observational content of our body of practical certainties would be significantly enlarged.

For simplicity, I have been speaking as if the law of thermal expansion were universal. It isn't, and most laws—at least, laws of this sort—have limited scope. The scope of the law is easy to build into this treatment. For a certain range of temperatures of a certain set of substances, distributions of values of their coefficients of thermal expansion are included as part of the data we have concerning the errors of measurement of temperature. For other temperatures and other substances, the law just doesn't apply. The full statement of the law in our language thus contains its limitations. Note that these limitations are *discovered* empirically: we do not confirm the law for each of a class of applications, but take it as universal, and discover its limitations.

Of course nowadays we know a lot more about the law of thermal expansion than is embodied in the special case I have been considering. This increasing sophistication is allowed for on the view of testing I am proposing. A simple illustration can be found in the gas law: At one point in time it is a feature of our scientific *language* that for relatively rarified gases, $PV = nRT$. Observations are to be construed as throwing light on the errors of measurement of pressure, volume, and temperature. (In fact, for high temperatures, this law does provide a means of measuring temperature by means of the gas thermometer.) At a later point in time it becomes a feature of our scientific *language* that for a wider range of instances, $(P + (a/v^2)) (v - b) = RT$, where v is specific volume, and a and b are characteristic of the gas in question. This is van der Waals's equation, and it is clear that for small values of a and b, the ideal gas law is a reasonable approximation to the truth. Now of course what is taken as a feature of our scientific language is yet more complicated; the ideal gas law and van der Waals's equation provide handy computational approximations (instruments?) but are strictly false, even as applied to the best old instances.

5. What does this approach have to say in general about the confirmation, disconfirmation, degree of credibility, and testing of scientific laws and theories? In one sense, it settles the whole problem—it reduces it to epistemology. We don't confirm laws, we don't test them, we don't

disconfirm them, because they are taken to be a priori features of our language. But we do have criteria, expressed in terms of a certain sort of content of our bodies of knowledge, for preferring a language with one a priori feature to a language with a different a priori feature.

Nevertheless something goes on that seems like the process characterized by the hypothetico-deductive model, or like confirmation by instances, or like Glymour's bootstrapping, or like something involving simplicity. Something should be said by way to explanation of the apparent relevance of these notions.

When we have a great deal of knowledge about the distributions of errors of measurement (direct *or indirect*) of the quantities that appear in a proposed law, we don't have to think very hard when we obtain a set of measurements that are widely at variance with the law. We forget about the law. Why? Because it seriously undermines our presumed knowledge of error distributions, and thereby impoverishes rather than enriches the predictive content of our body of knowledge. To show in detail how this comes about is nontrivial. I have made some efforts along these lines in my article "The Justification of Deduction in Science" (Kyburg, forthcoming). I have also argued there that in a very weak sense the successful testing of a law (the establishment of approximately confirming instances) does enhance its "acceptability"—instance confirmation may well help, particularly when indirect measurement is involved, to reduce the dispersion of our errors of measurement.

Nevertheless the general progress of scientific inquiry is contrary to that suggested by the deductive model. As the accuracy of measurement improves in a given domain, the discrepancies between our laws and our experimental results become more pronounced. *Successful* laws are progressively *dis*confirmed.

There are some interesting connections between Glymour's notion of bootstrapping and the approach I have been exploring. One is the fact that on Glymour's reconstruction, part of a theory may be needed to compute a parameter in a hypothesis, or a theoretical quantity. This is illustrated in my case by the calculation of the coefficient C in the example I belabored. It is easy to see, however, that one can only *accept* in his rational corpus an inequality like $C_1 < C < C_u$. But that is as it should be, and it is exactly what is needed to yield new predictive content.

Another connection lies in the valuable notion of a *computation*. Glymour shows computations going only one way, from data to parameters. It is important to note that error already enters here: we must begin with observation reports, embodying error, and trace the distribution of error through the calculation to arrive at a distribution of error for the quantity or parameter in question. But in applications of science—even in the case of simple predictions—the computation goes both ways. We compute the (credible) values of theoretical quantities on the basis of observation reports, then pass through yet further computations of theoretical quantities (keeping track of error distributions as we go along), and then back down to observation *sentences* (e.g., the angle almost certainly lies between 130.15° and 130.25°), and thence, through errors of measurement again, to a prediction (e.g., the angle will be reported to be between 129.5° and 130.5°).[7] What strikes me as most important and valuable about the notion of a computation is that it allows us to calculate what I would like to call a *trajectory of error* from observation report to observation report, as well as the trajectory of error from observation report to observation sentence to parameter to theoretical quantity.

Let me turn to questions of generality and simplicity. One aspect of generality and simplicity has already been touched upon. We discover statistical limitations in the applications of laws that lead us to prefer a language in which the scope of a law is limited: in other words, we make our (simple) conventions as general as we can get away with making them. But there are other aspects. One concerns the reduction of errors and the elimination of "discrepancies." In addition to extending the useful range of applicability of a law or theory by improving our techniques of measurement—and the distribution of errors provides a direct criterion of "improvement"—we may have a theory according to which a certain quantity is *mainly* a function of something else. For example, the orbit of a planet is mainly due to forces acting between the planet and the sun, but also, strictly speaking, a function of the forces exerted by other celestial bodies. In such a case the theory itself may provide an account of the deviant probabilistic difference between prediction sentence and observation report. And at the same time, it can tell us how to go about taking account of these systematic errors. It can tell us (if only roughly) how to "subtract the effect of the sun" on the orbit of the moon, or in what ways it would be *legitimate* to falsify and idealize the data to be used in calculating theoretical parameters or quantities. But the theory can function in this way only if it is construed as an a priori feature of our language.

There is another way in which a theory can be used to reduce the component of error in predictive sentences. This may be relevant to the discussion of Ptolemaic and Copernican astronomy. Either theory, it has been pointed out, can be made to fit the phenomena reasonably well. To do so involves determining parameters, and this determination is subject to error. On the Ptolemaic theory we have just one set of parameters characterizing the motion of each planet relative to the earth. On the Copernican theory this relative motion is characterized by *two* sets of parameters: one characterizing the motion of the earth about the sun (call this set E) and one characterizing the motion of the ith planet about the sun (call this P_i). On the Ptolemaic theory, observation of a given planet helps to reduce the uncertainty in the parameters characterizing its movement about the earth, independently of observations of other planets. On the Copernican theory, there are more parameters to worry about (E as well as P_i), but observations on the ith planet help to reduce the errors in E—and hence in the apparent motions of other planets—as well as P_i. Generality can thus contribute to the control of error. And of course one's concerns are prospective as well as immediate: the prospect of better control of error in the future, particularly as regards systematic error, has a bearing on our choice of theoretical languages.

I suppose that in some sense or other I am a victim of East Coast Holism: Here I am not only saying that our body of scientific knowledge must be treated as a single ball of wax, but that scientific change involves the very language in which we do science. But it is one thing to solemnly intone "All in One," and something else again to provide enough analysis of the Great White Unity to allow one to tinker with one part at a time and to choose rationally between two alternative balls of wax that differ only in specifiable limited ways. It is the latter sort of thing that I think worth pursuing, for the sake of the improvement of the understanding. But I confess that I have no more than sketched the beginnings of an approach.

I have attempted to provide some prima facie plausibility for my three preposterous claims: A, that our body of scientific practical certainties need be neither consistent nor deductively closed; B, that scientific laws need not be construed as having empirical content, nor as being confirmed by their instances; and C, that we need not regard any observations at all as incorrigible. I should also remark, though, that the deductive model is to the inductive model as the ideal gas law to van der Waals's equation: When we have adopted van der Waals's equation, the ideal gas law has no instances, but it is a handy approximation.

The analogy yields a further implication: when we have yet more insight into science, the inductive model I have sketched will probably have no instances either, just as van der Waals's equation has no instances today. But I think the inductive model will even then serve as a calculational approximation over a wider range of situations than the deductive model does.

Notes

1. There has in recent years been a stir about replacing the consideration of formal languages by models. Although there are advantages in so doing—we needn't bother about unintended models of our languages, we can replace the notion of deducibility by the tighter notion of model-theoretic entailment—these advantages are not important for present purposes. And we shall have occasion to refer to certain actual sentences—observation *reports*—of the language of science anyway.

2. It is a set-theoretical truth that most subsets of a given set resemble it with respect to the proportion of objects having a given property. This is an assertion without empirical content. Yet it may nevertheless be useful, when, for example, a specific subset—a sample that we have drawn—is a *random member* of the set of *n*-membered subsets of a given set. Similarly, it is vacuous to say that a fair die, tossed fairly, has an equal chance of exhibiting each face; but this can nonetheless be a useful piece of information if we can be *practically certain* that a given setup involves a (practically) fair die being tossed (nearly) fairly.

3. We must *specify* the high-level corpus, for it may be that there are two corpora of level higher than .9 relative to which the statement has different probabilities. For more details about probability, see Kyburg (1974).

4. We speak here of *meta*corpus since the argument concerns frequencies of error among sentences.

5. In the case of individuals, for example, we might suppose a maximum lifespan of a hundred years, and a maximum number of kinds of discrimination expressible in a given vocabulary (say, 10,000), and a minimum time to make an observation (say, a tenth of a second), and conclude that the upper bound on the contents of a rational corpus of an individual is $100 \times 365 \times 24 \times 60 \times 60 \times 10 \times 10,000 = 3.1 \times 10^{14}$ items. Here is where we might build in a cost function for both computation and for observation. *Here* is where "logical omniscience" becomes constrained.

6. More detail will be found in Kyburg (1979).

7. The distinction, introduced by Mary Hesse (1974), between observation statements and observation reports, is a handy one. It is somewhat elaborated on in Kyburg (forthcoming).

References

Glymour, Clark. 1980. *Theory and Evidence*. Princeton: Princeton University Press.

Hesse, Mary. 1974. *The Structure of Scientific Inference*. Berkeley: University of California Press.

Kyburg, Henry E. Jr. 1974. *The Logical Foundations of Statistical Inference*. Dordrecht: Reidel.

Kyburg, Henry E. Jr. 1979. Direct Measurement. *American Philosophical Quarterly*. 16: 259-272.

Kyburg, Henry E. Jr. Forthcoming. The Justification of Deduction in Science.

Lakatos, Imre. 1970. Falsification and the Methodology of Scientific Research Programmes. In *Criticism and the Growth of Knowledge*, ed. Imre Lakatos and Alan Musgrave, Cambridge: Cambridge University Press, 1970.

Popper, Karl. 1959. *The Logic of Scientific Discovery*. London: Hutchinson and Co.

VI. TESTING PARTICULAR THEORIES

Retrospective Versus Prospective Testing
of Aetiological Hypotheses
in Freudian Theory

I. Introduction

The repression-aetiology of neurotic disorders is the cornerstone of the psychoanalytic theory of unconscious motivations. Repressions are held to be not only the pathogens of the psychoneuroses but also the motives of dream construction, and the causes of various sorts of bungled actions ("parapraxes") in which the subject is normally successful (e.g., slips of the tongue or pen). Thus even in the case of "normal" people, Freud saw manifest dream content and various sorts of "slips" as the telltale symptoms of (temporary) *mini*neuroses, engendered by repressions.

Having modified Breuer's cathartic method of investigation *and* therapy, Freud arrived at the purported sexual aetiologies of the psychoneuroses, as well as at the supposed causes of dreams and parapraxes, by lifting presumed repressions via the patient's allegedly "free" associations. At the same time, excavation of the pertinent repressed ideation was to remove the pathogens of the patient's afflictions. Hence scientifically, Freud deemed the psychoanalytic method of investigation to be both heuristic *and* probative, over and above being a method of therapy.

Yet he and his disciples have not been alone in maintaining that the *clinical* setting of the psychoanalytic treatment sessions provides cogent epistemic avenues for testing his aetiologies *retrospectively*. The scrutiny of this thesis has highly instructive import for adjudicating the general polemic waged by critics against the clinical validation that Freudians have claimed for their theory. But the examination of the purported cogent testability likewise helps us assess the merits of prospective versus retrospective testing of aetiologic hypotheses in psychiatry. Hence I shall focus this essay on the scrutiny of the alleged clinical testability of the Freudian aetiologies.

II. Statement of the Controversy as to Clinical Testability

Hans Eysenck (1963) has maintained that "we can no more test Freudian hypotheses 'on the couch' [used by the patient during psychoanalytic treatment] than we can adjudicate between the rival hypotheses of Newton and Einstein by going to sleep under an apple tree." (p. 220) And, in Eysenck's view, only suitably designed *experimental* studies can perform the *probative* role of *tests*, while the clinical data from the couch may be heuristically fruitful by suggesting hypotheses. As against this denial of clinical testability, Clark Glymour (1974) has argued that "the theory Sigmund Freud developed at the turn of the century was strong enough to be tested [cogently] on the couch." (p. 304) And Glymour proposes to illuminate Eysenck's disparagement of clinical data but then to discount it by the following dialectical give-and-take:

> It stems in part, I think, from what are genuine drawbacks to clinical testing; for example, the problem of ensuring that a patient's responses are not *simply* the result of suggestion or the feeling, not without foundation, that the "basic data" obtained from clinical sessions—the patient's introspective reports of his own feelings, reports of dreams, memories of childhood and adolescence—are less reliable than we should like. But neither of these considerations seems sufficient to reject the clinical method generally, although they may of course be sufficient to warrant us in rejecting particular clinical results. Clinicians can hopefully be trained so as not to elicit by suggestion the expected responses from their patients; patients' reports can sometimes be checked independently, as in the case of memories, and even when they cannot be so checked there is no good reason to distrust them generally. But I think condemnations like Eysenck's derive from a belief about clinical testing which goes considerably beyond either of these points: the belief that clinical sessions, even cleansed of suggestibility and of doubts about the reliability of patients' reports, can involve no rational strategy for testing theories.
>
> I think that Eysenck's claim is wrong. I think there is a rational strategy for testing important parts of psychoanalysis, a strategy that relies almost exclusively on clinical evidence; moreover, I think this strategy is immanent in at least one of Freud's case studies, that of the Rat Man. Indeed, I want to make a much bolder claim. The strategy involved in the Rat Man case is essentially the same as a strategy very frequently used in testing physical theories. Further, this strategy, while simple enough, is more powerful than the hypothetico-deductive-falsification strategy described for us by so many philosophers of science. (p. 287-88)

Despite this epistemological tribute to Freud's couch, Glymour issues a *caveat:*

> I am certainly not claiming that there is good clinical evidence for Freud's theory; I am claiming that if one wants to test psychoanalysis, there is a reasonable strategy for doing so which can be, and to some degree has been, effected through clinical sessions. (p. 288)

Most recently, Glymour (1980) told us more explicitly why we should countenance the rationale that animated Freud's clinical investigation of psychoanalytic hypotheses during the treatment of his Rat Man patient Paul Lorenz.[1] Glymour points to at least three important specific episodes in the history of physical science in which he discerns just the logical pincer-and-bootstrap strategy of piecemeal testing that he also teased out from Freud's analysis of Paul Lorenz. Thus he says, "unlikely as it may sound . . . the major argument of the Rat Man case is not so very different from the major argument of Book III of Newton's *Principia*." (Glymour 1980, p. 265) And he stresses that this argument employs a logical *pincer* strategy of more or less *piecemeal* testing *within* an overall theory, instead of the completely global theory-appraisal of the hypothetico-deductive method, which altogether abjures any attempt to rate different components of the theory individually as to their merits in the face of the evidence.

Besides commending Freud's clinical study of the Rat Man for its rationale, Glymour likewise attributes a fair degree of scientific rigor to a *few* of Freud's other investigations. But he couples these particular appreciative judgments with a largely uncomplimentary overall evaluation, after deploring the very uneven logical quality of the Freudian corpus. Indeed, Glymour thinks he is being "gentle" (p. 265) when he deems Freud's 1909 case study of Little Hans to be "appalling." And he finds that "on the whole Freud's arguments for psychoanalytic theory are dreadful," marred by the frequent—though by no means universal—substitution of "rhetorical flourish" for real argument, and of a "superabundance of explanation" for cogent evidence. (p. 264) Yet clearly these quite fundamental dissatisfactions with Freud's all too frequent lapses do not militate against Glymour's espousal of the clinical testability of such central parts of psychoanalytic theory as the specific aetiologies of the psychoneuroses, at least in the aetiological versions that Freud enunciated before 1909.

Just this championship of the *probative* value of data from the analytic treatment sessions is philosophical music to the ears of those who echo Freud's own emphatic claim that the bulk of psychoanalytic theory is well founded empirically. For, as Ernest Jones reminded everyone in his Editorial Preface to Freud's *Collected Papers,* the clinical findings are "the real basis of Psycho-analysis. All of Professor Freud's other works and theories are essentially founded on the clinical investigations." (Jones 1959, vol. 1, p. 3) Thus most advocates of this theoretical corpus regard the analyst's many observations of the patient's interactions with him in the treatment sessions as the source of findings that are simply *peerless* not only heuristically but *also* probatively. We are told that during a typical analysis, which lasts for some years, the analyst accumulates a vast number of variegated data from each patient that furnish evidence relevant to Freud's theory of personality no less than to the dynamics and outcome of his therapy. The so-called psychoanalytic interview sessions are claimed to yield genuinely probative data because of the alleged real-life nature of the rich relationship between the analyst and the analysand. Even an analyst who just declared it to be high time that Freudians "move from overreliance on our hypothetical discoveries to a much needed validation of our basic theoretical and clinical concepts" (Kaplan 1981, p. 23) characterizes "the naturalistic observations within the psychoanalytic treatment situation" as "the major scientific method of psychoanalysis" (p. 18). Hence the clinical setting or "psychoanalytic situation" is purported to be the arena of *experiments in situ,* in marked contrast to the contrived environment of the psychological laboratory with its superficial, transitory interaction between the experimental psychologist and his subject. Thus the analysts A. M. Cooper and R. Michels (1978) tell us that "increasingly this [psychoanalytic] inquiry has recognized the analytic situation itself as paradigmatic for all human interactions." (p. 276) Indeed, the psychoanalytic method is said to be uniquely suited to first eliciting some of the important manifestations of the unconscious processes to which Freud's *depth* psychology pertains.

This superior *investigative value* of the analyst's clinical techniques is thus held to make the psychoanalytic interview at once the prime testing ground and the heuristic inspiration for Freud's theory of personality as well as for his therapy. Some leading orthodox analytic theoreticians have been concerned to *exclude* the so-called metapsychology of Freud's psychic energy model, and a fortiori its erstwhile neurobiological trappings from the avowed purview of clinical validation. Therefore it is to be understood that the term "psychoanalytic theory of personality" is here construed to *exclude* the metapsychology of psychic energy with its cathexes and anticathexes. In any case, most analysts have traditionally been quite sceptical, if not outright hostile, toward attempts to test Freudian theory experimentally *outside* the psychoanalytic interview.

Just such an assessment was enunciated again quite recently by Lester Luborsky and Donald Spence (1978). They do issue the sobering caveat that "psychoanalysts, like other psychotherapists, literally *do not know* how they achieve their results." (p. 360) But they couple this disclaimer with the tribute that analysts "possess a unique store of clinical wisdom." Moreover, Luborsky and Spence emphasize that *"far more is known now* [in psychoanalysis] *through clinical wisdom than is known through quantitative* [i.e., controlled] *objective studies."* (p. 350, italics in original) In short, they claim that—in this area—clinical confirmation is presently superior to experimentally obtained validation. And they deem findings from the psychoanalytic session to have such epistemic superiority not only therapeutically but also in the validation of Freud's general theory of unconscious motivations. (pp. 356-57) Similarly for the purported validation of Heinz Kohut's currently influential variant of psychoanalysis, which supplants Freud's Oedipal conflict by the child's *pre*-Oedipal narcissistic struggle for a cohesive self as the major determinant of adult personality structure. (Ornstein 1978)

Despite their strong differences, both of the parties to the above dispute about the probative value of *clinical* data for the empirical appraisal of psychoanalytic theory do agree that at least part of the Freudian corpus is indeed cogently testable by empirical findings of *some* sort: The Freudians have the support of Glymour, for example, in contending that observations made within the confines of the treatment setting do or can afford sound testability, while such anti-Freudian protagonists as Eysenck make the contrary claim that well-conceived tests are possible, at least in principle, but *only* in the controlled environment of the laboratory or in other *extra*-clinical contexts. And clearly the assumption of empirical testability shared by the disputants is likewise affirmed by someone who maintains that *both* clinical and extra-clinical findings are suitable, at least in principle, for testing psychoanalysis.

Yet this shared assumption of testability has again been denied simpliciter by Popper as recently as in his 1974 reply to his critics (Popper 1974, vol. 2, pp. 984-985). There he reiterates his earlier claim that Freud's theory, as well as Adler's, are "simply non-testable, irrefutable. There was no conceivable human behaviour which would contradict them." (Popper 1962, p. 37) It is then a mere corollary of this thesis of nontestability that *clinical* data, in particular, likewise cannot serve as a basis for genuine empirical tests.

III. Can Popper's Indictment of Freudian Theory Be Sustained?

In earlier publications, I have argued that neither the Freudian theory of personality nor the therapeutic tenets of psychoanalysis is untestable in Popper's sense. (Grünbaum 1976, 1977, 1979a) Furthermore, there I contended in detail that Popper's portrayal of psychoanalysis as a theory that is entitled to claim good *inductivist* credentials is predicated on a caricature of the requirements for theory validation laid down by such arch inductivists as Bacon and Mill. Thus I pointed out that Freud's theory is replete with *causal* hypotheses and that the evidential conditions that must be met to furnish genuine inductive support for *such* hypotheses are very demanding. But I emphasized that precisely these exacting inductivist conditions were pathetically *unfulfilled* by those Freudians who claimed ubiquitous confirmation of the psychoanalytic corpus, to Popper's fully justified consternation.

In Grünbaum (1984, chap. 1, sec B), the reader will find my most recent detailed critique of Popper's charge of unfalsifiability against all of psychoanalytic theory. Hence it will suffice here to give mere illustrations of the textually untutored and logically slipshod character of this wholesale charge.

Even a casual perusal of the mere *titles* of Freud's papers and lectures in the *Standard Edition* yields two examples of falsifiability. And the second is a case of acknowledged falsification to boot. The first is the paper "A Case of Paranoia Running Counter to the Psychoanalytic Theory of the Disease" (S.E. 1915, 14:263-272); the second is the Lecture "Revision of the Theory of Dreams" (S.E. 1933, 22:7-30, esp. pp. 28-30). Let us consider the first.

The "psychoanalytic theory of paranoia," which is at issue in the paper, is the hypothesis that *repressed* homosexual love is *causally necessary* for one to be afflicted by paranoid delusions. (S.E. 1915, 14:265-66). The patient was a young woman who had sought out a lawyer for protection from the molestations of a man with whom she had been having an affair. The lawyer suspected paranoia when she charged that her lover had gotten unseen witnesses to photograph them while making love, and that he was now in a position to use the photographs to disgrace her publicly and compel her to resign her job. Moreover, letters from her lover that she had turned over to the lawyer deplored that their beautiful and tender relationship was being destroyed by her unfortunate morbid idea. Nonetheless, aware that truth is sometimes stranger than fiction, the lawyer asked Freud for his psychiatric judgment as to whether the young woman was actually paranoid.

The lover's letters made "a very favorable impression" on Freud, thereby lending some credence to the delusional character of the young woman's complaints. But, assuming that she was indeed paranoid, Freud's initial session with her led to a theoretically disconcerting conclusion: "The girl seemed to be defending herself against love for a man by directly transforming the lover into a persecutor: there was no sign of the influence of a woman, no trace of a struggle against a homosexual attachment." (S.E. 1915, 14:265) If she was indeed delusional, then this seeming total absence of repressed homosexuality "emphatically contradicted" Freud's prior hypothesis of a homosexual aetiology for paranoia. Thus, he reasoned: "Either the theory must be given up or else, in view of this departure from our [theoretical] expectations, we must side with the lawyer and assume that this was no paranoiac combination but an actual experience which had been correctly interpreted." (S.E. 1915, 14:266) And furthermore: "In these circumstances the simplest thing would have been to abandon the theory that the delusion of persecution invariably depends on homosexuality." (p. 266) In short, Freud explicitly allowed that if the young woman *is* paranoid, then her case is a *refuting* instance of the aetiology he had postulated for that disorder. Alternatively, he reckoned with the possibility that she was not paranoid.

As it turned out, during a second session the patient's report on episodes at her place of employment not only greatly enhanced the likelihood of her being afflicted by delusions, but also accorded with the postulated aetiology by revealing a conflict-ridden homosexual attachment to an elderly woman there. But the point is that the psychoanalytic aetiology of paranoia is empirically falsifiable (disconfirmable) *and* that Freud explicitly recognized it. For, as we saw, this hypothesis states that a homosexual psychic conflict is causally necessary for the affliction. And empirical indicators can bespeak the absence of homosexual conflict as well as the presence of paranoid delusions.

Hence this example has an important general moral: Whenever empirical indicators can warrant the *absence* of a certain theoretical pathogen P as well as a differential diagnosis of the *presence* of a certain theoretical neurosis N, then an aetiologic hypothesis of the strong form "P is causally necessary for N" is clearly empirically falsifiable. It will be falsified by any victim of N who had not been subjected to P. For the hypothesis *predicts* that anyone not so subjected will be spared the miseries of N, a prediction having significant prophylactic import. Equivalently, the hypothesis *retrodicts* that any instance of N was also a case of P. Hence, if there are empirical indicators as well for the *presence* of P, then this retrodiction can be empirically instantiated by a person who instantiates both N and P.

Being a strict determinist, Freud's aetiological quest was for *universal* hypotheses. (S.E. 1915, 14:265) But he believed he had empirical grounds for holding that the development of a disorder N after an individual I suffers a pathogenic experience P depended on I's hereditary vulnerability. Hence his universal aetiologic hypotheses typically asserted that exposure to P is *causally necessary* for the development of N, *not* that it is causally sufficient.

Indeed, by claiming that P is the *"specific"* pathogen of N, he was asserting not only that P is causally necessary for N, but also that P is never, or hardly ever, an aetiologic factor in the pathogenesis of any other nosologically distinct syndrome. (S.E. 1895, 3:135-139) Robert Koch's specific aetiology of tuberculosis, i.e., the pathogenic tubercle bacillus, served as a model. (S.E. 1895, 3:137) By the same token, Freud pointed to the tubercle bacillus to illustrate that a pathogen can be quite explanatory, although its mere presence does not guarantee the occurrence of the illness. (S.E. 1896, 3:209) And Freud was wont to conjecture *specific* aetiologies for the various psychoneuroses until late in his career. (S.E. 1925, 20:55) Hence, as illustrated by the above example of paranoia, these aetiologies evidently have a high degree of empirical falsifiability, whenever empirical indicators can attest a differential diagnosis of N, as well as the absence of P. For the hypothesis that P is the specific pathogen of N entails the universal prediction that every case of non-P will remain a non-N, and equivalently, the universal retrodiction that any N suffered P, although it does not predict whether a given exposure to P will issue in N. Thus Glymour's account (1974) of Freud's case history of the Rat Man makes clear how Freud's specific aetiology of the Rat Man's obsession was falsified by means of disconfirming the retrodiction that Freud had based on it.

Let us return to our paranoia example. As I pointed out in an earlier paper (Grünbaum 1979a, pp. 138-139), the aetiology of paranoia postulated by psychoanalysis likewise makes an important "statistical" prediction that qualifies as "risky" with respect to any rival "background" theory that denies the aetiologic relevance of repressed homosexuality for paranoia. And, by Popper's standards, the failure of this prediction would count against Freud's aetiology, whereas its success would corroborate it.

To be specific, Freud originally hypothesized the aetiology of male paranoia (Schreber case) along the following lines. (SE. 1911, 12:63) Given the social taboo on male homosexuality, the failure to repress homosexual impulses may well issue in feelings of severe anxiety and guilt. And the latter anxiety could then be eliminated by converting the love emotion "I love him" into its opposite "I hate him," a type of transformation that Freud labeled "reaction formation." Thus the pattern of reaction formation is that once a dangerous impulse has been largely repressed, it surfaces in the guise of a far more acceptable *contrary* feeling, a conversion that therefore serves as a *defense* against the *anxiety* associated with the underlying dangerous impulse. When the defense of reaction formation proves insufficient to alleviate the anxiety, however, the afflicted party may resort to the further defensive maneuver of "projection" in which "I hate him" is converted into "He hates me." This final stage of the employment of defenses is then the full-blown paranoia. Thus this rather epigrammatic formulation depicts reaction formation and projection as the repressed defense mechanisms that are actuated by the postulated *specific* pathogen of homosexuality. But if repressed homosexuality is indeed the specific aetiologic factor in paraonia, then the decline of the taboo on homosexuality in our society should be accompanied by a decreased incidence of (male) paranoia. And, by the same token, there ought to have been relatively less paranoia in those ancient societies in which male homosexuality was condoned or even sanctioned. For the reduction of strong anxiety and repression with respect to homosexual feelings would contribute to the removal of Freud's *conditio sine qua non* for this syndrome.

Incidentally, as Freud explains (S.E. 1915, 14:265), before he enunciated universally that homosexuality is the specific pathogen of paranoia, he had declared more cautiously in his earlier publication that it is "perhaps an invariable" aetiologic factor. (S.E. 1911, 12:59-60 and 62-63, esp. p. 59) When I first drew the above "statistical" prediction from Freud's aetiology (Grünbaum 1979a, p. 139), I allowed for Freud's more cautious early formulation. For there I predicated the forecast of decreased incidence as a concomitant of taboo decline on the *ceteris paribus* clause that no other potential causes of paranoia become operative. But, by making repressed homosexuality the *conditio sine qua non* of the syndrome, Freud's specific aetiology clearly enables the prediction to go through *without* any such proviso.

On the other hand, even assertions of pathogenic causal relevance that are logically *weaker* than the specific aetiologies can be empirically disconfirmable. For they can have testable (disconfirmable) predictive import, although they fall short of declaring P to be causally necessary for N. Thus, when pertinent empirical data fail to bear out the prediction that P positively affects the incidence of N, they bespeak the causal *irrelevance* of P to N. Consequently the currently hypothesized causal relevance of heavy cigarette smoking to lung cancer and cardiovascular disease is disconfirmable. And so is the alleged causal relevance of laetril to cancer remission, which was reportedly discredited by recent findings in the United States.

The aetiology that Freud conjectured for one of his female homosexual patients furnishes a useful case in point. He states its substance as follows:

> It was just when the girl was experiencing the revival of her infantile Oedipus complex at puberty that she suffered her great disappointment. She became keenly conscious of the wish to have a child, and a male one; that what she desired was her *father's* child and an image of *him*, her consciousness was not allowed to know. And what happened next? It was not *she* who bore the child, but her unconsciously hated rival, her mother. Furiously resentful and embittered, she turned away from her father and from men altogether. After this first great reverse she forswore her womanhood and sought another goal for her libido. (S.E. 1920, 18:157)

But later on he cautions:

> We do not, therefore, mean to maintain that every girl who experiences a disappointment such as this of the longing for love that springs from the Oedipus attitude at puberty will necessarily on that account fall a victim to homosexuality. On the contrary, other kinds of reaction to this trauma are undoubtedly commoner. (S.E. 1920, 18:168)

Thus he is disclaiming the predictability of lesbianism from the stated pubescent disappointment *in any one given case*. Yet the frustration does have disconfirmable predictive import, although its causal relevance is not claimed to be that of a specific pathogen. For by designating the stated sort of disappointment as *an* aetiologic factor for lesbianism, Freud is claiming that occurrences of such disappointment *positively affect* the incidence of lesbianism.

This predictive consequence should be borne in mind, since Freud's case history of his lesbian patient occasioned his general observation that the aetiologic explanation of an already existing instance of a disorder is usually not matched by the predictability of the syndrome *in any one given case*. (S.E. 1920, 18:167-68) And an apologist for Popper was thereby led to conclude that the limitation on predictability in psychoanalysis thus avowed by Freud is tantamount to generic nonpredictability and hence to nondisconfirmability. But, oddly enough, this apologist is not inclined to regard the causal relevance of heavy smoking to cardiovascular disease as wholly nonpredictive or nondisconfirmable, although chain smoking is not even held to be a specific pathogen for this disease, let alone a universal predictor of it.

The comments I made so far in response to Popper's aforecited 1974 statement focused largely on Freud's 1915 paper on paranoia, whose very title announces an instance of empirical falsifiability. Besides, Freud's 1933 "Revision of the Theory of Dreams" presents an acknowledged falsification by the recurrent dreams of war neurotics.

These are mere illustrations of falsifiability. But I trust they have prepared the ground for my claim that Popper's refutability criterion is too insensitive to reveal the genuinely egregious epistemic defects that indeed bedevil the clinical psychoanalytic method and the aetiologies based upon it. And, as I shall argue, time-honored inductivist canons for the validation of causal claims do have the capability to exhibit these cognitive deficits.

IV. Does Neo-Baconian Inductivism Sanction the Testability of Psychoanalytic Aetiologies by *Clinical* Data?

I should remind the reader that "clinical data" are here construed as findings coming from *within* the psychoanalytic treatment sessions. When I am concerned to contrast these data from the couch with observational results secured from *outside* the psychoanalytic interview, I shall speak of the former as "*intra*-clinical" for emphasis. Freud gave a cardinal epistemological defense of the psychoanalytic method of clinical investigation that seems to have gone entirely unnoticed in the literature, as far as I know, until I recently called attention to its significance. I have dubbed this pivotal defense "The Tally Argument" in earlier publications (Grünbaum 1979b, 1980), and I shall use this designation hereafter.

It will be important for the concerns of this paper to have a statement of some of the theses for which I have argued (Grünbaum 1983, 1983b).

1. It was Freud's "Tally Argument"—or its bold lawlike premise—that was all at once his basis for five claims, each of which is of the first importance for the legitimation of the central parts of his theory. These five claims are the following:

(i) The denial of an irremediable epistemic contamination of clinical data by suggestion.

(ii) The affirmation of a crucial difference, in regard to the *dynamics* of therapy, between psychoanalytic treatment and all rival therapies that actually operate entirely by suggestion.

(iii) The assertion that the psychoanalytic method is able to validate its major causal claims—such as its specific sexual aetiologies of the various psychoneuroses—by essentially *retrospective* methods without vitiation by post hoc ergo propter hoc, and without the burdens of prospective studies employing the controls of experimental inquiries.

(iv) The contention that favorable therapeutic outcome can be warrantedly attributed to psychoanalytic intervention *without* statistical comparisons pertaining to the results from untreated control groups.

(v) The avowal that, once the patient's motivations are no longer distorted or hidden by repressed conflicts, credence can rightly be given to his or her introspective self-observations, because these data then do supply probatively significant information. (Cf. Waelder 1962, pp. 628-29)

2. The epistemological considerations that prompted Freud to enunciate his Tally Argument make him a sophisticated scientific methodologist, quite better than is allowed by the appraisals of relatively friendly critics like Fisher and Greenberg (1977) or Glymour (1980), let alone by very severe critics like Eysenck.

Yet evidence that has been accumulating in the most recent decades makes the principal premise of the Tally Argument well-nigh empirically untenable, and thus devastatingly undermines the conclusions that Freud drew from it. Indeed, no empirically plausible alternative to that crucial discredited premise capable of yielding Freud's desired conclusions seems to be in sight.

3. Without a viable replacement for Freud's Tally Argument, however, there is woefully insufficient warrant to vindicate the intraclinical testability of the cardinal tenets of psychoanalysis—especially its ubiquitous causal claims—a testability espoused traditionally by analysts, and more recently by Glymour on the strength of the pincer-and-bootstrap strategy. This unfavorable conclusion is reached by the application of neo-Baconian inductivist standards whose demands for the validation of causal claims can generally not be met intraclinically, unless the psychoanalytic method is buttressed by a powerful substitute for the defunct Tally Argument. And in the absence of such a substitute, the epistemic decontamination of the bulk of the patient's productions on the couch from the suggestive effects of the analyst's communications appears to be quite utopian.

4. Insofar as the credentials of psychoanalytic theory are currently held to rest on clinical findings, as most of its official spokesmen would have us believe, the dearth of acceptable and probatively cogent clinical data renders these credentials quite weak. And without a viable alternative to the aborted Tally Argument having comparable scope and ambition, the future validation of Freudian theory, if any, will have to come very largely from *extra*-clinical findings.

5. Two years before his death, Freud invoked the *consilience* or convergence of clinical inductions (in the sense of William Whewell) to determine the probative cogency of the patient's assent or dissent in response to the interpretations presented by the analyst. (S.E. 1937, 23:257-269) But such a reliance on consilience is unavailing unless and until there emerges an as yet unimagined trustworthy method for epistemically decontaminating each of the *seemingly* independent consilient pieces of clinical evidence. For the methodological defects of Freud's "fundamental rule" of free association (S.E. 1923, 18:238; 1925, 20:41; 1940, 23:174) ingress *alike* into the interpretation of several of these prima facie independent pieces of evidence (e.g., manifest dream content, parapraxes, waking fantasies). And this multiple ingression renders the seeming consilience probatively spurious.

6. Given the aforestated dismal inductivist verdict on clinical testability, that traditional inductivist methodology of theory appraisal no more countenances the *clinical* validation of psychoanalysis than Popper does. (1962, p. 38, fn. 3) Hence the specifically clinical confirmations claimed by many Freudians but abjured as spurious by inductivist canons are unavail-

able as a basis for Popper's charge of undue permissiveness against an inductivist criterion of demarcation between science and nonscience. And I have already argued in section III above that the actual falsifiability of psychoanalysis undercuts his reliance on Freud's theory as a basis for claiming greater stringency for his criterion of demarcation.

Here I shall focus solely on the issue posed by the *third* of these six theses. In Grünbaum 1983a I argued in detail that there are fundamentally damaging flaws in the actual *clinical* arguments given by Freud and his disciples for the very foundation of his entire edifice: the theory of repression. Thus I claim to have shown there that the actual clinical evidence adduced by Freudians provides no cogent basis for the repression-aetiology of neuroses, for the compromise-model of manifest dream content, or for the causal attribution of parapraxes ("slips") to repressed ideation. And in Grünbaum 1983, sec. III, 3 I canvassed solid evidence for the considerable epistemic contamination and hence lack of probative value on the part of three major kinds of clinical findings that Freud deemed either initially exempt from such adulteration or certifiably not marred by it because of due precautions: the products of "free" association, the patient's assent to analytic interpretations that he (she) had initially resisted, and memories recovered from early life.

Indeed, the epistemic adulteration I documented there seems to be *ineradicable* in just those patient responses that are supposed to lay bare his repressions and disguised defenses after his resistances have been overcome. And yet Freud attributed pride of place to these very data in the validation of his theory of repression, a doctrine that is avowedly "the cornerstone on which the whole structure of psychoanalysis rests . . . the most essential part of it." (S.E. 1914, 14:16) Thus, generally speaking, clinical findings—in and of themselves—forfeit the probative value that Freud had claimed for them, although their potential heuristic merits may be quite substantial. To assert that the contamination of intraclinical data is *ineradicable* without extensive and essential recourse to *extra*-clinical findings is *not*, of course, to declare the automatic falsity of any and every analytic interpretation which gained the patient's assent by means of prodding from the analyst. But it *is* to maintain—to the great detriment of intraclinical testability!—that, in general, the epistemic devices confined to the analytic setting cannot reliably *sift* or decontaminate the clinical data so as to *identify* those that qualify as probative. And clearly all of these liabilities apply fully as much to Heinz Kohut's "self-psychology" version of psychoanalysis, which places the major determinants of adult personality structure into an even earlier phase of childhood than Freud did. (Cf. Meyers 1981; Basch 1980, chapter XI)

One must admire the strenuous and ingenious efforts made by Freud to legitimate his psychoanalytic method by arguing that it had precisely such an identifying capability. As I pointed out in Grünbaum 1983, these efforts included the attempt to vouchsafe the probity of free associations by secluding their *contents* in the bastion of *internal* causal relatedness. And Freud's dialectical exertions culminated in the generic underwriting of clinical investigations by the Tally Argument. But the empirical untenability of the cardinal premise of the Tally Argument that I have documented elsewhere (Grünbaum 1980, sec. II) has issued in the latter's collapse, leaving intraclinical validation defenseless against all of the sceptical inroads from the substantial evidence for the distortion and tailoring of its data by such mechanisms as suggestion.

Hence it is *unavailing* to take contaminated findings from the psychoanalytic interview more or less at face value, and then to try to employ them probatively in some testing strategy whose *formal* structure is rational enough as such. All the more so, since no viable substitute for the Tally Argument appears to be in sight. Indeed, the seeming ineradicability of epistemic contamination in the clinical data adduced as support for the cornerstones of the psychoanalytic edifice may reasonably be presumed to doom any prospects for the cogent intraclinical testing of the major tenets espoused by Freud.

These considerations can now be brought to bear on the scrutiny of Glymour's defense of clinical testability, which was outlined in our Section I above.

Glymour gives an illuminating reconstruction of Freud's account of the Rat Man case by means of the logical pincer-and-bootstrap strategy that Glymour had teased out of that account. I have no reason to gainsay this strategy in general, as far as it goes. But I shall now argue that—with or without it—strong reasons militate against the intraclinical testability of the specific aetiologic hypothesis that was at issue in the case of the Rat Man, Paul Lorenz, who suffered from an obsessional fear of rats.

At the time of the Rat Man case, Freud had postulated that premature sexual activity such as excessive masturbation, subjected to severe repression, is the specific cause of obsessional neurosis. As will be recalled from Section III, in his carefully defined usage of "specific cause," the claim that

X is the specific cause of Y entails unilaterally that X is causally *necessary* for Y. And the latter, in turn, unilaterally entails that all cases of Y were X's. Thus if *this particular consequence* of the conjectured sexual aetiology is to get confirmation from Lorenz's psychoanalysis, the intraclinical data yielded by it need to be able to certify the following: Lorenz, who was an adult victim of obsessional neurosis, engaged in precocious sexual activity, which was then repressed. Hence let us inquire first whether intraclinical data produced by the adult patient can *reliably* attest the actual occurrence of a childhood event of the stated sort. But, as I shall argue, even if the answer to this question were positive, this much would be quite insufficient to support Freud's aetiologic hypothesis that repressed precocious sexual activity is *causally relevant* to adult obsessional neurosis.

As Glymour (1980) notes, "Freud had . . . arrived at a retrodicted state of affairs, namely, the patient's having been punished by his father for masturbation." (p. 272) And indeed, "the crucial question is whether or not Lorenz was in fact punished by his father for masturbation." (p. 273) But Freud's specific aetiology of adult obsessional neurosis as such calls only for an early childhood event in which precocious sexual activity was repressed. Why then should it be probatively "crucial" whether it was the patient's *father* who was involved in the sexual event required by the hypothesized aetiology?

As is clear from Freud's account, the elder Lorenz's involvement became probatively weighty, because of the unconscious significance that psychoanalytic theory assigns to the patient's recollection of recurring fears of his father's death, at least after the age of six. While having these fears, the child Paul bore his father deep conscious affection. Freud derived the presumed unconscious origin of the fears from a theoretical postulate of so-called precise contrariety, which he took pains to communicate to the patient, who then became "much agitated at this and very incredulous." (S.E 1909, 10:180) Freud explains both his reasoning and revealingly relates his indoctrination of the patient:

> . . . he was quite certain that his father's death could never have been an object of his desire but only of his fear.—After his forcible enunciation of these words I thought it advisable to bring a fresh piece of theory to his notice. According to psycho-analytic theory, I told him, every fear corresponded to a former wish which was now repressed; we were therefore obliged to believe the exact contrary of what he had asserted. This would also fit in with another theoretical requirement, namely, that the unconscious must be the precise

contrary of the conscious.—He was much agitated at this and very incredulous. He wondered how he could possibly have had such a wish, considering that he loved his father more than any one else in the world. . . . I answered that it was precisely such intense love as this that was the necessary precondition of the repressed hatred. (S.E. 1909, 10:179-180)

Having thus theoretically inferred the patient's deep childhood grudge against his father from the recurring fears of losing the father, Freud also conjectured that the grudge remained so durably unconscious only because it was a response to the father's interference with the patient's sensual gratification.

And this conclusion was, then, serendipitous in suggesting that there had been an early event satisfying the specific aetiology that Freud had hypothesized for Lorenz's obsessional neurosis. Since this aetiology required precocious masturbation events, Freud retrodicted that the patient had been punished by his father for masturbation "in his very early childhood . . . before he had reached the age of six." (S.E. 1909, 10:183) Clearly the actual occurrence of an event having these attributes would *simultaneously* satisfy the initial condition of the postulated aetiology and explain Lorenz's early dread of his father's death via Freud's principle of precise contrariety.

Let us now suppose, just for argument's sake, that Freud's avowedly *well-coached* adult patient had actually reported having a memory of the very early childhood event that Freud had retrodicted. Then I ask: Could such a clinical event have reliably attested the actual occurrence of the distant event? I have framed this question hypothetically, because it so happened that Lorenz actually had no *direct* memory of any physical punishment by his father, let alone of a punishment for a *sexual* offense. He did remember having been *told* repeatedly by his *mother* that there had been *one* incident of angry conflict with his father at age three or four, when he was beaten by him. And when the mother was consulted whether this beating had been provoked by a misdeed of a sexual nature, her answer was negative. Furthermore, this was apparently the *only* beating the child had ever received from the father.

But for the purpose of our inquiry, we are positing that, at some point in his analysis, the patient had claimed to remember just the kind of early childhood event that Freud had retrodicted via his specific aetiology of obsessional neurosis. Then I am concerned to show that, taken by itself, such a finding would be quite insufficient to lend any significant support to the hypothesized aetiology of obsessional neurosis. And my reasons for this claim will then enable me to argue quite generally for the following conclusion: Given the demise of the Tally Argument, the intraclinical testing of the causal assertions made by Freud's specific aetiologies of the psychoneuroses, and by his ontogenetic developmental hypotheses, is *epistemically quite hopeless.*

Let "N" (neurosis) denote a psychoneurosis such as the syndrome of obsessional neurosis, while "P" (pathogen) denotes the kind of sex-related antecedent event that Freud postulated to be the specific cause of N. Thus I shall say that a person who had a sexual experience of the sort P "is a P." And if that person was then afflicted by N, I shall say that he was both a P and an N, or just "a PN." It is taken for granted, of course, that *there are* both N's and non-N's, as well as P's and non-P's. To support Freud's aetiologic hypothesis that P is causally necessary for N, evidence must be produced that being a P *makes a difference* to being an N. But such causal relevance is *not* attested by *mere* instances of N that were P's, i.e., by patients who are both P's and N's. For even a large number of such cases does not preclude that just as many *non*-P's would also become N's, if followed in a horizontal study from childhood onward! Thus instances of N that were P's may just *happen* to have been P's, whereas being a P has no aetiologic role at all in becoming an N.

A telling, sobering illustration of this moral is given by the following conclusion from a review of forty years of research (Frank, 1965):

> No factors were found in the parent-child interaction of schizophrenics, neurotics, or those with behavior disorders which could be identified as unique to them or which could distinguish one group from the other, or any of the groups from the families of the [normal] controls (p. 191).

Hence it is insufficient evidence for causal relevance that any N who turns out to have been a P does instantiate the retrodiction "All N's were P's," which is entailed by Freud's specific aetiology. Thus, to provide evidence for the causal relevance claimed by Freud, we need to *combine* instances of N that were P's with instances of non-P who are *non*-N's. And indeed, since he deemed P to be causally necessary for N—rather than just causally relevant—his aetiology requires that the class of non-P's should not contain *any* N's whatever, whereas the class of P's is to have a positive (though numerically unspecified) incidence of N's.

One can grant that since "All N's are (were) P's" is logically equivalent to "All non-P's are (will be) non-N's," any case of an N who was a P will support the latter to whatever extent it supports the former. But this fact is unavailing to the support of Freud's aetiology. For the issue is *not* merely to provide evidential support for "All non-P's are (will be) non-N's," or for its logical equivalent, by some instances or other. Instead the issue is to furnish evidential support for the (strong kind of) *causal relevance* claimed by Freud. And, for the reasons I have given, the fulfillment of that requirement demands that there be cases of non-P's that are non-N's no less than instances of N's that were P's. Yet *at best*, the Rat Man could furnish only the *latter* kind of instance. In other words, if we are to avoid committing the fallacy of post hoc ergo propter hoc, we cannot be content with instances of N's that were P's, no matter how numerous. Analogously, suppose it were hypothesized that drinking coffee is causally relevant to overcoming the common cold. And consider the case of a recovered cold-sufferer who retrospectively turns out to have been drinking coffee while still afflicted with the cold. Then such an instance, taken by itself, would hardly qualify as *supportive* of the hypothesized causal relevance.

Psychoanalytic theory and therapy have encouraged the disregard and even flouting of the elementary safeguards against the pitfalls of causal inference familiar since the days of Francis Bacon, not to speak of J. S. Mill. Yet even informed laymen in our culture are aware that such safeguards are indeed heeded *in medicine* before there is public assent to the validity of such aetiologic claims as "heavy tobacco smoking causes cardiovascular disease." This double standard of evidential rigor in the validation of aetiologic hypotheses even makes itself felt in current criminal law. For legal prohibitions—and so-called expert psychiatric testimony in courts of law—are sometimes predicated on such hypotheses even when their credentials are no better than that blithe repetition has turned them into articles of faith. The recently publicized reiteration of the purported pathogenicity of child molestation in opposition to its decriminalization is a case in point.

In our society, such sexual molestation is often alleged to be pathogenic, even when it is affectionate and tender rather than violent. And this allegation has been invoked to justify making it illegal and fraught with substantial penalties. Yet recently a number of sexologists have maintained that very young children should be allowed, and perhaps even encouraged, to have sex with adults, unencumbered by interference from the law. In their view, such activity itself is harmless to the child and becomes harmful only when parents raise a fuss about it. Indeed, *some* of these advocates have made the daring and quite unfashionable aetiologic claim that unless children do have early sex, their psychological development will go awry. And even the less daring champions of harmlessness are opposed to jailing affectionate pedophiles.

Reasons of elementary prudence and also of humaneness make it a good policy, in my view, to put the burden of proof on those who maintain that affectionate and tender child molestation is *not* even distressing to the child, let alone pathogenic. But a cautionary basis for a legal prohibition is a far cry from the confident assertion of demonstrated pathogenicity. And the difference between mere caution and authenticated causation of neurosis may, of course, be relevant to the severity of the punishment appropriate for violations of the interdiction.

In a recent issue of *Time* magazine, John Leo inveighs aetiologically *against* the demand to legalize tender pedophilia, which he sees as a thinly disguised manifesto for child molesters liberation. And the justification offered by him for his indictment is as follows:

> Unfortunately, few responsible child experts have reacted . . . so far to the radical writing on child sex. One who has is Child Psychiatrist Leon Eisenberg of Children's Hospital Medical Center, Boston: "Premature sexual behavior among children in this society almost always leads to psychological difficulties because you have a child acting out behavior for which he is not cognitively or emotionally ready. . . .
> Psychotherapist Sam Janus, author of a new book, *"The Death of Innocence,"* says that people who were seduced early in life "go through the motions of living and may seem all right, but they are damaged. I see these people year after year in therapy." U.C.L.A. Psychiatrist Edward Ritvo also says that much of his work is with children who have been involved in catastrophic sexual situations. His conclusion: "Childhood sexuality is like playing with a loaded gun." (September 7, 1981, p. 69)

But the aetiologic reasoning of those whom *Time* cites to document the pernicious effects of child molestation is just as shoddy as the causal inferences of those advocates of pedophilia who claim dire psychological consequences from the *failure* of infant boys to act on their erections, and of infant girls to utilize their vaginal lubrications. For the findings adduced by *Time* do not answer either of the following two questions:

(1) Is the occurrence of childhood seduction not equally frequent among those who are well enough never to see a psychotherapist? In the parlance of John Stuart Mill, this question calls for the use of the *joint* method of agreement and difference, rather than just the heuristic method of agreement.

(2) Would a representative sample of those who were *not* seduced in childhood have a significantly *lower* incidence of adult neurosis than those who *were* seduced? By the same token, we must ask those who claim seduction to be *beneficial* psychologically to show that those who were indeed seduced *fared better* psychologically than those who were not sexually active in this way. Without the appropriate answers to these questions, the respective assertions of causal relevance remain gratuitous.

Thus we must ask those who *condemn* childhood seduction the foregoing questions, because it may be that childhood seduction just *happens* to be quite common among neurotics, even though it has no aetiologic role in the production of neurosis. In that case, the same people would have become neurotics anyway, in the absence of early seduction. Without answers to these questions, the evidence given by those whom *Time* invokes as authorities merely suggests the bare *possibility* that childhood seduction is pathogenic. By the same token, psychoanalysts have overlooked the fact that repressed homosexual feelings cannot be shown to be the pathogen of adult paranoia, by merely pointing to the frequency of homosexually-tinged themes in the associative output of paranoiacs during analysis. This finding does not tell us whether homosexual themes would not likewise turn up to the same extent in the so-called free associations of nonparanoiacs who lead well-adjusted lives and never see a therapist. Here no less than in the case of the Rat Man, the invocation of J. S. Mill's heuristic method of agreement is not enough to lend support to the hypothesis of aetiologic relevance.

Hence even if the Rat Man did in fact have the sexually repressive experience P retrodicted via Freud's aetiology of obsessional neurosis, this alone would hardly qualify as evidential support for that aetiology. And there is a further reason for concluding that even if the child Paul Lorenz had actually been punished by his father for masturbating, as retrodicted via Freud's aetiology, this putative occurrence would confer little, if any, support on this aetiology. For, as Ronald Giere has remarked (in a private communication), the occurrence of this sort of event is to be routinely expected in the Victorian child-rearing culture of the time on grounds *other than* psychoanalytic theory.

Moreover, Freud had made the adult Rat Man patient well aware, as we saw, of the inferences that Freud had drawn about his childhood via psychoanalytic theory. Given the substantial evidence I adduced for the notorious docility of patients in analysis (Grünbaum 1983), I submit that one ought to discount Lorenz's *putative* early childhood memory as too contaminated to attest reliably the actual occurrence of the retrodicted early event.

It can be granted, of course, that requirements of consistency or at least overall coherence do afford the analyst *some* check on what the patient alleges to be bona fide memories. But Freud's own writings attest to the untrustworthiness of purported adult memories of early childhood episodes that had presumably been repressed in the interim and then retrieved by the analysis. (Cf. the documentation in Grünbaum 1980, p. 353) Indeed, the malleability of adult memories from childhood is epitomized by a report from Jean Piaget (Loftus 1980, pp. 119-121), who thought he remembered vividly an attempt to kidnap him from his baby carriage along the Champs Elysées. He recalled the gathered crowd, the scratches on the face of the heroic nurse who saved him, the policeman's white baton, the assailant running away. However vivid, Piaget's recollections were false. Years later the nurse confessed that she had made up the entire story, which he then internalized as a presumed experience under the influence of an authority-figure.

The discounting of the Rat Man's putative early childhood memory is hardly a general derogation of the reliability of adult memories in ordinary life. But in the clinical context, the *posited* memory is simply not sufficiently dependable to qualify as evidence for the retrodicted event. Thus the retrospective intra-clinical ascertainment of the actual occurrence of the retrodicted distant event is just too unreliable. And, in general, the patient's memory may simply fail to recall whether the pertinent event did occur, as Freud himself stressed. (S.E. 1920, 18:18; 1937, 23:265-66) Indeed, even in survey studies of lung cancer patients who are asked about their prior smoking habits, and of heroin addicts who are questioned about previous use of marijuana, the retrospective ascertainment of the actual occurrence of the suspected causal condition is epistemically flawed. (Giere 1979, pp. 216, 265)

Have I provided adequate grounds for maintaining that long-term *prospective* studies, which employ control groups and spring the clinical confines of the usual psychoanalytic setting, must supplant the *retrospective* clinical testing of aetiology defended by Glymour? Not just yet. For suppose that analysts could secure reasonable numbers of patients who, though presumed to need analysis for some neurotic affliction or other, are certifiably free of the *particular* neurosis N (say, obsessional neurosis) whose aetiology is currently at issue. Since neuroses usually occur in mixed rather than pure form, this is a generous assumption. All the same, let us operate with it. Then if we are given such patients who, though neurotic, are non-N, Freud's pertinent specific aetiology does *not* retrodict whether patients of *this* sort were P's or non-P's. For his hypothesized pathogenesis allows given non-N's to have been P's no less than to have been non-P's, although it does require any non-P to become a non-N. Now postulate, for argument's sake that, though retrospective, psychoanalytic inquiry *were* typically able to ascertain *reliably* whether a given case of non-N was indeed a non-P or a P. If so, then non-N's who putatively turn out to have been P's would merely be compatible with Freud's aetiologic hypothesis instead of supporting it, since this hypothesis allows these instances without requiring them.

But what of patients who are *neither* N's nor P's? Would such people, together with other persons who are both N's and P's, jointly bespeak that P is pathogenic for N (*obsessional* neurosis) within the class of all persons?

Note that the clinical testing design I have envisaged for scrutiny is *confined* to the class of neurotics. For even the non-N's of this design are presumed to be afflicted by some neurosis other than N. The reason is that persons who have practically no neuroses of any sort are hardly available to analysts in sufficient numbers to carry out the putative retrospective determination of whether they were non-P's or P's. But, as Mr. Blake Barley has noticed, the confinement of this retrospective clinical determination to the class of neurotics has the following consequence: Even if every observed non-N (non-obsessive neurotic) is a non-P while every observed N is a P, these combined instances lend credence only to the hypothesis that, *within* the class of neurotics, P is aetiologically relevant to N. But these putative combined instances do not support the Freudian claim of such aetiologic relevance within the wider class of persons.

In short, the Freudian clinical setting does *not* have the epistemic resources to warrant that P is *neurotogenic*. And this unfavorable conclusion emerges even though it was granted, for argument's sake, that the retrospective methods of psychoanalytic inquiry can determine *reliably* whether adult neurotics who are non-obsessives were non-P's in early life.

But is it reasonable to posit such reliability? It would seem not. For clearly, even if the patient believes to have the required memories, the retrospective clinical ascertainability of whether a given non-N was actually a non-P is epistemically on a par with the psychoanalytic determination of whether a given N was a P. And, as we saw, the latter is unreliable. Moreover, as Freud himself acknowledged, "the patient cannot remember the whole of what is repressed in him, and what he cannot remember may be precisely the essential part of it." (S.E. 1920, 18:18)

Now contrast the stated epistemic liabilities of the retrospective psycho-analytic inference that a given adult patient was or was not a P during his early childhood with the assets of *prospective* controlled inquiry: A *present* determination would be made, under suitably supervised conditions, that children in the experimental and control groups are P's and non-P's respectively; again, during long-term follow-ups, later findings as to N or non-N would also be made in what is then the present.

Recently experimental validations of therapeutic efficacy have been carried out by using the response history of single individuals *without* control groups drawn from other individuals. (Hersen and Barlow 1976; Kazdin 1981; Kazdin, in press) Thus in these validations, the *causal* claims inherent in the pertinent assertions of therapeutic efficacy have been validated by single-case experimental designs. Hence it behooves us to ask whether these "*intra*subject" validations could become prototypic for using a given analysand to test *intra*clinically the causal assertions made by the long-term aetiologic hypotheses of psychoanalytic theory and by such claims of efficacy as are made for its avowedly slow therapy. To answer this question, let us first look at situations in physics in which the *probative equivalent* of controlled experiments is furnished by other means.

When a billiard ball at rest on a billiard table suddenly acquires momentum upon being hit by another billiard ball, we are confident that the acceleration of the first ball is due to the impact of the second. And even more importantly, astronomers made sound causal claims about the motions of planets, binary stars, etc. before they were able to manipulate artificial earth satellites, moon probes, or interplanetary rockets. What took the probative place of control groups in these cases? In the case of the billiard ball, Newton's otherwise well-supported first law of motion gives us background knowledge as to the "natural history" of an object initially at rest that is not exposed to external forces: Such an object will remain at rest. And this information, or the law of conservation of linear momentum, enables us to point the finger at the moving second billiard ball to furnish the cause of the change in the momentum of the first. A similar reliance on otherwise attested background knowledge supplies the probative equivalent of experimental controls in the astronomical cases.

Turning to the *single*-case validations of therapeutic efficacy, they pertain to the following sort of instance:

> . . . a seven-year-old boy would beat his head when not restrained. His head was covered with scar tissue and his ears were swollen and bleeding. An extinction procedure was tried: the boy was allowed to sit in bed with no restraints and with no attention given to his self-destructive behavior. After seven days, the rate of injurious behavior decreased markedly, but in the interim the boy had engaged in over ten-thousand such acts, thus making the therapists fearful for his safety. A punishment procedure was subsequently introduced in the form of one-second electric shocks. In a brief time, the shock treatment dramatically decreased the unwanted behavior. (Erwin 1978, pp. 11-12)

Here the dismal prospects of an untreated autistic child are presumably known from the natural history of other such children. And, in the light of this presumed background knowledge, the dramatic and substantial behavior change ensuing shortly after electric shock allowed the attribution of the change to the shock without control groups. For, under the circumstances, the operation of *other* causal agencies seems very unlikely. More generally, the *paradigmatic* example of an *intra*subject clinical validation of the causal efficacy of a given intervention is furnished by the following *variant* of using the single patient as his own "historical" control: (i) The natural history of the disorder is presumably otherwise known, *or* (ii) the therapist intervenes only in on-off fashion, and this intermittent intervention is found to yield alternating remissions and relapses with dramatic rapidity.

Can the causal validation designs employed in these intrasubject clinical tests of therapeutic efficacy become prototypic for using an individual analysand to validate Freud's *long*-term aetiologic hypotheses, or to furnish evidence that an analysis whose typical duration extends over several years deserves credit for any therapeutic gain registered by the patient after, say, four years? To ask the question is to answer it negatively. The natural history of a person *not* subjected to the experiences deemed pathogenic by Freudian theory is *notoriously* unknown! And as for crediting therapeutic gain to analytic intervention on the basis of an intrasubject case history, how could such an attribution possibly be made in the face of Freud's own acknowledgment of the occurrence of *spontaneous* remissions (S.E. 1926, 20:154)? At best, Freudians can hope to show that the incidence of therapeutic gain in groups of patients who undergo analysis exceeds the spontaneous remission rate in untreated control groups belonging to the same nosologic category (Rachman and Wilson 1980). In short, the stated intrasubject validation by means of dramatic therapeutic gains can hardly be extrapolated to underwrite the prospective single-case evaluation of slow analytic therapy, let alone to vindicate the *retrospective* testing of a Freudian aetiology in the course of an individual analysis.

Although Freud's specific aetiologies did nòt specify numerically the percentage of P's who become N's, it is noteworthy that only prospective investigation can yield the information needed for such a statistical refinement. For let us suppose that retrospective data confirm the retrodiction of Freud's specific aetiology that the incidence of P's within the sample group of N's is 100 percent. Then this percentage incidence clearly does not permit an inference as to the percentage incidence of N's within the class of P's. Yet such information is clearly desirable, if only in order to estimate the probability that a child who was subjected to P will become an N. More generally, when P is not deemed causally necessary for N, but merely causally relevant, retrospective data simply do not yield any estimates of P's degree of causal effectiveness. (Giere 1979, pp. 274, 277)

Our inquiry into the Rat Man case so far has operated with a *counterfactual* posit in order to discuss the reliability of clinical data by reference to this case. The *hypothetical* clinical datum we used was that the patient *had* reported having a memory of the early childhood event retrodicted by Freud. As against Glymour's generic thesis that the specific psychoanalytic aetiologies can be cogently tested "on the couch," I have argued that, at least typically, such testing is epistemically quite hopeless. And hence it would seem that Paul Lorenz's psychoanalysis would have completely failed to furnish evidential support for the *aetiologic relevance* of childhood sexual repression to obsessional neurosis, even if Paul's father had reliably reported having repeatedly punished his young son for masturbation.

Incidentally, when Waelder defended the clinical confirmation of the psychoanalytic aetiologies, he overlooked precisely that their substantiation requires evidential support for the *causal relevance* of the purportedly pathogenic experience, and not merely the historical authentication of the bare occurrence of that experience. (1962, pp. 625-626)

Let us return to Glymour's account of the testing strategy in Paul Lorenz's analysis, which was predicated on Lorenz's failure, in actual fact, to have any *direct* recall of receiving a punishment from his father, let alone a castigation for a sexual offense. Therefore let us now see how Glymour evaluated the probative significance of this finding. I shall be concerned to stress the scope that Glymour does give to *essential* reliance on *extra*clinical data for probative purposes. Indeed, it will turn out that the entire testing procedure in the Rat Man case comes out to be probatively *parasitic* on an extraclinical finding. And hence I wonder how Glymour can see himself as rebutting Eysenck's denial of intraclinical testability, although he does succeed in impugning the demand that all extraclinical disconfirmation be *experimental*.

By Glymour's own account of the Rat Man case, the probatively "crucial" data came from the *extra*clinical testimony of the patient's mother. On Glymour's reading of Freud, at the time of Lorenz's analysis, Freud still postulated *actual* rather than fancied early sexual experiences to be the pathogens of obsessional neurosis. (Glymour 1980, pp. 274-275) And as Glymour explains lucidly, what made Lorenz's case a *counterexample* to this aetiology was *not* the mere failure of the patient to recall the event retrodicted by Freud. Instead, it was the *extra*clinical testimony from the *mother* that had this negative probative import. (p. 273) For it was her testimony that supplied the probatively crucial datum by contravening Freud's retrodiction, when she answered the question that Glymour characterized as "the crucial question." And he himself characterizes "the memory of an adult observer"—in this case that of the mother—as "the most reliable available means" for answering this decisive question as to the character of the offense for which the child Paul had been punished. (p. 273) How, then, in the face of the *extra*clinical status of the *decisive* datum, can Glymour justify his description of the testing rationale used in the Rat Man case as "a strategy that relies almost exclusively on clinical evidence"? (Glymour 1974, p. 287)

It is true enough that, as we know from the case history of the Wolf Man, Freud regarded stories told by other members of the family to the patient about his childhood to be generally "absolutely authentic" and hence as admissible data. (S.E. 1918, 17:14 and fn. 2) But Freud completes this assertion by cautioning that responses by relatives to pointed inquiries from the analyst—or from the patient while in analysis—may well be quite contaminated by misgivings on their part:

> So it may seem tempting to take the easy course of filling up the gaps in a patient's memory by making enquiries from the older members of his family; but I cannot advise too strongly against such a technique. Any stories that may be told by relatives in reply to enquiries and requests are at the mercy of every critical misgiving that can come into play. One invariably regrets having made oneself dependent upon such information; at the same time confidence in the analysis is shaken and a court of appeal is set up over it. Whatever can be remembered at all will anyhow come to light in the further course of analysis. (S.E. 1918, 17:14, fn. 2)

Even if one were to discount Freud's caveat, several facts remain: (i) it is misleading to claim intraclinical testability if, as in the Rat Man case, the avowedly crucial datum does *not* come from "the couch." (ii) What makes the reliance on extra-clinical devices important is that, far from being marginal epistemically, its imperativeness derives from the typically present probative defects of the analytic setting, defects that are quite insufficiently acknowledged by Glymour. And in my view, it does not lessen the liabilities of intraclinical testing that the compensations for its deficits from *outside* the clinical setting *may occasionally* be available in situ (e.g. from family records) and thus do not necessarily have to require the experimental laboratory. For even when supplemented by such nonexperimental, extra-clinical devices, the thus enlarged "clinical" testing procedure is not adequate or epistemically autonomous. For example, when it becomes necessary to resort to extraclinical information for the sort of reason that was operative in the Rat Man case, it will be a matter of mere happenstance whether suitable relatives are even available, let alone whether they can *reliably* supply the missing essential information. Why then dignify as a "clinical testing strategy" a procedure of inquiry that depends on such contingent good fortunes and hence, when luck runs out, cannot dispense with experimental information? (iii) The real issue is whether the clinical setting *typically*—rather than under contingently favorable circumstances—does have the epistemic resources for the cogent validation of the aetiology at issue in the Rat Man case, and of other analytic aetiologies. In dealing with that issue, Glymour's otherwise illuminating account has not demonstrated the existence of a cogent intraclinical testing strategy, even if he succeeded in showing that extraclinical compensations for its lacunae need not be wholly experimental.

Indeed, the extent of his essential epistemic reliance on extraclinical findings can now be gauged from his view of the effect that Freud's modifications of the specific sexual aetiology of obsessional neurosis (and of other neuroses) had on the *testability* of these evolving aetiologic hypotheses. Glymour (1980) recounts this evolution:

> After the turn of the century and before 1909, . . . there is no statement of the view that sexual phantasies formed in childhood or subsequently, having no real basis in fact, may themselves serve *in place of* sexual experiences as etiological factors. . . . Yet after the Rat Man case the view that either infantile sexual experiences *or* phantasies of them may equally serve as etiological factors became a standard part of Freud's theory. In *Totem and Taboo*, four years after the Rat Man case appeared, Freud emphasized that the guilt that obsessional neurotics feel is guilt over a happening that is physically real but need not actually have occurred [fn. omitted]. By 1917 Freud not only listed phantasies themselves as etiological factors alternative to real childhood sexual experiences, but omitted even the claim that the former are usually or probably based on the latter [fn. omitted]. The effect of these changes is to remove counterexamples like that posed by the Rat Man case, but at the cost of making the theory less easily testable. For whereas Freud's theories, until about 1909, required quite definite events to take place in the childhood of a neurotic, events that could be witnessed and later recounted by adults, Freud's later theory required no more than psychological events in childhood, events that might well remain utterly private. (pp. 276-27)

Thus Glymour attributes the diminishing testability of Freud's modified aetiologies quite rightly to the lessening *extra*clinical ascertainability of the sorts of events that Freud successively postulated as aetiologic. But if the testability of the psychoanalytic aetiologies is in fact "almost exclusively" intraclinical, as Glymour told us, why should it be *vital* for their testability that the aetiologic events required by Freud's later theory are just mental states of the patient to which only the patient himself and his analyst become privy?

Incidentally, the problem of testing Freud's sexual aetiology of the neuroses—either clinically or extraclinically—became less well defined after he gave up the quest for qualitatively *specific* pathogens of nosologically distinct psychoneuroses in favor of a generic Oedipal aetiology for all of them. In fact, he used the analogy of explaining the great qualitative differences among the chemical substances by means of quantitative variations in the proportions in which the same elements were combined. But having thus dissolved his prior long-standing concern with the problem of "the choice of neurosis," he was content to leave it at vague metapsychological remarks about the constant intertransformation of "narcissistic libido" and "object libido." (S.E. 1925, 20:55-56)

What of Glymour's reliance on *intra*clinical data? In that context, he seems to have taken much too little cognizance of even the evidence furnished by analysts that intraclinically the suggestibility problem is radically unsolved, if not altogether insoluble because there is no viable substitute for the defunct Tally Argument. Can we place any stock in Glymour's aforecited aspiration that clinicans can be "trained so as not to elicit by suggestion the expected responses from their patients"? In view of the evidence for the *ineradicability* of suggestive contamination, it would now seem that this hope is sanguine to the point of being quite utopian. In an Afterword to his (1974) for a second edition of the Wollheim volume in which it first appeared, Glymour (1982) has reacted to some of these particular doubts as follows:

> I do not see . . . that the experimental knowledge we now have about suggestibility requires us to renounce clinical evidence altogether. Indeed, I can imagine circumstances in which clinical evidence might have considerable force: when, for example, the clinical proceedings show no evident sign of indoctrination, leading the patient, and the like; when the results obtained fall into a regular and apparently law-like pattern obtained independently by many clinicians; and when those results are contrary to the expectation and belief of the clinician. I do not intend these as *criteria* for using clinical evidence, but only as indications of features which, in combination, give weight to such evidence.

To this I say the following:

(1) I do *not* maintain that any and all clinical data are altogether irrelevant probatively. Instead, I hold that such findings cannot possibly bear the probative burden placed upon them by those who claim, as Glymour did, that psychoanalysis can TYPICALLY be validated or invalidated "on the couch," using a clinical testing strategy that is mainly confined to the analytic setting.

(2) The existence of *some* circumstances under which we would be warranted in not renouncing clinical evidence "altogether" is surely not enough to sustain clinical testing as a largely cogent and essentially autonomous scientific enterprise. And as for Glymour's illustrations of such circumstances, I cannot see that absence of evident indoctrination, or regular concordance among the results obtained independently by many clinicians exemplify circumstances under which "clinical evidence might have considerable force." For—apart from the arguments I gave against these illustrations, if only à propos of "free" association (Grünbaum 1983)—it seems to me that their utopian character as a step toward solving the compliance problem is epitomized by the following sobering results, which are reported by the analyst Marmor (1962):

> . . . depending upon the point of view of the analyst, the patients of each [rival psychoanalytic] school seem to bring up precisely the kind of phenomenological data which confirm the theories and interpretations of their analysts! Thus each theory tends to be self-validating. Freudians elicit material about the Oedipus Complex and castration anxiety, Jungians about archetypes, Rankians about separation anxiety, Adlerians about masculine strivings and feelings of inferiority, Horneyites about idealized images, Sullivanians about disturbed interpersonal relationships, etc." (p. 289)

(3) I do not deny at all that *now and then* clinical results "are contrary to the expectations and belief of the clinician." But as a step toward vindicating clinical inquiry qua epistemically autonomous testing strategy, I can only say "One swallow does not a summer make."

What seems to me to emerge from Glymour's interesting reconstruction is that, on the whole, data from the couch *acquire* probative significance when they are independently corroborated by extraclinical findings, or are inductively consilient with such findings in Whewell's sense. Thus I do not maintain that any and all clinical data are altogether irrelevant probatively. But this much only conditionally confers *potential* relevance on intraclinical results beyond their heuristic value. And surely this is not enough to vindicate testability on the couch in the sense claimed by its Freudian exponents, and countenanced by Glymour.

Note

1. Sigmund Freud, Notes upon a case of obsessional neurosis. In *Standard Edition of the Complete Psychological Works of Sigmund Freud*, trans. J. Strachey et al. (London: Hogarth

Press, 1955), 10:155-318. This paper first appeared in 1909. Hereafter any references given to Freud's writings in English will be to this *Standard Edition* under its acronym "S.E." followed by the year of first appearance, the volume number, and the page(s). Thus the 1909 paper just cited in full would be cited within the text in abbreviated fashion as follows: S.E. 1909, 10:155-318.

References

Basch, M. 1980. *Doing Psychotherapy*. New York: Basic Books.

Cooper, A. M. and Michels, R. 1978. An Era of Growth. In *Controversy in Psychiatry*, J. P. Brady and H. K. H. Brodie, ed. Philadelphia: W. B. Saunders, pp. 369-385.

Erwin, E. 1978. *Behavior Therapy*. New York: Cambridge University Press.

Eysenck, H. 1963. *Uses and Abuses of Psychology*. Baltimore: Penguin.

Fisher, S. and Greenberg, R. P. 1977. *The Scientific Credibility of Freud's Theory and Therapy*. New York: Basic Books.

Frank, G. H. 1965. The Role of the Family in the Development of Psychopathology. *Psychological Bulletin* 64:191-205.

Giere, R. 1979. *Understanding Scientific Reasoning*. New York: Holt, Rinehart and Winston.

Glymour, C. 1974. Freud, Kepler and the Clinical Evidence. In *Freud*, R. Wollheim, ed. New York: Anchor Books, pp. 285-304.

Glymour, C. 1980. *Theory and Evidence*. Princeton: Princeton University Press.

Glymour, C. 1982. Afterword to Glymour (1974). In *Philosophical Essays on Freud*, R. Wollheim and J. Hopkins, ed. New York: Cambridge University Press, pp. 29-31.

Grünbaum, A. 1976. Is Falsifiability the Touchstone of Scientific Rationality? Karl Popper Versus Inductivism. In *Essays in Memory of Imre Lakatos*, ed. R. S. Cohen, P. K. Feyerabend, and M. W. Wartofsky, pp. 215-229. Boston Studies in the Philosophy of Science, vol. 38. Dordrecht: D. Reidel.

Grünbaum, A. 1977. How Scientific is Psychoanalysis? In *Science and Psychotherapy*, R. Stern *et al.*, ed. New York: Haven Press, pp. 219-254.

Grünbaum, A. 1979a. Is Freudian Psychoanalytic Theory Pseudo-Scientific by Karl Popper's Criterion of Demarcation. *American Philosophical Quarterly* 16:131-141.

Grünbaum, A. 1979b. Epistemological Liabilities of the Clinical Appraisal of Psychoanalytic Theory. *Psychoanalysis and Contemporary Thought* 2:451-526.

Grünbaum, A. 1980. Epistemological Liabilities of the Clinical Appraisal of Psychoanalytic Theory. *Noûs* 14:307-385. This is an *enlarged* version of Grünbaum (1979b).

Grünbaum, A. 1983. The Foundations of Psychoanalysis. In *Mind and Medicine: Explanation and Evaluation in Psychiatry and the Biomedical Sciences*. ed. L. Laudan. Pittsburgh Series in Philosophy and History of Science, vol. 8. Berkeley and Los Angeles: University of California Press.

Grünbaum, A. 1983a. Logical Foundations of Psychoanalytic Theory. In *Festschrift for Wolfgang Stegmüller*, eds. W. K. Essler and H. Putnam, Boston: D. Reidel. To appear likewise in *Erkenntnis*.

Grünbaum, A. 1983b. Freud's Theory: The Perspective of a Philosopher of Science. Presidential Address to the American Philosophical Association (Eastern Division, 1982). *Proceedings and Addresses of the American Philosophical Association* 57, No. 1 (in press).

Grünbaum, A. 1984. *The Foundations of Psychoanalysis: A Philosophical Critique*. Berkeley and Los Angeles: University of California Press.

Herson, M. and Barlow, D. H. 1976. *Single Case Experimental Designs*. New York: Pergamon Press.

Jones, E. 1959. Editorial Preface. S. *Freud, Collected Papers*, vol. 1. New York: Basic Books.

Kaplan, A. H. 1981. From Discovery to Validation: A Basic Challenge to Psychoanalysis. *Journal of the American Psychoanalytic Association* 29:3-26.

Kazdin, A. 1981. Drawing Valid Inferences from Case Studies. *Journal of Consulting and Clinical Psychology* 49:183-192.

Kazdin, A. Forthcoming. Single-Case Experimental Designs. In *Handbook of Research Methods of Clinical Psychology,* ed. P. C. Kendall and J. N. Butcher. New York: John Wiley and Sons.

Loftus, E. 1980. *Memory*. Reading, Mass.: Addison-Wesley.

Luborsky, L. and Spence, D. P. 1978. Quantitative Research on Psychoanalytic Therapy. In *Handbook of Psychotherapy and Behavior Change,* S. L. Garfield and A. E. Bergin, ed. 2nd ed. New York: John Wiley and Sons, pp. 331-368.

Marmor, J. 1962. Psychoanalytic Therapy as an Educational Process. In *Psychoanalytic Education,* ed. J. Masserman. Science and Psychoanalysis, vol. 5. New York: Grune and Stratton, pp. 286-299.

Meyers, S. J. 1981. The Bipolar Self. *Journal of the American Psychoanalytic Association* 29: 143-159.

Ornstein, P. H., ed. 1978. *The Search for the Self; Selected Writings of Heinz Kohut: 1950-1978,* 2 vols. New York: International Universities Press.

Popper, K. R. 1962. *Conjectures and Refutations*. New York: Basic Books.

Popper, K. R. 1974. Replies To My Critics. In *The Philosophy of Karl Popper,* ed. P. A. Schilpp, Book 2. LaSalle, IL: Open Court, pp. 961-1197.

Rachman, S. J. and Wilson, G. T. 1980. *The Effects of Psychological Therapy*. 2nd enlarged ed. New York: Pergamon Press.

Waelder, R. 1962. Review of *Psychoanalysis, Scientific Method and Philosophy,* S. Hook, ed. *Journal of the American Psychoanalytic Assn.,* vol. 10, pp. 617-637.

Subjectivity in Psychoanalytic Inference: The Nagging Persistence of Wilhelm Fliess's Achensee Question

An alternative subtitle to this essay, which my non-Freudian Minnesota colleagues urged upon me, would have been, "Whose mind does the mind-reader read?" To motivate discussion of a topic not deemed important by some today, consider the story of the last "Congress" between Freud and Fliess, the rupture of their relationship at Achensee in the summer of 1900—the last time the two men ever met, although an attenuated correspondence continued for a couple of years more. Setting aside the doubtless complex psychodynamics, and the prior indications (from both content and density of correspondence) that the relationship was deteriorating, I focus on the intellectual content of the final collision. Fliess had attacked Freud by saying that Freud was a "thought reader" who read his own thoughts into the minds of his patients. Freud correctly perceived that this choice of content for the attack was deadly, that it went for the jugular. Freud's letter to Fliess after the meeting (Freud 1954) indicates that Fliess had written, apparently to soften the blow of the criticism, something about "magic," which Freud again refused to accept and referred to as "superfluous plaster to lay to your doubts about thought reading." (p. 330) A year later Freud is still focusing on the thought-reading accusation, and writes, "In this you came to the limit of your penetration, you take sides against me and tell me that 'the thought-reader merely reads his own thoughts into other people,' *which deprives my work of all its value* [italics added]. If I am such a one, throw my everyday-life [the parapraxis book] unread into the wastepaper basket." (p. 334) In a subsequent letter Freud quotes himself as having exclaimed at Achensee, "But you're

Expansion of a paper read at the Confirmation Conference; an earlier version was prepared for and presented in the Lindemann Lecture Series celebrating the 25th anniversary of the founding of the Institute for Advanced Psychological Studies at Adelphi University (March 11, 1977).

undermining the whole value of my work." (p. 336) He says that an interpretation of Fliess's behavior made the latter uncomfortable, so that he was "ready to conclude that the 'thought-reader' perceives nothing in others but merely projects his own thoughts into them... *and you must regard the whole technique as just as worthless as the others do.*" (p. 337) (Italics added)

Not to belabor the point, it seems that when Fliess wanted to hurt, he knew precisely what was the tender spot, and so did Freud. So that in addressing myself to this vexed topic of the subjectivity of psychoanalytic inference, I am at least in good company in thinking it important. Surely it is strange that four-fifths of a century after the publication of the *Interpretation of Dreams* it is possible for intelligent and clinically experienced psychologists to reiterate Fliess's Achensee question, and it is not easy to answer it.

One has the impression that the epistemology of psychoanalytic inference is less emphasized today than it was in Freud's writings, or in the discussions as recorded in the minutes of the Vienna Psychoanalytic Society. Despite the scarcity of psychoanalytic tapes and protocols (relative to, say, Rogersian and rational-emotive modes), and the lack of any verbatim recordings from the early days, it seems safe to say that the kinds of inferences to unconscious *content* and *life history episodes* that so fascinated Freud, and played the dominant role in his technique, are much less emphasized today. We cannot ignore the fact that Freud considered the dream book his best book. Why is there less emphasis upon discerning the hidden meaning, whether in the restrictive sense of "interpretation" or the more complicated sense of a "construction," than there used to be? I suppose one reason is the tendency among analysts to say, "Well, we don't worry as much about it, because we know the answer." The trouble with that is that there are *two* groups in American psychology who think we now "know the answer," and their answers are very different, consisting of the Freudian answers and the non-Freudian answers. Nor are the non-Freudian answers found only among experimentalists or behaviorists or dust-bowl psychometrists. They are found widely among practitioners and psychotherapy teachers.

One source of the lessened attention to psychoanalytic evidence is the long-term shift—especially complicated because of Freud's never having written the promised treatise on technique—from the original Breuer-Freud abreaction-cartharsis under hypnosis, to the pressure technique focusing upon specific symptoms, to the more passive free association (but still emphasizing the content of the impulse defended against or the memory repressed), to resistance interpretation and, finally, the heavy focus on interpretation of the transference resistance. So that today a large part of analytic intervention is directed at handling the momentary transference, aiming to verbalize the patient's current transference phenomenology with interpretations that are hardly distinguishable from a Rogersian reflection during Rogers's "classical nondirective" period. Such sessions sound and read uninteresting to me. My first analyst was Vienna trained in the late twenties and my second was a product of the Columbia Psychoanalytic Clinic under Rado's aegis, and both spent quite a bit of effort on a variety of interpretations and constructions, the Radovian very actively.

Perhaps the seminal papers of Wilhelm Reich on character analysis—despite Reich's own objection to analysts simply "floating in the patient's productions" and "permitting the development of a chaotic situation" or as Fenichel somewhere puts it, "communicating intermittently and unselectively various thoughts that occur as they listen"—nevertheless had the long-term effect, because they focused on resistance and specifically on the characterological resistances as interferences with obedience to the Fundamental Rule, of narrowing interpretive interventions almost wholly to varying forms of the question, "How are you feeling toward me right now?"

The playing down of the importance of old-fashioned interpreting and constructing I see, perhaps wrongly, as related to an oddity in the views expressed by some well-known institute-trained analysts who, though in good standing with the American Psychoanalytic Society, adopt strange positions. Take Dr. Judd Marmor, whose views are expressed in the preface to the huge tome he edited *Modern Psychoanalysis: New Directions and Perspectives* (Marmor 1968). Before touching gingerly on the topic of nonmedical analysts and considering ambivalently the nature and purpose of the training analysis, he has told us that modern psychoanalysis builds upon the great work of that genius Freud, whose followers we are, and who discovered for the first time a powerful and truly scientific way of investigating the human mind. But we are also told that of course today the classical psychoanalytic technique is not used much because it doesn't work, and that the constructs in Freud's psychoanalytic theory need not be taken very seriously. It is clear that Dr. Marmor is jealous of the designation "psychoanalyst" and "psychoanalysis," but I find it hard to see why. An imaginary analogy: Suppose I tell you that I am a microscopist, that I stand in the succession of that great genius, the founder of true scientific microscopy, Jan van Leeuwenhoek, upon whose discoveries, made by means of the microscope, we contemporary microscopists build our work. Nobody practices microscopy, or is entitled to label himself "a microscopist," who has not attended one of our van Leeuwenhoekian night schools. Of course we no longer use the microscope, since it doesn't work as an instrument; and the little animals that van Leeuwenhoek reported seeing by the use of this device do not exist. What would we think of such a position? It seems to me incoherent. One can argue that if our practice consists almost entirely of handling the moment-to-moment transference phenomenology that occurs during interviews, then the study of Freud's writings is largely a waste of time in preparation for practice of psychoanalytic therapy, and should be classed along with requirements that one study brain physiology or correlational statistics, as tribal educational hurdles for the coveted Ph.D. or M.D. degree that must be met, pointlessly, by would-be psychotherapists! One recalls Carl Rogers's famous view, pushed by him toward the end of World War II, that it takes only a few weeks of intensive training in client-centered therapy to become skillful at it and that most of what therapists study in medical school or graduate school is a waste of time.

I have taken somewhat too long on these preliminary remarks, but I wished to sketch the historical and current sociological context in which Fliess's question is, I think wrongly, often set aside. I must also mention four matters I am *not* considering here, although all of them have great interest and importance. First, I am not concerned to discuss the therapeutic efficacy of classical analysis or psychoanalytically oriented therapy, on which my views are complicated, especially since in recent years I have been doing quite a bit of modified RET (Rational Emotive Therapy), and only recently returned to mixing RET with a modified psychoanalytic approach. Rational emotive therapy and behavior modification are probably the treatments of choice for 80 or 90 percent of the clientele. This view is not incompatible with a view I hold equally strongly, that if you are interested in learning about your mind, there is no procedure anywhere in the running with psychoanalysis. Second, I am not going to address myself to the validation of metapsychological concepts, even though I think that a view like Marmor's is in need of clarification. Third, I'm not going to talk about the *output* aspect of analytic interpretation, i.e., the timing and wording of interventions, but only abut the *input* (cognitive) side, i.e., the way one construes the material, whatever he decides to do with it, including the usual decision to wait. Finally, I shall set aside entirely the experimental and statistical studies, not because I think them unimportant but because I am not very familiar with them except through summaries such as the recent paper by Lloyd H. Silverman in the *American Psychologist* (1976). I believe a person would not become convinced of the truth or falsity of the first-level theoretical corpus of psychoanalysis solely on the basis of the experimental and statistical studies, and that most psychologists who are convinced that there is a good deal in psychoanalytic theory have become convinced mainly by their own experience with it as patient and therapist.

It goes without saying, for anyone familiar with current philosophy of science, that there is a complicated two-way relationship between facts and theories. On the one hand, we can say that clinical experience with psychoanalytic material provides some sort of prior probability when we come to evaluate the experimental and correlational evidence, both when it's positive and when it's adverse; and, on the other hand, whatever general principles can come from the study of either human beings or animals, even of the somewhat attenuated, distantly related kind reported in Robert R. Sears's (1943) classic survey on objective studies of psychoanalytic concepts, can in turn give support to inferences made during the psychoanalytic session itself. My own view, despite my Minnesota training, is that if you want to find out what there is to psychoanalysis, the psychoanalytic hour is still the best place to look. It would be strange—although not logically contradictory—to say that we have in that hour a set of important discoveries that a certain man and a few of his coworkers hit upon while listening to what patients said about their dreams and symptoms and so on, under special instruction for how to talk (and even a prescribed physical posture while talking); but that setting is not a good one for investigating the matters allegedly brought to light! So : The rest of my remarks will deal wholly with the subjectivity in inferences reached from the verbal and expressive behavior the patient displays during the psychoanalytic hour.

I realize that I have not given a clear statement of the problem, and it's not easy. The patient speaks; I listen with "evenly hovering attention" (and background reliance upon my own unconscious). From time to time I experience a cognitive closure that, its content worded, characterizes an inferred latent psychological state or event in the patient's mind. For example, it occurs to me that the patient is momentarily afraid of offending me, or that the dream he reported at the beginning of the session expresses a homoerotic wish, or that the Tyrolean hat in the manifest content is connected with his uncle who wore one, and the like. The essence of psychoanalytic listening is listening for that which is not manifest (Reik 1948), for an inferred (and theoretical) entity in the other's mind that has then imputed to it a causal status. Fliess's Achensee question is: "What credentials does this kind of alleged knowledge bring? When you listen to a person talk, you can cook up all sorts of plausible explanations for why he says what he says. I accuse you, therefore, of simply putting your own thoughts into the mind of the helpless patient."

The epistemological scandal is that we do not have a clear and compelling answer to this complaint eighty years after Fliess voiced it, and a century after Josef Breuer discovered his "chimney sweeping" hypnocathartic technique on Anna O. We may motivate the topic as one of great theoretical interest, which I confess is my main one, as it was Freud's. From the clinical standpoint, however, we "mind-healers" do have the long-debated question as to whether, and how, processes labeled "insight," "uncovering," and "self-understanding" work therapeutically. Non-Freudian therapists like Joseph Wolpe, Albert Ellis, and Carl Rogers have argued plausibly that psychoanalytic efficacy (marginal as it is) is an incidental byproduct of something other than what the analyst and the patient think they are mainly doing. Whatever the merits of these explanations, it is difficult to answer questions about whether and how a correct interpretation or construction works, if one has no independent handle on the epistemological question, "How do we know that it is correct?" In putting it that way I am of course referring to its content correctness and not to its technical correctness, not to the output aspect of the analyst's interpretation. It would be strange, would it not, if we were able to investigate the technically relevant components of the output side of analytic interpretation without having some independent test of its cognitive validity? That is, how would I research the question whether, for instance, summary interpretations at the end of the session are, on the average, more efficacious than tentative ones dropped along the way, (cf. Glover 1940), if the problem of inexact interpretation or totally erroneous "barking up the wrong tree" were wholly unsettled? I do not advance the silly thesis that one must know *for sure* whether the main content of an interpretation is cognitively valid before investigating these other matters. But without some probabilistic statement as to content correctness, it is hard to imagine an investigation into the comparative therapeutic efficacy of the two interpretative tactics. The usual statement that an interpretation is "psychologically valid" when it results in a detectable dynamic and economic change may be all right as a rule of thumb, but it does not satisfy a Fliessian critic, and I cannot convince myself that it should. Although there occur striking experiences on the couch or behind it in which the quality, quantity, and temporal immediacy of an effect will persuade all but the most anti-Freudian skeptic that something is going on, these are not the mode. Furthermore, "Something important happened here" is hardly the same as "*What* happened here is that a properly timed and phrased interpretation also had substantive validity and hence the impulse-defense equilibrium underwent a marked quantitative change."

There are few phenomena—and I do mean *phenomena*, that is, virtually uninterpreted raw observations of speech and gesture, not even first-level thematic inferences—that are so persuasive to the skeptic when he is himself on the couch, or so convincing (even when related without tape recordings or verbatim protocol) to clinical students, as the *sudden* and *marked* alteration in some clearly manifested mental state or ongoing behavior immediately following an analytic interpretation. For readers without psychoanalytic experience, I present a couple of brief examples.

When I was in analysis, I was walking about a half block from the University Hospitals to keep my analytic appointment and was in a more or less "neutral" mood, neither up nor down and with no particular line of thought occupying me, but rather observing the cars and people as they passed. I perceived approaching me a man and woman in their late thirties, both with distinctly troubled facial expressions and the woman weeping. The man was carrying a brown paper sack and over his arm a large Raggedy Ann doll. It is not, of course, in the least surprising (or requiring any special psychodynamic interpretation) that the thought occurred to me from their behavior, the doll, and the fact that they were leaving the University Hospital, that a child was very ill or possibly had just died. It would not be pathological for a person of ordinary human sympathy, and especially a parent, to feel a twinge of sympathetic grieving at such a sight. That is not what befell me on this occasion, however. I was suddenly flooded with a deep and terrible grieving and began to weep as I walked. I don't mean by that that I was a little teary; I mean that I had difficulty restraining audible sobs as I passed people, and that tears were pouring down my face. I told myself this was absurd, I must be reacting to something else, and so on and so forth, none of which self-talk had the slightest discernible effect. On the elevator to go up to my analyst's office were two of our clinical psychology trainees who looked at me somewhat embarrassedly, saying "Good morning, Dr. Meehl," vainly trying to appear as if they had not noticed the state I was in. Even under those circumstances, in an elevator full of people, I literally could not control the weeping, including muffled sobbing sounds. I did not have to wait more than a minute or two for my analyst to appear. Trying to ignore the puzzled expression of a psychiatric social worker whose hour preceded mine, I went in, lay down, and at that point began to sob so loudly that I was unable to begin speaking. After acquiring enough control to talk, I described briefly the people I had met, whereupon my analyst (who, while he had had analysis with Helene Deutsch and Nathan Ackerman, had been exposed to strong Radovian influences in his training institute) intercepted with the brief question, "Were you harsh with Karen [my five-year-old daughter] this morning?" This question produced an immediate, abrupt, and total cessation of the inner state and its external signs. (I had spoken crossly to Karen at the breakfast table for some minor naughtiness, and remembered leaving the house, feeling bad that I hadn't told her I was sorry before she went off to kindergarten.) I emphasize for the nonclinical reader, what readers who have had some couch time will know, that the important points here are the *immediacy* and the disappearance of any problem of *control*—no need for counterforces, "inhibition" of the state, or its overt expression. That is, the moment the analyst's words were perceived, the affective state immediately vanished. I don't suppose anyone has experienced this kind of phenomenon in his own analysis without finding it one of the most striking direct behavioral and introspective evidences of the concepts of "mental conflict," "opposing psychic forces," and "unconscious influences"—the way in which a properly timed and formulated interpretation (sometimes!) produces an immediate dynamic and economic change, as the jargon has it.

Comparable experiences when one is behind the couch rather than on it, usually carry less punch. The reason is not that analysis is a "religious experience," as my behaviorist friends object when I point it out, but that the analysand is connected with his inner events more closely and in more modalities than the analyst is, which fact confers an evidentiary weight of a different qualitative sort from what is given by the analyst's theoretical knowledge and his relative freedom from the patient's defensive maneuvers. True, it is generally recognized that we see considerably fewer "sudden transformations" today than apparently were found in the early days of the analytic movement. We do not know to what extent this reduced incidence of sudden lifting of repression with immediate effects, especially dramatic and permanent symptomatic relief, is attributable to the cultural influence of psychoanalytic thinking itself (a development Freud predicted in one of his prewar papers). There are doubtless additional cultural reasons for changes in the modal character neurosis. There was perhaps some clinical peculiarity (that still remains to be fathomed) in some of the clientele studied during the early days, such that true "Breuer-Freud repression," the existence of a kind of "cold abcess in the mind" that could be lanced by an analytic interpretation-cum-construction that lifted the repression all at once, was commoner in the 1890s than today. These are deep questions, still poorly understood. But it remains true that from time to time symptomatic phenomena that have been present for months or years, and have shown no appreciable alteration despite the noninterpretative adjuvant and auxiliary influences of the therapeutic process (e.g. reassurance, desensitization, and the mere fact that you are talking to a helper) do occur and help to maintain therapist confidence in the basic Freudian ideas.

I recall a patient who had among her presenting complaints a full-blown physician phobia, which had prevented her from having a physical examination for several years, despite cogent medical reasons why she should have done so. She was a professionally trained person who realized the "silliness" of the phobia and its danger to her physical health, and attributed the phobia—no doubt rightly, but only in part—to the psychic trauma of a hysterectomy. Her efforts to overcome it were unsuccessful. Repeatedly she had, after working herself up to a high state of drive and talking to herself and her husband about the urgency of an examination, started to call one or another physician (one of whom was also a trusted personal friend who knew a lot about her) but found herself literally unable to complete even the dialing of the telephone number. Now, after seventy-five or eighty sessions, during which many kinds of material had been worked through and her overall anxiety level markedly reduced, the doctor phobia itself remained completely untouched. From themes and associations, I had inferred, but not communicated, a specific experience of a physical examination when she was a child in which the physician unearthed the fact of her masturbation, which had unusually strong conflictful elements because of the rigid puritanical religiosity of her childhood home (and of the physician also). During a session in which fragments of visual and auditory memory and a fairly pronounced intense recall of the doctor's examining table and so on came back to her, and in which she had intense anxiety as well as a feeling of nausea (sufficient to lead her to ask me to move a wastebasket over in case she should have to vomit), she recalled, with only minimal assistance on my part, the physician's question and her answer. This occurred about ten minutes before the end of the hour. She spent the last few minutes vacillating between thinking that she had been "docile," that I had implanted this memory, but then saying that she recalled clearly enough, in enough sense modalities, to have a concrete certainty that it was, if imperfectly recalled, essentially accurate. As one would expect in a sophisticated patient of this sort, she saw the experience as the earlier traumatic happening that potentiated the effect of the adult hysterectomy and led to her doctor phobia. She called me up the following morning to report cheerily, although a bit breathlessly, that she had refrained from making a doctor's appointment after the session yesterday, wondering whether her feeling of fear would return. But when, on awakening in the morning, she detected only a faint anxiety, she found it possible without any vacillation to make a phone call, and now reported that she was about to leave for her appointment and was confident that she would be able to keep it. I think most fairminded persons would agree that it takes an unusual skeptical resistance for us to say that this step-function in clinical status was "purely a suggestive effect," or a reassurance effect, or due to some other transference leverage or whatever (75th hour!) rather than that the remote memory was truly repressed and the lifting of repression efficacious.

Some argue simply that "clinical experience will suffice to produce conviction in an open minded listener." We are entitled to say, with Freud, that if one does not conduct the session in such and such a way, then he will very likely not hear the kind of thing that he might find persuasive. But the skeptic then reminds us of a number of persons of high intelligence and vast clinical experience, who surely cannot be thought unfamiliar with the way to conduct a psychoanalytic session, who subsequently came to reject sizable portions of the received theoretical corpus, and in some instances (e.g., Wilhelm Reich, Albert Ellis, Melitta Schmideberg, and Kenneth Mark Colby) abandoned the psychoanalytic enterprise. Nobody familiar with the history of organic medicine can feel comfortable simply repeating to a skeptic, "Well, all I can say is that my clinical experience shows. . . ."

The methodological danger usually labelled generically "suggestion," that of "imposing theoretical preconceptions" by "mind-reading one's own thoughts into the patient's mind," is itself complex. An experimental or psychometric psychologist (I am or have been both) can distinguish four main sources of theory-determined error in the psychoanalytic process. First, *content implantation*, in which memories, thoughts, impulses, and even defenses are explicitly "taught" to the patient via interpretation, construction, and leading questions. Second, *selective intervention*, in which the analyst's moment-to-moment technical decisions to speak or remain silent, to reflect, to ask for clarifications, to call attention to a repetition, similarity, or allusion, to request further associations, to go back to an earlier item, etc., can operate either as *differential reinforcement* of verbal behavior classes (a more subtle, inexplicit form of implantation!) or as a *biased evidence-sifter*. By this latter I mean that even if the patient's subsequent verbalizations were uninfluenced by such interventions, what the analyst has thus collected as *his* data surely has been. Third, on the "input" side, there is the purely perceptual-cognitive aspect of subjectivity in discerning the "red thread," the thematic allusions running through the material. (As my Skinnerian wife says, we want the analyst to *discern* the "red thread," we don't want him to spin it and weave it in!) Fourth, supposing the theme-tracing to be correct, we make a *causal* inference; and what entitles us to infer the continued existence and operation of an unconscious background mental process *guiding* the associations (Murray's *regnancy*)? Such a construction does not follow immediately from correct detection of a theme *in* the associations. I focus the remainder of my remarks almost wholly on the third of these dangers, the subjective (critics would say "arbitrary") construing of what the verbal material *means*, "alludes to," "is about."

I do not trouble myself to answer superbehaviorist attacks, such as those that say that science can deal only with observables, hence an unconscious fantasy is inherently an illicit construct; since these attacks, besides being dogmatic, are intellectually vulgar, historically inaccurate, and philosophically uninformed. *The crunch is epistemological, not ontological.* The problem with first-level psychoanalytic constructs is not that they are not observable as test-item responses or muscle twitches or brain waves, but that their inferential status, the way in which they are allegedly supported by the data base of the patient's words and gestures, is in doubt.

I also reject, in the most high-handed manner I can achieve, the typical American academic psychologists' objection that psychoanalysis is not "empirical," which is based upon a failure to look up the word "empirical" in the dictionary. There is, of course, no justification for identifying the empirical with the quantitative/experimental other than either behaviorist or psychometric prejudice, nor to identify the quantitative/experimental with the scientific, nor to identify any of these with what is, in some defensible sense, "reasonable to believe." These mistaken synonymies involve such elementary errors in thinking about human knowledge generally, and even scientific knowledge in particular, that I refuse to bother my head with them.

My late colleague Grover Maxwell used to ask me why I think there is a special problem here, once we have shed the simplistic American behaviorist identification: reasonable = empirical = quantitative/experimental = scientific. Do we not recognize the intellectual validity of documentary disciplines like law, history, archeology, and so on, despite the fact that they (with interesting exceptions, such as the cliometrists) proceed essentially as we do in psychoanalysis? Or, for that matter, what about all of the decisions, judgments, and beliefs we have in common life, such as that we could probably lend money to our friend Smith, or that our wife is faithful to us, or that one Swedish car is better than another?

An analogy between psychoanalytic inference and decision making or beliefs adopted in "ordinary life" is defective for at least three reasons, and probably more.

Most "ordinary life" beliefs do not involve high-order theorizing, but concern fairly simple connections between specific happenings, easily and reliably identified. Herewith a list of ten circumstances that affect the degree of confidence or skepticism with which nonquantified impressions from clinical experience should be assessed:

1. Generalized observations about a variate that is of a "simple, physical, quickly-decidable" nature are more to be trusted on the basis of common experience, clinical impressions, anecdotal evidence, "literary psychology," or the fireside inductions generally (Meehl 1971) than claims about variates, however familiar to us from common life or sophisticated clinical experience, that are not "observable" in a fairly strong and strict sense of that slippery word. Thus shared clinical experience that persons in a grand mal epileptic seizure fall down is more dependable than the

(equally shared) experience that schizotypes readily act rejected by a clumsy therapist's remark in an interview. The fact that an experimental animal stops breathing is a more trustworthy protocol, absent solid data recording, than the "fact" that an experimental animal shows "anxious exploring behavior." We would think it odd if somebody published an article proving, with scientific instrumentation and significance tests, that if you hold a bag of kittens under water for an hour, they will be dead!

2. Contrasted with the preceding are three main categories of not simply observable variates that are readily inferred by us, both in common life and in clinical practice: (a) Inferred inner states or events ("anxiety," "dependency," "hostile," "seductive," "manipulative," "passive-aggressive," "anhedonic," "guarded," "paranoid"); (b) clusters, composites, behavior summaries—more generally *traits*, a trait being conceived as an empirically correlated family of content-related dispositions; and (c) inferred external events and conditions, either current but not actually under the clinician's observation (e.g., "patient is under work stress") or historical (e.g., "patient is obviously from a lower-class social background").

3. Even simple physical observables, however, can sometimes be distorted by theory, prejudice, or otherwise developed habits of automatic inference. A classic example is Goring's study of the English convict in which estimated heights of foreheads were positively correlated with intelligence as estimated by prison personnel, although the estimated intelligence (like the estimated forehead height) did not correlate with *measured* forehead heights! The interesting methodological point here is that guards and prison officials could agree quite reliably on how bright a man is, intelligence being a socially relevant property and one that we know (from many data in educational, military, and industrial settings) has an interjudge reliability of .50 or better, so that pooled judgments can have a high reliability; but because these persons shared the folklore belief that high forehead goes with brains, they apparently "perceived" a prisoner's forehead as higher when they thought the prisoner was bright.

4. If the event being correlated is something strikingly unusual, such as an occurrence or trait that deviates five standard deviations from the mean in populations with which the observer is accustomed to dealing, it is obviously going to be easier to spot relationships validly.

5. States, events, or properties that fluctuate spontaneously over time are hard to correlate with causative factors such as intervention, compared

to those that do not fluctuate much "spontaneously" (that is, absent intervention) over time.

6. States, properties, or dimensions that normally move monotonically in a certain direction over time (e.g., patients usually get progressively sicker if they have untreated pernicious anemia) are more easily relatable than those that show numerous spontaneous "ups and downs" (e.g., spontaneous remissions and exacerbations in diseases such as schizophrenia or multiple sclerosis). The long list of alleged beneficial treatments for multiple sclerosis that have been pushed enthusiastically by some clinicians and have subsequently been abandoned as enthusiasm dies out or controlled quantitative studies are performed, is due not only to the urgency of trying to help people with this dread illness, but also to the normal occurrence of spontaneous remissions and exacerbations, the considerable variability in interphase times, in their severity, and in the functional scarring following an acute episode of this illness (see Meehl 1954, p. 136).

7. The more causal influences are operative, the harder it will be to unscramble what is operating upon what. In the case of cross-sectional correlation data in the social sciences, the intractability of this methodological problem is so great as to have resulted in a special methodological approach, known as path analysis, disputes about whose conceptual power in unscrambling the causal connections still persist to the extent that some competent scholars doubt that it has any widespread validity.

8. If a causal influence shows a sizable time lag to exert its effect, which is often true in medical and behavioral interventions, it is harder to correlate validly than if its effect, when present, is immediate. The other side of this coin is the tendency of minimal effects in behavior intervention to fade out with time. Differences in the impact of an educational procedure—such as the difference between two methods of teaching fractions to third graders —if it is barely statistically significant but not of appreciable size immediately after learning, the chances are good that the children's ability to do fractions problems two years later (let alone as adults) will not be different under the two teaching methods. Yet in opposition to this admitted tendency of interventions to fade out, we have some tendency— claimed but not well documented, if at all—for long-term influences of successful psychotherapeutic intervention to escape detection in immediate post treatment assessment. Whatever the influence of these opposed

tendencies, the point is that the existence of the first and the probability (at least in some cases) of the second greatly increase the difficulty of ascertaining an effect.

9. If the time lag between an influence and its consequence, whatever its average size, is highly variable among individuals (or over different occasions for the same individual), a valid covariation is harder to discern.

10. If there are important feedback effects between what we are trying to manipulate and the subsequent course of our manipulation, the relationships are harder to untangle, especially because there are likely to be sizable differences in the parameters of the feedback system.

More generally, a complicated and controversial topic deserving more discussion than the present context permits, we still do not have an adequate methodological formulation as to the evidentiary weight that ought rationally to be accorded the "clinical experience" of seasoned practitioners when it is not as yet corroborated by quantitative or experimental investigation that meets the usual "scientific" criteria for having formal "researched status." This problem is troublesome enough when the situation is that of practitioners asserting something on the basis of their clinical experience, which, when pressed, they can document only by what amount essentially to an educated guesser's anecdotes, whereas the "anecdotal method" is repudiated as an unacceptable method in any sophomore general psychology course! The problem is made worse when purportedly scientific research on the clinician's claims has been conducted and seems to be unfavorable to his generalization.

On the one side, it must be admitted that some laboratory or even field survey studies of clinical hypotheses are clumsy, naive, and unrealistic in one or more ways, so that one cannot fault a good clinician for dismissing them as irrelevant to what he intended to say. I think, for instance, of a silly study by some academic psychologist (whose familiarity with psychoanalysis must have been confined to reading one or two tertiary sources) who published a paper in a psychology journal alleging to refute Freud's idea of the Oedipal situation because a simple questionnaire item administered to college undergraduates asking whether they preferred their father to their mother, showed that both boys and girls preferred their mother but the latter more so. One can hardly blame Freud for not spending much time monitoring the journals of academic psychology in the 1920s if this is the kind of production they were coming up with. I think it appropriate, being myself both an experienced practitioner *and* a psychometric and experimental psychologist, to venture an opinion as to the main sources of this "pseudoscientific unrealism" on the part of some academics attempting to study a clinical conjecture. (I think it also fair to say that it happens less frequently today than it did, say, between World Wars I and II, during which time very few academic social scientists had any real firing-line experience with mental patients or with intensive psychotherapy.) First, the nonclinician literally fails to understand the clinician's theoretical conjecture with sufficient precision and depth to know what would constitute a reasonable statistical or experimental test of it. Of course, sometimes this is partly the fault of the clinician for not troubling to expound the theory with even that minimal degree of scientific rigor that the state of the art permits. Second, one can understand the essential features of the theory but make simplistic auxiliary assumptions, the most tempting of which is the reliance upon instruments that the clinician would probably not trust (the above undergraduate questionnaire being a horrible example). Third is the possibility that, although the instruments employed are adequate for the purpose, the particular psychological state of affairs is not qualitatively of the same nature or quantitatively as intense as that which the clinician had in mind; as, for example, paradigm studies of psychotherapy in which, rather than having a full-blown clinical phobia brought in by a suffering patient, one has a small-scale "artificial phobia" generated in the psychology laboratory. Fourth, clinicians are likely to do an inadequate job of characterizing the clientele, so that a selection of individuals from a population may yield an incidence of something whose base rate in that population is so different from that of the clinic that a statistically significant result is hard to achieve with only moderate statistical power. (How the clinician can have detected something here that the statistician can't is such a complicated question that I must forego discussion of it here, but it deserves an article in its own right.)

Against all of these proclinical points must be a simple, clear, indisputable historical fact: In the history of the healing arts, whether organic medicine or psychological helping, there have been numerous diagnostic and therapeutic procedures that fully trained M.D.s or Ph.D.s, who were not quacks and who were honorable and dedicated professionals, have passionately defended, that have subsequently been shown to be inefficacious or even counterproductive. No informed scholar disputes this. I cannot see the following as anything but a form of intellectual dishonesty or carelessness: A person with a doctorate in psychology advocates a certain interpretation of neurosis on the basis of his clinical experience. He is challenged by another seasoned practitioner who has, like himself, interviewed, tested, and treated hundreds or maybe thousands of patients, and who is familiar with the conceptual system of the first clinician, but persists in not believing it, and denies the causal relations that the first clinician alleges. The first clinician persists in repeating "Well, of course, I *know* from my clinical experience that . . . ."

I suspect that this kind of cognitive aberration occurs partly because introductory psychology courses no longer emphasize the classic studies on the psychology of testimony, the psychology of superstitions, and the inaccuracy of personnel ratings from interviews and the like, which used to be a staple part of any decent psychology course when I was an undergraduate in the 1930s. It is absurd to pretend that because I received training in clinical psychology, I have thereby become immunized to the errors of observation, selection, recording, retention, and reporting, that are the universal curse on the human mind as a prescientific instrument. Nobody who knows anything about the history of organic medicine (remember venesection!) should find himself in such a ridiculous epistemological position as this. Fortunately, we do find statements about certain classes of patients agreed on by almost all clinicians (it is perfectionistic to require *all*, meaning "absolutely every single one") provided they have adequate clinical exposure and do not belong to some fanatical sect, *their diagnostic impressions being shared despite marked differences in their views on etiology and treatment*. Although consensus of experienced practitioners is strictly speaking neither a necessary nor sufficient condition for the truth—it would be as silly to say that here as to say it about consensus of dentists, attorneys, engineers or economists—presumably something that practitioners trained at different institutions and holding divergent opinions about, say, a certain mental disorder, agree upon is *on the average* likely to be more trustworthy as a clinical impression than something that only a bare majority agrees on, and still less so something that is held by only a minority. One says this despite realizing from history, statistics, or general epistemology that a group that's currently in a small minority may, in the event, turn out to have been out right after all.

Suppose, for example, that two psychologists have each spent several thousand hours in long-term intensive psychotherapy of schizophrenics, either psychotic or borderline. They may disagree as to the importance of genetic factors or the potentiating impact of the battle-ax mother. But it would be hard to find *any* experienced clinician, of whatever theoretical persuasion, who would dispute the statement that schizophrenics have a tendency to oddities of thought and associated oddities in verbal expression. Even a clinician who has bought in on the labeling-theory nonsense and who doesn't think there is any such mental disease as schizophrenia (if there are some funny-acting people in the mental hospitals, they surely don't have anything wrong with their brains or genes—a view that it takes superhuman faith or ignorance of the research literature to hold at the present time), will hardly dispute that one of the main things that leads *other* wicked practitioners to attach the label "schizophrenia" is a fact that he himself has observed *in the people so labeled;* to wit, that they have funny ways of talking and thinking, and it's a kind of funniness that is different from what we hear in neurotics or people who are severely depressed or psychopathic or mentally deficient.

But there simply isn't any way of getting around the plain fact that individuals in the healing arts are not immune from overgeneralization and are sometimes recalcitrant in the presence of refuting evidence, even when the statistical or experimental study cannot be faulted on any of the clinical grounds given above. Everyone knows that this is true in the history of organic medicine (where, by and large, we expect the clinical phenomena to be relatively more objective and easier to observe than in a field like psychotherapy), so that for a long time physicians practiced venesection or administered medicinal substances that we now know have no pharmaceutical efficacy. Surely this should lead an honest psychotherapist to face the possibility that he might *think* that he is helping people or—the main question before us in the present paper—that he is making more correct than incorrect inferences from the patient's behavior, even though in reality he is not doing so, and is himself a victim of a large-scale institutionalized self-deception. Even a practitioner like myself who finds it impossible to really believe this about say, a well-interpreted dream, ought nevertheless to be willing to say in the metalanguage that it *could* be so. His inability to believe it belongs in the domain of biography or "impure pragmatics" rather than in science or inductive logic.

Several thousand people are today totally blind because they developed the disease called retrolental fibroplasia as a result of being overoxygenated as premature newborns. For twenty years or so, obstetricians and pediatricians debated hotly the merits of this allegedly "prophylactic" procedure. It was only when an adequate statistical analysis of the material was conducted by disinterested parties that the question was finally resolved. It is incredible to me that psychotherapists familiar with this kind of development in organic medicine nevertheless counter objections to psychotherapeutic interpretations by doggedly reiterating, "My clinical experience proves to my satisfaction that. . . ."

It is not easy to convey to the nonclinician reader how a seasoned experienced practitioner who has had plently of diagnostic and therapeutic exposure to a certain clientele could come into collision with another one, without giving examples outside psychoanalysis. Consider, for instance, the widely-held view that the battle-ax mother (often called by the theory-laden term "schizophrenogenic mother," despite rather feeble quantitative support for her causal relevance) has a great deal to do with determining the psychopathology of schizophrenia and, perhaps, even its very occurrence. The point is that the clinicians who are convinced of her etiological importance have not made up the raw data, and this explains why other equally experienced clinicians skeptical of the schizophrenogenic mother hypothesis don't find that their (similar) clinical experience convinces them of the same etiological view. I have not talked to any experienced practitioner, whatever his views of schizophrenia or its optimal treatment, who disputes certain "observations" about the way schizophrenics talk when they get on the subject of their mother or subsequent mother figures. But collecting these chunks of verbal behavior about battle-ax mothers is several steps removed inferentially from the common (American) clinicians' conclusion that this patient is psychotic mainly because of the way his battle-ax mother treated him. There are half a dozen plausible factors tending to generate this kind of verbal behavior, and they are not incompatible, so that when taken jointly, it is easy to construct a statistical-causal model that will explain the widespread extent of this clinical experience by practitioners without assuming even the tiniest causal influence of the battle-ax mother syndrome upon the subsequent development of a schizophrenia (see Meehl 1972, pp. 370-371). For a more general discussion of the relationship of clinical or anecdotal generalizations to criticism and the difficulty of assessing the relative weight to be given to it in relation to more scientifically controlled studies, see Meehl 1971, especially pages 89-95.

Even those ordinary-life conclusions that are not themselves explicitly statistical nevertheless are often *based upon* experimental or statistical findings by somebody else. Sometimes these findings are known to us, sometimes we rely on reports of them because we have previously calibrated the authorities involved. Thus, for instance, in buying life insurance we rely upon actuarial tables constructed by insurance statisticians, and we also know that the law constrains what an insurance company can charge for a given type of policy on the basis of these statistics. The actuarial table's construction and interpretation is a highly technical business, beyond the insured's competence to evaluate. But he does not *need* to understand these technicalities in order (rationally) to buy life insurance.

In most ordinary-life examples, one is forced to make a decision by virtue of the situation, whereas a psychology professor is not forced to decide about psychoanalytic theory. If somebody replies to this by saying that the *practitioner* is forced to decide, that's not quite true, although it has a valid element. The practitioner is not forced to decide to proceed psychoanalytically in the first place; and, pushing the point even farther back, the psychologist was not forced to be a psychotherapist (rather than, say, an industrial psychologist).

For these three reasons, the easy analogy of a psychoanalytic inference about an unconscious theme, or mechanism, or whatever, with those less-than-scientific, action-related inferences or assumptions we require in ordinary life, is weak, although not totally without merit.

Suppose one drops "ordinary life" as the analogy and takes instead some other nonexperimental, nonstatistical but technical scholarly domain, such as law or, usually, history. The evidence in a law court is perhaps the closest analogue to psychoanalytic inference, but inferences in history from fragmentary data and empirical lawlike statements that cannot be experimented on are also a good comparison. The analogy breaks down somewhat in the case of law, however, in that the application of legal concepts is not quite like an empirical theory, although the lawyer's inferences as to the fact situation are epistemologically similar to those of psychoanalytic inference.

A difficulty in relying on Aristotle's dictum about "precision insofar as subject matter permits" is that this rule doesn't tell us whether, or why, we ought to have a high intellectual esteem for a particular subject matter. If the subject matter permits only low-confidence, nonquantitative, impressionistic inferences, operating unavoidably in a framework not subjected to experimental tests in the laboratory, or even file-data statistical analysis, perhaps the proper conclusion is simply that we have a somewhat shoddy and prescientific discipline. Doubtless some physical scientists would say that about disciplines such as history, or the old-fashioned kind of political science, as well as psychoanalysis. In either case one cannot take very much heart from the analogy. It is, I fear, really a rather weak defense of psychoanalytic inference unless we can spell it out more.

However, an interesting point arises here in connection with the flabbiness of statistical significance testing as a research method. I think it can be shown—but I must leave it for another time—that the use of either a Popperian or a Bayesian way of thinking about a criminal case gives stronger probabilities than those yielded by null hypothesis testing. This is an important point: The fact of "explicit quantification," i.e., that we have a procedure for mechanically generating *probability numbers*, won't guarantee strong inference or precision, although it looks as if it does, and most psychologists seem to think it must. I am convinced that they are mistaken. The "precision" of null hypothesis testing is illusionary on several grounds, the following list being probably incomplete:

1. Precision of the tables used hinges upon the mathematical exactness of the formalism in their derivation, and we know that real biological and social measures don't precisely fill the conditions. That a statistical test or estimator is "reasonably robust" is, of course, no answer to this point, when precision is being emphasized.

2. Random-sampling fluctuations are all that these procedures take into account, whereas the most important source of error is systematic error because of the problem of validity generalization. Almost nothing we study in clinical psychology by the use of either significance tests (or estimation procedures with an associated confidence belt) is safely generalizable to even slightly different populations.

3. The biggest point is the logical distance between the statistical hypothesis and the substantive theory. For a discussion of this see Meehl (1978, pp. 823-834; Meehl 1967/1970; Lykken 1968; and Meehl 1954, pp. 11-14, Chapter 6, and passim).

As to the analogy of psychoanalytic inference with highly theoretical interpretive conjectures in history (e.g., hypotheses about the major factors leading to the fall of Rome, of which there are no fewer than seven, including that the elite all got lead poisoning from drinking wine out of those pewter vessels!), I find critics are as skeptical toward these as they are toward psychoanalysis. So that one doesn't get us very far, at most taking us past the first hurdle of a simplistic insistence that nothing can be "reasonable" or "empirical" unless based upon laboratory experiments or statistical correlations. We have to admit to the critic that not all psychoanalytic sessions are understandable, and even a session that is on the whole comprehensible has many individual items that remain mysterious, as Freud pointed out. We must also grant the point that physicians and psychologists who have certainly had the relevant clinical experiences—thousands of hours in some cases—have "fallen away" from the Freudian position, and claim to have rational arguments and evidence *from their clinical experience* for doing so. And finally, as a general epistemological point, we have to confess that human ingenuity is great, so that if you have loose enough criteria you can "explain anything."

It may help to ask, "Just *why* is there a problem here?" Why, in particular, do some of us find ourselves caught in the middle between, on the one hand, people who think there is no problem, that we know how to interpret, and that any seasoned practitioner has his kit of tools for doing so; and on the other, those skeptical people (e.g., Sir Karl Popper) who think our situation is conceptually hopeless because of a grossly uncritical methodology of inference? I can highlight the dilemma by examining Freud's jigsaw-puzzle analogy. The plain fact is that the jigsaw-puzzle analogy is false. It is false in four ways. There are four clearcut tests of whether we have put a jigsaw puzzle together properly. First, there must not be any extra pieces remaining; second, there must not be any holes in the fitted puzzle (Professor Salmon has pointed out to me that although this requirement holds for jigsaw puzzles, *provided all the pieces are present*, it is too strong for such cases as a broken urn reconstructed by an archeologist, since a few "holes" due to unfound pieces do not appreciably reduce our confidence in the reconstruction); third, each piece has to fit cleanly and easily into a relatively complicated contour provided by the adjacent pieces; finally, when a piece is fitted "physically," that which is painted on it also has to fit into a meaningful gestalt as part of a picture. Now the first two of these do not apply to the great majority of psychoanalytic sessions; and the second two apply only with quantitative weakening. Combining this point with the variation among psychoanalytic sessions (ranging from sessions almost wholly murky, in which neither patient or analyst can even be confident about a generic theme lurking behind the material, to a minority in which it seems as though "everything fits together beautifully"), we recognize a statistical problem of *selective bias,* emphasizing theoretically those sessions that are, so to say, "cognitively impressive." A good Skinnerian will remind us that the interpreter of psychoanalytic material is on an intermittent reinforcement schedule and that therefore his verbal behavior and his belief system will be maintained, despite numerous extinction trials that constitute potential refuters. The statistical problem presented here is that when, in any subject matter, a large number of arrangements of many entities to be classified or ordered is available, and some loose (although not empty) criterion of "orderliness" has been imposed, then we can expect that even if the whole thing were in reality a big random mess, *some expected subset of sequences will satisfy the loose ordering criteria.* This is one reason why we worry about significance testing in the inexact sciences, since we know that articles with significant t-tests are more likely to be accepted by editors than articles that achieve a null result, especially with small samples.

If you wonder why this problem of selecting an orderly-looking subset of cases from a larger mass arises especially in psychoanalysis, my answer would be that it does not arise there more than it does in other "documentary" disciplines, in which the interpreter of facts cannot manipulate variables experimentally but must take the productions of individuals or social groups as they come, as "experiments of nature," to use Adolf Meyer's phrase for a mental illness. My conjecture is that this is a far more pervasive and threatening problem than is generally recognized. It appears distressingly frequent, to anyone who has once become alerted to it, in diverse domains other than psychodiagnosis, including the nonsocial life sciences, the earth sciences, and in almost any discipline having an important "historical" dimension. To digress briefly, lest this point be misunderstood for want of nonpsychoanalytical examples, I give two.

Philosophers of science have recognized in recent years that the largely nonhistorical, nonempirical, "armchair" approach of the Vienna Circle can be misleading. Most of us would hold that a proper philosophy of science must (as Lakatos has emphasized) combine critical rational reconstruction with tracing out the historical sequence of the growth of knowledge as it actually occurred. Hence we find an increasing use of evidence for or against various philosophies of science from historical examples. I find it odd that so few agree with me that this problem involves a constant danger of selecting one's examples tendentiously. Thus, for instance, my friend

Paul Feyerabend loves to talk about Galileo and the mountains on the moon in order to make a case against even a sophisticated falsificationism; whereas almost any article by Popper can be predicted to contain a reference to the quick slaying (with Lakatos's "instant rationality") of the Bohr-Kramers-Slater quantum theory by a sledgehammer falsification experiment. I don't think I am merely displaying the usual social scientist's liking for doing chi squares (on tallies of practically anything!) when I say that this seems to me an inherently statistical problem, and that it cannot be settled except by the application of formal statistical methods.

A second example occurs in paleontology. Almost any educated person takes for granted that the horse series, starting out with little old terrier-size, four-toed *Eohippus*, is chosen by paleontologists for effective pedagogy, but that there are countless similar examples of complete, small-step evolutionary series in the fossil record—which in fact there are not. I think the evaluation of the fossil evidence for macroevolution is an inherently statistical problem, and is strictly analogous to such documentary problems in copyright law as whether a few bars of music should be considered plagiarized, or a few words of verbal text lifted. (See Meehl 1983.) In the evolution case, given a very large number (literally hundreds of thousands) of species of animals with hard parts that existed for various periods of time; and given the heavy reliance on index fossils for dating rocks (because in most instances neither the radium/lead clock nor purely chemical and geological criteria suffice); it stands to reason that some subset of kinds of animals should, *even if the historical fact of evolution had never occurred at all*, give an appearance of evolutionary development as found in the famous horse series.

We deal here with what I learned in elementary logic to call "the argument from convergence." In the light of a hypothesis H we can see how certain facts f_1, f_2, f_3 would be expected (ideally, would be deduced—although in the biological and social sciences that strict deducibility is rare). So when we are presented with f_1, f_2, f_3 we say that they "converge upon" the hypothesis H, i.e., they give it empirical support. I believe it is generally held by logicians and historians of science (although the late Rudolf Carnap was a heavyweight exception, and I never managed to convince him in discussions on this point) that the argument from convergence in inductive logic (*pace* Popper!) is inherently weaker than the argument from prediction, given the same fact/theory pattern. That is to say, if the facts that support hypothesis H are f_1, f_2, and f_3, let's contrast the two situations. First we have pure convergence, in which the theorist has concocted hypothesis H in the presence of f_1, f_2, f_3 and now presents us with the pure argument from convergence; second, we have a mixed argument from convergence *and prediction*, in which the theorist has concocted hypothesis H in the presence of facts f_1, f_2 and then predicted the third fact f_3, which was duly found. Every working scientist (in any field!) that I have asked about this says that Carnap was wrong and Popper is right. That is, the second case is a stronger one in favor of the hypothesis, despite the fact that precisely the same data are present in both instances at the time of the assessment, and their "logical" relationship to the theory is the same. If the logicians and philosophers of science cannot provide us with a rational reconstruction of why scientists give greater weight to a mixed argument from convergence and prediction than to a pure argument from convergence (given identical fact/theory content), I think they had beter work at it until they can.

The danger of content-implantation and the subtler, more pervasive danger of differential reinforcement of selective intervention, combine here with the epistemological superiority of prediction over (after-the-fact) convergence to urge, "Wait, don't intervene, keep listening, get more uninfluenced evidence." But our recognition of the factor of resistance often argues the other way, as, e.g., to get a few associations to a seemingly unconnected passing association, especially when the patient seems anxious to get past it. We simply won't *get* certain thoughts if we never intervene selectively, and those never-spoken thoughts may be crucial to our theme-tracing. The technical problem posed by these countervailing considerations is unsolved.

In my own practice, I usually follow a crude rule of thumb to avoid an intervention (whether requesting further associations to an item or voicing a conjecture) until I receive at least two fairly strong corroborator associations. If the corroborators are weak or middling, I wait for three. *Clinical example:* The manifest content of a male patient's dream involves reference to a urinal, so one conjecture, doubtless at higher strength in my associations because of his previous material, is that the ambition-achieve-ment-triumph-shame theme is cooking. (Cf. Freud 1974, p. 397, index item "Urethral erotism.") Half-way through the hour he passingly alludes to someone's headgear and suddenly recalls an unreported element of the dream's manifest content, to wit, that hanging on a wall peg in the urinal was a "green hat." This recalls to my mind, although not (unless he is editing) to his mind, a reference several weeks ago to a green hat. The patient had an uncle of whom he was fond and who used to be an avid mountaineer, given to recounting his mountain-climbing exploits to the boy. Sometimes when the uncle was a bit in his cups, he would don a green Tyrolean hat that he had brought back from Austria. The uncle had several times told the boy the story about how Mallory, when asked why he wanted so much to conquer the Matterhorn, responded, "Because it's there." The uncle would then usually go on to say that this answer showed the true spirit of the dedicated mountain climber, and that it should be the attitude of everybody toward life generally. We may choose to classify the passing allusion to a green hat as belonging to the same thematic cluster as this material. Later in the session, if it doesn't emerge spontaneously by a return to that element in the associations, we may decide (how?) to ask the patient to say more about the hat, ascertaining whether he says it was a Tyrolean hat and, even better, a Tyrolean hat "such as my uncle used to wear." I call this a strong corroborator for the obvious reason that the base rate of green-hat associations for patients in general, and even for this patient, is small. That generalization isn't negated by the fact that he *once before* had this thought. Once in scores or hundreds of hours is still a pretty low base rate. But more important is the fact that the sole previous mention is what enables us to link up a green hat with the achievement motive.

On the other hand, the presence of alternative and competing hypotheses tends to lower the corroborative power of our short-term prediction. How much it is lowered depends on how many competitors there are, how good a job they do of subsuming it, and, especially, on the antecedent or prior probability we attach to them. This prior probability is based upon general experience with persons in our clinical clientele but also, of course, upon the base rate for the particular patient. For example, in the present instance the patient, although not an alcoholic, has reported having a minor drinking problem; he has also revealed—although it has not been interpreted—a linkage between alcohol and the homoerotic theme. The uncle's tendency to tell this story when in his cups, and his further tendency to get a little boy to take a sip of beer, produces an unwelcome combination of competing hypotheses.

We have also the possibility, frequently criticized by antipsychoanalytic skeptics as a form of "fudging," that what appear at one level of analysis to be competing hypotheses are, at another level (or one *could* say, "when properly characterized thematically"), not competitors but aspects or facets of a core theme. In the present instance, at one level one might view the major determiner of manifest content about urinals and a Tyrolean hat as being *either* ambition or homoeroticism, and unfortunately the cluster of memories concerning the uncle can allude to both of these competing thematic hypotheses (and hence be useless for predictive corroboration). But we may without artificiality or double talk point out that achievement, especially that which involves marked features of competition with other males, has a connection with the theme of activity vs. passivity, strength vs. weakness, masculinity vs. femininity, the latent fear in males of being aggressively and/or homosexually overpowered by other males. (One thinks here of the ethologists' observations on our primate relatives, in which a "subordinate" male wards off threatening aggressive behavior from a dominant male by adopting the "sexual presenting" posture of females.) It could easily be that the additional associative material in the session is useful to us primarily as a means of separating these linked themes of homosexuality and achievement but, although we recognize their thematic linkage and overlapping dynamic sources, is primarily useful in differentiating the aspects of that common cluster as to what the "predominant" emphasis is at the moment. We do not need to force an arbitrary and psychologically false *dis*connection between the ambition theme and defense against passive feminine wishes and fears in order to ask and, in probability, answer the question, "Is the regnant [Murray 1938, pp. 45-54] wish-fear aspect of the theme today, and in the creation of the dream, that of homosexual passivity or that of competitive achievement ["male aggressiveness"]?"

The mathematical psychoanalysis of the utopian future would be plugging in values, or at least setting upper and lower bounds on values, of three probability numbers—none of which we know at the present time, but which are in principle knowable. The first probability is the prior probability of a particular theme, whether from patients in general or from the patients in a particular clinic or therapist clientele or, as presumably the most accurate value, this particular patient. The second probability number would be the conditional probability going from each competing theme to the associative or manifest content element taken to corroborate it. (It boggles the mind to reflect on how we would go about ascertaining that one! Yet it has some objective value, and therefore it should be possible to research it.) Third, we want the probability of the associative or manifest content item without reference to a particular dynamic source. The reason we want that is that we need such a number in the denominator of Bayes' Formula, and the only other way to ascertain that number is to know the probability of each of those elements on the whole set of competing dynamic or thematic hypotheses, a quantity that is as hard to get at as the second one, and maybe harder.

But there is worse to come. Even the pure argument from convergence gets its strength when the facts that converge upon the hypothesis are numerical point predictions, i.e., facts having low prior probability, an argument that can be made from either a Popperian standpoint of high risk-taking or from the non-Popperian standpoint of the Bayesians—in this respect, the two positions come to the same thing. We have to admit that the deductive model in which H strictly entails facts f_1, f_2, f_3 is not satisfied by psychoanalytic inference, although we can take some comfort from the fact that it is not satisfied in other documentary disciplines either. Freud points out, after discussing the dream work, that it would be nice if we had rules for actually constructing the manifest content from the latent dream thoughts arrived at by interpretation, but that we cannot do this. Hundreds, or in fact thousands, of alternative manifest contents could be generated from the list of latent dream thoughts, plus knowledge of the precipitating event of the dream day that mobilized the infantile wish, plus the stochastic nomologicals (if I may use such a strange phrase; or see Meehl 1978, pp. 812-814, "stochastologicals") that we designate by the terms condensation, displacement, plastic representation, secondary revision, and symbolism. The situation is rather like that of a prosecutor making a case for the hypothesis that the defendant killed the old lady with the axe, when he tries to show that there was a motive, an opportunity, and so forth. That the defendant decides to kill her does not tell us on which night he will do so, which weapon he will use, whether he will walk or drive his automobile or take a taxi, and the like. Whether a Hempel deductive model can be approximated here by designating a suitably broad *class* of facts as what is entailed, and by relying upon probabilistic implication (a notion regrettably unclear), I shall not discuss.

There is the further difficulty that the mind of the interpreter plays a somewhat different role in the argument, either from prediction or convergence, than it does in the physical and biological sciences. In order to "see how" a dream element or an association "alludes to" such and such a theme, one makes use of his own psyche in a way that most of us think is qualitatively different from the way in which we solve a quadratic equation. This is, of course, a deep and controversial topic, and one thinks of the nineteenth-century German philosophers of history who emphasized the qualitative difference between *Naturwissenschaften* and *Geisteswissenschaften*, or the famous thesis of Vico—which sounds so strange to contemporary behaviorist and objectivist ears—that man can understand history in a way he cannot understand inanimate nature, because of the fact that history, being human actions under human intentions, is of his own making, whereas the physical world is not! Whether or not Brentano was correct in saying that intentionality is the distinctive mark of the mental, I think we can properly say that there is a role played in psychological understanding of words and gestures that involves so much greater reliance upon the interpreter's psychological processes and content—the fact of the similarity between his mind and that of the other person—that it would be a case of "quantity being converted into quality." In recent years, the business of intentionality has led some philosophers, especially of the ordinary language movement, to deny that purposes can be causes and especially that reasons can be causes, a view I consider to be a mistake—as I am certain Freud would. (See my paper against Popper on "Clouds and Clocks," Meehl 1970; cf. also Feigl and Meehl 1974.)

In this connection, it is strange that one common objection to idiographic psychoanalytic inferences is that they seem too clever, an objection Fliess once made to Freud. I call attention to an unfair tactic of the critic, a "heads I win tails you lose" approach—unfair in that he relies on the relative weakness of the pure argument from convergence when the facts are limited in number and especially when the facts are qualitatively similar (like replicating the same experiment in chemistry five times, instead of doing five different *kinds* of experiments, which everybody recognizes as more probative); but when presented with a more complicated network in which this epistemological objection is not available, he then objects on the grounds that the reconstruction *is* too complicated! It's a kind of pincers movement between epistemology and ontology such that the psychoanalytic interpreter can't win. If the thing seems fairly simple, it doesn't converge via multiple strands (and hence not strongly), or it can be readily explained in some "nonmotivational" way, as by ordinary verbal habits and the like; if the structure is complex, so that many strands and cross connections exist, tending to make a better argument from convergence than the simple case, he thinks it unparsimonious or unplausible, "too pat," "too cute," to "too clever" for the unconscious to have done all this work.

My response to this pincers movement is to blunt the second, that is, the ontological pincer. I do not accept the principle of parsimony, and I am in good company there because neither does Sir Karl Popper, no obscurantist or mystic he! I see no reason to adopt the "postulate of impoverished reality" (as an eminent animal psychologist once called it). I see no reason to think that the human mind must be simpler than the uranium atom or the double helix or the world economy. Furthermore, the critic contradicts himself, because when he says that the interpreter is attributing to the subject's mind something too complicated for the mind to concoct, he is of course attributing that complicated a concoction to the interpreter himself! It doesn't make any sense to say, "Oh, nobody's head works like that," when the subject matter has arisen because Freud's head, for one, obviously *did* work like that. With this rebuttal, the second pincer is blunted and can be resharpened only by saying that the conscious mind does it; but obviously the unconscious couldn't be that complicated. I cannot imagine the faintest ground for such a categorical denial about a matter of theoretical substance; and it seems to me obvious that there are rich and numerous counter-examples to refute it.

In the experimental literature there is a vast body of non-Freudian research dealing with the establishment of mental sets, with the superior strength of thematic similarity over sound similarity in verbal conditioning, and so forth. Or, for that matter, take the righting reflex of the domestic cat. It was not until the advent of modern high speed photography, permitting a slow motion analysis of the movement, that this amazing feline talent was understood from the standpoint of physics. The cat's nervous system, wired in accordance with what must be an awesome complication in DNA coding, embodies, so to speak, certain principles of mechanics concerning the moment of inertia of a flywheel. As the cat begins to fall, he extends his rear limbs and adducts his forelimbs so that his front section has a small radius of gyration and hence a small moment of inertia, whereas his rear is large in these respects. The torque produced by appropriate contractions of the midriff musculature consequently produces a greater rotation of the forebody, so that the rear twists only a little and the front a great deal. When he is nearly "head up" from this first half of the maneuver, he then extends the forelimbs and adducts the rear limbs and again uses his muscles to apply opposed torques, but now his front has a greater moment of inertia and therefore twists less than the rear. This is an extraordinarily complicated behavior sequence, and nobody supposes that the cat is familiar with differential equations for the laws of mechanics. I therefore meet the second prong of the criticism with sublime confidence, because I know the critic cannot possibly demonstrate, as a general thesis of mammalian behavior, either (a) that the nervous system is incapable of complex computings or (b) that all of its complex computings are introspectable and verbally reportable.

As I pointed out in a previous paper on this subject (Meehl 1970), however, answering a silly objection is merely answering a silly objection, and does not suffice to make an affirmative case. Recognizing the weakness of the jigsaw analogy, and recognizing that complicated inference in the other documentary disciplines, while it might reassure us to know that we are in the same boat with historians and historical geologists and prosecuting attorneys and journalists is not, upon reflection, terribly reassuring in the face of methodological criticism; what are the possibilities for reducing the subjectivity of psychoanalytic inference? I hope it is clear that in putting this question I am not focusing on the possible development of some cookbook mechanical "objective" procedure to be employed in the interview. (See Meehl 1956, 1957.) I still try to distinguish between Hans Reichenbach's two contexts, i.e., the context of discovery ("how one comes to think of something") and the context of justification ("how one rationally supports whatever it is he has come to think of"). A caveat is imperative here, however, if we are to be intellectually honest: Distinguishing between these two contexts, important as it is epistemologically, should not lead to the mistaken idea that there is a clean qualitative distinction between a high probability (and hence, in some sense, "cookbook" process?) relied on, and a low probability in a more idiographically creative one. Freud himself relies upon both; witness his acceptance of Stekel's view that there are more or less standard dream symbols and scattered remarks that claim universality. For example, when the patient says, "I would not have thought of that," the remark can be taken as a quasi-definitive confirmation of the interpretation; or when he says, "My mind is a blank, I am not thinking of anything now," this is almost invariably a violation of the Fundamental Rule with respect to transference thoughts. With this warning, I focus then upon research procedures that aim to reduce the subjectivity, in the special sense that they ought to be rationally persuasive to a fair-minded skeptic who himself has not experienced or conducted analyses, leaving open to what extent such research might add to the list of high-probability "rules of thumb" that already exist in psychoanalytic lore. I list five approaches, without evaluating their merits and without claiming that they are entirely distinct, which they aren't.

First, I am convinced that sheer laziness, aggravated by excessive faith in the received tradition (plus the fact that most practitioners are not research-oriented, plus the lamentable dearth of psychoanalytic protocols), has prevented the application of simple and straightforward nonparametric statistics to test some basic, "first-level" psychoanalytic concepts (not the metapsychology yet). To take my favorite example because of its simplicity, when in 1943 I was seeing a patient (as a graduate student, before my own analysis) and proceeding in a modified Rogersian fashion, the patient reported a dream about firemen squirting water on a burning building. Almost all of the rest of the interview dealt with ambition as a motive and its correlated affects of triumph or shame, which struck me forcibly because it just happened that I had been reading one of Freud's long footnotes on the puzzling "urethral cluster" of urinary function, pyromania, and ambition. Being in those days extremely skeptical of psychoanalytic thinking, I resolved to take note in the future of exceptions to such a crude rule. Now, some forty years later, I can report that I have as yet never found one single exception *among male patients* to the induction that, if the manifest content of a dream deals with fire and water, the dominant theme of the rest of the session will be in the domain of Murray's *n Recognition* (or its aversive correlate *n Infavoidance*) and the associated affects of triumph and elation on the one side or shame and embarrassment on the other. Now such a finding (which I cannot record in the research literature because, stupidly enough, I haven't been keeping affirmative tallies but only waiting vainly to find the first exception) does not exist as a "statistic" because nobody—myself included—has bothered to analyze it systematically. And what I am claiming about this relationship is that it yields a fourfold table with one empty cell—a thing we almost never find in the behavioral sciences. I believe that simple chi squares applied to such intermediate levels of psychoanalytic inference should be computed, and that no new techniques need to be developed for this purpose.

Second, it is possible that the application of existing statistical techniques of various complexity, such as factor analysis and cluster analysis, might be helpful. (Cf. Luborsky's factor analysis of time-series correlations based on Cattell's P-technique, in Luborsky, 1953.) I am inclined to doubt the value of these approaches because I am not persuaded that the factor-analytic model is the appropriate formalism for this subject matter. I think it fair to say that what few studies have been done along these lines have not been illuminating.

A third possibility is the application of new formalisms, kinds of mathematics with which social scientists do not customarily operate, chosen for their greater structural appropriateness to the problems of the psychoanalytic hour. I am particularly open to this one because my own current research is in taxometrics and has convinced me that a psychologist with even my modest mathematical competence can come up with new search techniques in statistics that are superior to those customarily relied on. The taxometric procedures I have developed over the last decade appear, so far, to be more sensitive and powerful than such familiar methods as factor analysis, cluster analysis, hierarchical clustering, and Stephenson's Q-technique. (I do not believe that interdisciplinary exchange between mathematical statisticians and psychoanalytic psychologists is likely to be fruitful unless each party possesses some real understanding of the other one's subject matter, a rare situation.) If someone were to ask me what novel formalisms I have in mind, I don't know enough mathematics to come up with good examples, although graph theory with its nonmetric theorems about paths, nodes, and strands of networks is urged by a colleague of mine. I do not suggest that a really new branch of mathematics needs to be invented, but I would not exclude even that, since I think the social sciences have had an unfortunate tendency to assume that their kind of mathematics has to look like the mathematics that has been so powerful in sciences like chemistry and physics. There have been kinds of mathematics that had no application to empirical scientific questions for long periods, the standard example in the history of science being Galois's invention of group theory in the 1820s, a branch of mathematics that found no application in any empirical science until somebody realized over a century later that it was useful in quantum theory.

A fourth possibility is along the lines of computer programming for complex content analysis that we find in the book by Stone, Dumphy, et al., *The General Inquirer* (1966). In my 1954 monograph on statistical prediction I tried to make a case against T. R. Sarbin and others who imagined that there could be a "mechanical" method for analyzing psychotherapeutic material, at least at any but the most trivial level. I think today we must be extremely careful in setting limits on the computer's powers. Ten years ago computers could play a pretty good game of checkers, and had been programmed to search for more parsimonious proofs of certain theorems in Russell and Whitehead's *Principia Mathema-*

tica; but no computer, although it could obey the *rules* of chess, could play even a passable chess game. Recently in Minneapolis two computers were entered as competitors in our Minnesota Chess Open Tournament, and they did rather well. In 1979 a computer got a draw with an international grand master. On the other hand, machine translation of foreign languages has turned out to be so intractable a "context" problem (so my colleagues in psycholinguistics inform me) that even the Russians have dropped it. I don't really want to push the computer as psychoanalyst. I merely warn you that it would be rash for people like me who do not possess computer expertise to say, "Well, whatever they program the damn things to do, it's obvious they will never be able to interpret dreams." Maybe they will, and maybe they won't.

No doubt one reason for persistence of the Achensee Question is that objectification of psychoanalytic inference, in any form that would be persuasive to psychometricians, behaviorists, and other social science skeptics who have scientifc doubts about the validity of the psychoanalytic interview as a data source, would presumably rely on well-corroborated *quantitative* lawlike statements from several disciplines that are themselves in their scientific infancy. Since psychoanalysis deals mainly with words, the most obvious example of such a related discipline would be psycholinguistics, whose conceptual and evidentiary problems (I am reliably informed by its practitioners) are not in much better scientific shape than psychoanalysis itself. But cognitive psychology more broadly —the psychology of imagery, the general psychology of motivation, the social psychology of the psychoanalyst as audience, not to mention the recent non-Freudian experimental work on the mental and cerebral machinery of dreams—would all have to be put together in some utopian integrated whole before it would be possible to write the sort of nomological or stochastological (Meehl 1978, pp. 812-814) relations that would generate any sort of plausible numerical probabilities for the inferences. I am not here lamenting lack of exact numerical values; I am referring rather to the difficulty of expressing crude forms of functional dependencies that would yield any numbers at all. Suppose the "objectifier" requires a kind of evidentiary appeal *more explicit* than, "Well, just look at these clusters and sequences of speech and gesture. Don't you agree that they sort of hang together, if you view them as under the control of a certain (nonreported) guiding theme?" I am not accepting here that it is imperative to meet such a standard of explicitness. I am simply reacting to the social fact that most objectifiers would not be satisfied with less. Now since we don't have a good *general* psychology of language, of imagery, of dreams, of short-term fluctuations in state variables like anxiety or anger or erotic impulse, (and obviously we cannot turn to psychoanalysis itself for that when the Achensee Question is before us), then it is almost pointless for anyone in the present state of knowledge to even speculate utopianly about how such a psychology might look. We do not even know whether the kinds of mathematics favored among social scientists would be appropriate. Thus, for instance, it may be that the mathematical formalisms of factor analysis (Harman 1976) or taxometrics (Meehl and Golden 1982) are much less appropriate for tracing themes in the psychoanalytic interview than, say, something like graph theory (Read 1972 pp. 914-958) or finite stochastic processes (Kemeny, Snell, and Thompson 1957). All of this is music of the future, and I shall not discuss it further.

Does that mean that we have to put the whole thing on the shelf for another century, pending satisfactory development and integration of these underpinning disciplines? One hopes not, and let me try to say why that *may* not be required (although it may!). When we think about the relationship of a mathematical psycholinguistics to a mathematical science of short-term motivational state variables, we are fantasizing a rather detailed prediction (or explanation) of the analysand's choice of words and their sequence. Not perhaps the specific word, which is likely to be beyond the power of utopian psycholinguistics, just as the prediction of precisely how a collapsed bridge falls today is beyond our advanced science of physics; but at least a kind of intermediate-level analysis of the verbal material. Suppose we can bypass that by permitting a somewhat more global categorization of the patient's discourse, allowing for a kind of global subsumption under motivational themes by some kind of partly objective, partly subjective procedure of the sort that I discuss in the research proposal below. That is, we aren't (pointlessly?) trying to predict exactly which verb, or even which class of active or passive verbs, will be emitted during a specified small time interval of the session, having been preceded by a narrowly specified number of allusions to the "mother theme," or whatever. What we are saying is that we have a small collection of rather broadly identified remarks (or gestures or parapraxes), and we hope to persuade the skeptic (assuming we have convinced ourselves) that it is more reasonable to construe this small set of happenings as produced by one dynamic state or event—I shall simply refer to *a theme* in what follows—rather than to reject that possibility in favor of multiple separate psychological hypotheses that are unrelated dynamically or thematically, but each of which might easily be capable of explaining the *particular* individual remark, gesture, or parapraxis that is treated as *its* explanandum.

In several places Freud likens the analytic detective work to that of a criminal investigation, and the analogy is a good one on several obvious counts. We do not consider the prosecutor's summation of evidence to the jury, or the judge's standardized instructions on how to assess this evidence, as somehow irrational or mystical or intuitive, merely because it is not possible to express the invoked probabilistic laws *in numerical form,* let alone ascribe a net joint "empirical support" *number* to the total evidence. As I have said elsewhere in this essay, there are other disciplines that we consider intellectually respectable and worth pursuing and, in certain important private and social decision makings, even deserving of our reliance in grave matters, despite the absence of an effective procedure (algorithm) for computing a numerical probability attachable to outcomes, or to alleged explanations of events that have already taken place.

I don't intend to get much mileage out of those nonpsychoanalytic examples. But it is important, before we proceed, to recognize that the domain of the rational and empirical is not identical with the domain of the statistical-numerical, as some overscientized psychologists and sociologists mistakenly believe. What reservations must be put by a rational skeptic on the force of such extrapsychoanalytic analogies I discuss elsewhere herein; and I agree with the skeptics that those reservations are discouragingly strong. But the point is that we can and do talk about evidence converging, of explanations as being parsimonious or needlessly complicated, and so forth, with an idealized inductive logic (if you don't like that phrase, an idealized empirical methodology), including one that accepts a particular *kind of inferential structure*—say, a Bayesian inference—even though, in the concrete domain of application, we cannot plug in any actual *probability numbers*. If we do hazard mention of numerical values—and this is an important point one should never forget in thinking about psychoanalysis and allied subjects—at most we have some upper and lower bounds, acceptable by almost all members of our clinical and scientific community, on the numerical values of expectedness, priors, and conditionals in Bayes's Formula. I think it is generally agreed that we can "think Bayesian" and, thinking Bayesian, can come up with some reasonable numerical bounds on likelihood ratios and posterior probabilities, without claiming to have determined relative frequencies empirically as the initial numbers that get plugged into Bayes's Theorem. (See Carnap's discussion of the two kinds of probability, Carnap 1950/1962, passim.) For someone who is uncomfortable even with that weak a use of a widely accepted formalism (which is, after all, a high-school truth of combinatorics and does not require one to be a "strong Bayesian" in the sense of current statistical controversy), at the very least one can point to a list of separate causal hypotheses to explain half a dozen interview phenomena and then to a single causal hypothesis—the psychoanalytic one—as doing the job that it takes half a dozen others to do as its joint competitors.

Example: A patient drops her wedding ring down the toilet. In speaking of her husband, Henry, she mistakenly refers to him as George, the name of an old flame of hers. An evening dining out in celebration of their wedding anniversary was prevented because the patient came down with a severe headache. Without any kinds of challenge to the contrary, she "spontaneously" makes four statements at different times in the interview about what a fine man her husband is, how fortunate she is that she married him, and so on. Now you don't have to go through any elaborate psycholinguistics or even any of that "within-safe-bounds" application of Bayes's Theorem, to argue that it may be simpler and more plausible to attribute these four phenomena to the unreported guiding influence of a single psychological entity, namely, some ambivalence about her husband, than to deal with the four of them "separately." In the latter case we would be, say, attributing the wedding ring parapraxis to nondynamic clumsiness (the patient happens to be at the low end of the O'Connor Finger Dexterity Test); the anniversary headache to insufficient sleep and oversmoking; the misnaming George-for-Henry to the fact that old George was in town recently and called her up; and the unprovoked overemphasis on her happy marriage to some recent observations on the unhappy wives that are her neighbors on either side. Setting aside the independent testing of those alternatives, it's basically a simple matter. We have four competing hypotheses whose separate prior probabilities are not much higher than that of the marital ambivalence hypothesis, although some of them might be a little higher and others a little lower. We think that the conditional probabilities are also roughly in the same ball park numerically as the conditional probability of each of the four observations upon the ambivalence hypothesis. The argument that one would make if he knew nothing about statistics, Bayesian inference, or inductive logic but did know legal

practice, or common sense, or diagnosing what's the matter with somebody's carburetor, would be: "We can easily explain these four facts with
one simple hypothesis, so why not prefer doing it that way?"

It is not difficult to tighten this example up a bit and make it semi-formal,
to such an extent that it is the skeptic who is put on the spot—provided, of
course, that he will accept certain reasonable bounds or tolerances on the
estimated numbers. Thus, suppose the average value of the four priors is
not greater than the prior on the ambivalence hypothesis; and suppose the
average value of the four conditionals required to mediate an explanation of
each of the four observations is not greater than the average conditional of
the four observations on the ambivalence theory. Since the "expectedness"
in the denominator of Bayes's Theorem is some unknown but determinate
true value (however we break it up into the explanatory components
associated with the possibilities), and since a dispersion of four probabilities yields a product less than the fourth power of their average, then when
we compute a likelihood ratio for the ambivalence hypothesis against the
conjunction of the separate four (assuming these can be treated as
essentially independent with respect to their priors, quite apart from
whether they are explanations of the four explananda), things cancel out,
and we have a ratio of the prior on the ambivalence hypothesis to the
product of the other four priors. If, as assumed above, the dynamic
hypothesis is at least as probable antecedently as the other four average
priors, a lower bound on this likelihood ratio is the reciprocal of the prior
cubed. So that even if the priors were all given as one-half—an unreason-
ably large value for this kind of material—we still get a likelihood ratio, on
the four facts, of around eight to one in favor of the psychodynamic
construction.

Before setting out my fifth (and, as I think, most hopeful) approach to
psychoanalytic theme-tracing, it will help clarify a proposed method to say
a bit more about theory and observation, at the risk of boringly repetitious
overkill. I think our characterization of the theory/observation relation is
especially important here because (a) both critics and defenders of
psychoanalytic inference have tended to misformulate the issue in such a
way as to prevent fruitful conversation, and (b) the pervasive influence of
the antipositivist line that all observations are theory-infected (Kuhn,
Feyerabend, and Popper) lends itself readily to obscurantist abuse in fields
like psychopathology. Having mentioned Popper in this connection, I
must make clear that I do not impute to him or his followers any such abuse;

and, as is well known, Popper himself is extremely skeptical about the alleged scientific status of psychoanalytic theory.

As a starter, let us be clear that there are two methodological truisms concerning the perceptions and subsumptions (I think here the line between these two need not be nicely drawn) of "experts," in which the expertise is partly "observational" and partly "theoretical." One need not be appreciably pro-Kuhnian, let alone pro-Feyerabendian, as I most certainly am not, to know that technical training, whether in the methods of historiography or electron microscopy or psychodiagnosis or criminal investigation, enables the trained individual to perceive (and I mean literally *perceive,* in a very narrow sense of that term that would have been acceptable even to Vienna positivism) things that the untrained individual does not perceive. Anybody who has taken an undergraduate zoology class knows this; it cannot generate a dispute among informed persons, whatever their views may be on epistemology or history of science. On the other hand, it is equally obvious, whether from scientific examples or everyday life, that people tend to see and hear things they expect to see and hear, and that, given a particular close-to-rock-bottom "perceptual report," the expert's theoretical conceptions will affect under what rubric he subsumes the observation. These truisms are so obvious that my philosophy-trained readers, who comfortably accept both, will be puzzled to learn that the situation is otherwise in psychology; but it is. Psychoanalytic clinicians have become accustomed to relying on analogies of psychoanalytic method to the microscope or telescope or whatever, together with the alleged combination of sensitization to unconscious material and reduced defensive interferences, supposedly producing *some* variant of "objectivity," consequent upon the analyst's personal analysis and the corrective experiences of his control cases. Nonpsychoanalytic clinicians, especially those coming from the behaviorist tradition, fault the analytic theme-detector for failing to provide "objective operational definitions" in terms of the behavior itself, of constantly going behind the data to concoct unparsimonious causal explanations, and the like. The same polarization is true for nonpractitioners, i.e., academic theoreticians with psychoanalytic versus antipsychoanalytic orientations. In these disputes, when the parties are not simply talking past each other (the commonest case), the difficulty is that each thinks that his methodological principle clashes with his opponent's methodological principle, which is almost silly on the face of it. Consider the following two statements:

M_1: "It takes a specially skilled, specially trained, clinically sensitized, and theoretically sophisticated observer to notice certain sorts of behavioral properties and to subsume them under a psychodynamically meaningful rubric."

M_2: "Training in a special kind of observation predicated upon a certain theory of (unobserved) states and events underlying the behavior being observed may sometimes have the result, and presents always the danger, of seeing things that aren't there, or subsuming them in arbitrary ways, or forcing a conceptual meaning upon them that has no correspondence to the actual causal origins of the behavior observed."

These two methodological assertions M_1 and M_2 are perfectly compatible with each other, as is obvious so soon as we state them explicitly. In fact, in large part each of them flows from the *same* imputed characteristics of the "sensitive" and "insensitive" observer! They just aren't in conflict with one another as assertions. Why would anyone have supposed they were, or (better), argued as if he supposed that?

The problem is that they are in a kind of pragmatic conflict, even though they are logically and nomologically compatible, in that there is a tension generated in anyone who accepts both of them, between his aim to discover the nonobvious (an aim pursued by reliance on M_1) and his aim to avoid theoretically generated self-deception or projection (an aim aided by remembering M_2). This epistemic tension between two aims that are both reasonable and legitimate is not, I suggest, any more puzzling than some better-understood cases, such as the fact in statistical inference that one has tension between his desire to avoid Type I errors (falsely inferring parameter difference from the observed statistical trend) and Type II errors (wrongly sticking with the null hypothesis of "no difference" when there is a difference in the state of nature.) There isn't any mathematical contradiction, and under stated and rather general conditions, one uses a single coherent mathematical model to compute the trade off and assign it numerical values. Nor is there any sort of philosophical collision or semantic confusion. It is simply a sad fact about the human epistemic situation, even for a fairly developed and rigorous science like mathematical statistics, that when we wish to apply it to such a simple task as detecting bad batches of shotgun shells, we experience a pragmatic tension, and that tension has, so to speak, a realistic (not a "neurotic") basis.

Similarly, in dealing with material of appreciable complexity such as the stream of speech and gesture produced by an analysand, if the observer (NB: *noticer, attender*), classifier, and interpreter lacks certain kinds of training and expertise, he won't be able to do the job. But if he does have those kinds of training and expertise, he may be seduced to do a job that is too good. So the first thing one has to do in thinking rationally about this problem is to get away from the pseudo-collision of Principles M_1 and M_2, to wholeheartedly and unreservedly accept both of them, recognizing that the two principles are logically consistent and even flow from the same facts about the human observer and interpreter and the effects of his training and experience. They lead to a pragmatic tension generated by our reasonable desire to avoid two kinds of errors which might be described as errors of omission versus commission, errors of "under-discovery" versus "over-belief," or even William James' famous errors of the tenderminded and the toughminded. One thinks of James' comment on William Kingdon Clifford's stringent "ethics of belief," to the effect that Professor Clifford apparently thought that the worst possible fate that could befall a man was to believe something that wasn't so!

Excursus on Observations

Because of some current tendency in psychology to rely on the Kuhn-Feyerabend "theory-laden" doctrine, I shall permit myself here a few general remarks on the controversy about meaning invariance and theory ladenness, because it seems to me that the situation in psychology is importantly different from the favorite examples employed by philosophers and historians of science in discussing this difficult and important issue. First, the theory-ladenness of observational statements and the associated meaning-variance is most clearly present and most important in what Feyerabend calls "cosmological theories," i.e., theories that say something about everything there is (as he once put it to me in discussion). I believe a psychologist can be seduced into attributing undue importance to the Kuhn-Feyerabend point if he takes it as a matter of course that philosophical arguments concerning meaning variance and theory-ladenness applying to the Copernican hypothesis or relativity theory or (less clearly?) quantum mechanics, apply equally strongly *and equally importantly* to rats running mazes or psychoanalytic patients speaking English on the couch. I recall a Ph.D. oral examination in which the candidate, a passionate and somewhat dogmatic Kuhn-Feyerabend disciple, was asked by one of my philosophy colleagues to explain just how the protocol "Rat number 7 turned right in the maze and locomoted to the right-hand goal box" was theory-laden; in what sense the experimenter would experience a perceptual gestalt-switch if the experiment converted him from being a Tolmanite to being a Skinnerian; and which of the words in the protocol sentence would undergo a meaning change as a result of his theoretical conversion? It was a good question, and our surprise was not at the candidate's utter inability to deal with it, but the extent to which he was prepared to go in a hopeless cause. What he said was that the very genidenity of rat number 7 from yesterday's trial run to today's test run, and the very fact that we called today's run a "test run," were theory-laden. (I think this is pathetic, but if there are readers who don't, I won't press the point.) Of course the "theory" that today's rat is genidentical with yesterday's rat, is not a psychological theory in any sensible or interesting use of the word "psychological," and even if it were, it is firmly believed (I should say presupposed) by both Tolman and Skinner as well as all of their followers, including those that are in transition to the opposing paradigm. If one wants to say that the ordinary common-sense world view that macro-objects as described in Carnap's "physical thing language" are spatio-temporally continuous constitutes a kind of theory, I have no strong objection to this, although I don't find it an illuminating way to speak. But the point is that if it is a theory, it is a theory that cuts across psychological theories of animal learning. And, of course, it's for that reason that the candidate was unable to point to any of the words in the sentence that would have undergone a meaning change as a result of the experimental psychologist's conversion from one theory of learning to another.

The question whether theory defines the concept "test day" is another piece of obscurantism, since, although the theory is what *leads us to make a certain test* (not in dispute), this fact is quite incapable of showing meaning variance or theory-ladenness of the behavioral terms describing what the rat *does* on the test. Witness the fact that an undergraduate with correct 20/20 vision and normal hearing, who is nonpsychotic and familiar with the macro-object English terms "right" and "left," could be safely trusted to make the relevant observations even if he knew nothing about the latent learning concept or about differences between the rat's condition (hungry? anxious?) this night and on previous nights. In most latent learning experiments, the question about test (or critical pre-test) occasions "How is this night different from all other nights?" has its answer either in the precedent *operations* of feeding (or fasting), or what, if anything, is present in the goalbox that wasn't there on previous occasions.

In psychological research on behavior, whether animal or human, the theory-dependence of observations and the theory-ladenness of operational meanings involve a cluster of questions that are related but clearly distinguishable, as follows:

(1) Instruments of observation are usually theoretically understood, whether in the inorganic or life sciences. (Not quite always, e.g., the Wasserman test!). Such instruments are used (a) to control all causally influential variables, (b) to extend the human sensorium, (c) to dispense with human perception by substituting a physical record for the human sense-report, and (d) to replace human memory by a more reliable record (storage and retrieval). In physics and astronomy (I do not know about chemistry, but I imagine there also) there is sometimes a theoretical question concerning the extent to which the instrument can be relied upon for a certain experimental purpose, because the laws of physics involved as auxiliaries in the test proposed are themselves "connected," via overlapping theoretical terms, with the conjectured laws (including dispositional, causal, and compostional properties attributed to conjectural entities) that the experiment is designed to test. There are problematic entities whose role in the nomological net would be altered importantly (for the experiment) if certain other adjacent regions of the net were altered in such-and-such respects. This may happen in psychology also, but it is not as common in good research as those who emphasize theory-ladenness seem to assume. Whether the observing instrument is the unaided human eye, the kymographic recording of a lever press, or a photo cell showing which alley the rat passed through, these are all instruments that rely on no currently competing theory of animal learning. For that matter, no reasonable person believes their deliverances are dependent upon a theory of how the rat's brain works in choosing which direction to turn in a T-maze.

(2) Auxiliary theories, and hence auxiliary particularistic hypotheses formulated in terms of these theories, may be problematic but essentially independent of the subject matter under study. An example would be an auxiliary theory about the validity of the Rorschach Ink Blot Test as a detector of subtle cognitive slippage in schizoidia. We might wish to rely on this in testing the dominant schizogene theory of that disorder, but it is an auxiliary theory that is highly problematic and, in fact, as problematic as

the major substantive hypothesis under test. (Meehl 1978, pp. 818-821) A less problematic auxiliary theory used in listening to psychoanalytic session discourse would be the vague (but not empty) "theory" that speech is determined by interactive causal chains between the current stimulating field (is the analyst moving restlessly in his chair?) and the subject's inner states and dispositions, whether those latter are characterized in brain language (not presently available except for very rough statements of certain systems or regions of the brain that are more relevant to verbal behavior than, say, to keeping one's balance when standing up) or in the molar language (whether behavioristic or phenomenological) again. We tacitly presuppose the "theory" that people cannot speak a language they have not learned. Suppose I know that a patient is a classics professor, and that his wife's name is Irene. Then the possibility that a dream about a "peace conference" conjoined with an association during the session about "knocking off a piece" is connected with his wife, is plausible only because of our tacit auxiliary theories about language. If he didn't know any Greek, he could still link in his unconscious the homophones "peace" and "piece," but he would not link that with the name of his wife because he wouldn't know the etymology of the female name "Irene." (NB: If the therapist doesn't know that amount of Greek, he won't be able to create the thematic conjecture either!)

(3) *Did* this particular investigator design the experiment in the light of the theory he was interested in testing (a biographical question of fact)?

(4) *Could* a seasoned, clever researcher have designed this experiment without the disputed learning theory in mind—perhaps without any learning theory in mind except the minimal nondisputed statement (hardly a technical doctrine of animal psychology) that organisms usually have a knack to "learn things?"

(5) Granted that a purely atheoretical, "blind inductive" ethologist *could* have designed a certain experiment, is it *likely* that he would have designed this particular one and chosen to observe the things he chose to observe, absent this theory? Ditto, absent some other theory about learning?

(6) Whatever has led, or plausibly could lead, to the designing of the experiment and the specification of what was to be observed on the test night, *could* a theoretically naive person with normal sensory equipment make and record the observations without the theory?

It is a grave source of confusion to lump all these questions together by the single seductive statement, "Observations are theory-laden." One can make entirely too much of the simple fact that people's visual and auditory perceptions, and their imposition of a spatial or other reference frame upon what comes to them through distance receptors, is always influenced by their implicit beliefs about the world (genidentity of macro-objects, etc.), their previous experiences, their verbal habits, and their culture's dominant interests and values. Nobody today holds—what one doubts even our philosophical ancestors held—that the favorite epistemological prescriptions of Vienna positivism can be founded upon theorems in formal logic, as if we were to pretend we did not know that we are organisms, occupying space-time, having distance receptors, wired somewhat similarly to one another, able to remember, to speak, and so forth. The agreement among scientists (or critical, skeptical, tough-minded contemporary nonscientists) observing a witchcraft trial as to which witnesses were believable, which sense modalities were generally reliable, could not be achieved without experience of the human mind and society. Does anybody dispute this? On the other hand, these kinds of minimal "epistemological basics" are part of our general theory of macro-objects, formulable in Carnap's "physical thing language," and our knowledge of human beings as observers, recorders, rememberers, and reporters.

These shared, common-sense, well-tested notions are *not* "theories" in the interesting and complex sense we have in mind when we talk about constructing a satisfactory psychology of perception, or an adequate psychology of cognition, or a sound descriptive (non-normative) theory of rat decision making. It is not perfectionist epistemology or vulgar positivism—nor, I think, even an antipathy to ghosts, leprechauns, and capricious deities (not to mention fortune tellers and other epistemological disreputables)—that we form the societal habit of employing instruments of observation and recording as often as we do in the sciences whether these sciences deal with organic or inorganic subject matter. I must confess I see nothing complicated, philosophically earthshaking, or especially interesting about the fact that since humans are not always accurate in noticing whether a visual or auditory stimulus occurred first, one has a problem in such cases about relying on human observation For that reason one substitutes nonhuman instruments whose subsequent deliverances to the human perceiver are of a different form, a form chosen by the instrument maker and the scientist who wanted the instrument built so as to be less ambiguous perceptually. (Incidentally, the notion that we cannot do any basic epistemology or formulate any workable methodology of science prior to having respectably developed sciences of perception, cognition, sociology of knowledge, etc., seems very odd to me when I reflect that astronomy, chemistry, physics, and physiology were well advanced sciences, relying on a developed "method" that made them so superior to pre-Galilean thought, long before psychology or sociology were even conceived as scientific disciplines.)

Do we really have to write a big book about this? Maybe for a critical epistemology of astrophysics, or quark theory; but not, I urge, for animal conditioning, or classical psychometrics, *or even psychoanalysis*. In N.R. Campbell's (today underrated) *Physics: The Elements* (1920/1957) he has a nice section (pp. 29-32) about judgments for which universal assent can be obtained. He is of course exaggerating somewhat when he refers to these judgments as being "universal," and one supposes that even if we eliminate known liars, psychotics, persons with aberrant vision and hearing—things which can be independently tested—we still might have a minuscule fraction of observers who would puzzlingly persist in making judgments out of line with the rest of us. Note that even in such extremely rare cases, it is usually possible without vicious ad hockery to give a satisfactory *causal* explanation of why the incurably deviant response is made. But let's admit that it is not always so. Nothing in the method of science requires that we should come to a quick decision in such cases. What happens is that (a) we get a steady accumulation of protocols from the reliable observers (calibrated in the past in other contexts); (b) we are able to give a satisfactory explanation, again without vicious ad hockery, of the discrepant observer's findings; so (c) we decide to "pull a Neurath" (Neurath 1933/1959) by simply refusing to admit the aberrant protocol into the corpus. We prefer to rest decision (c) on a conjunction of (a) and (b), but if (a) is sufficiently extensive and varied, and the theory forbidding the aberrant protocol is doing well enough, we will dispense with (b)—while keeping our hopes up. Thus, e.g., physicists dealt "Neurath-style" with the irksome Dayton C. Miller ether-drift protocol for some 30 years before Shankland and colleagues provided an acceptable (b). But I think it unfortunate that psychologists do not at least read and reflect upon Campbell's discussion or that of an undervalued contributor—C.R. Kordig (1971a, 1971b, 1971c, 1973) before jumping on the popular Kuhn-Feyerabend bandwagon. Campbell sets up three kinds of perceptual judgments on which universal assent could be attained among normal calibrated observers—to wit, judgments of simultaneity, consecutiveness, and betweenness in time; judgments of coincidence and betweenness in space; and judgments of number. Now this is a pretty good list. Everyone knows that the plausibility of Eddington's famous statement "science consists of pointer readings" (wrong, but not a stupid remark at the time) arises from the very large extent to which pointer readings of instruments substituted for the human eye and ear at the first interface with the phenomenon under study have replaced the human eye and ear as Aristotle and Pliny used them. Again, I don't see anything mysterious about this, and it flows directly from Campbell's principles. That a pointer lies *somewhere between* the line marked "7" and the line marked "8" on a dial is an example of spatial betweenness; that this may be an observation made *during* a few seconds interval during which a single tone was being sounded is an example of (local grammar) temporal coincidence. Now whenever we find that most people's theoretical notions unduly affect consensus in perceiving alleged N-rays (or auras, or levitations)—not to mention "anxiety" in the rat or "latent hostility" in the psychiatric patient—we have recourse to some strategem in which we try to achieve an equivalent of the physicist's pointer reading *without throwing away perceptual input that is relevant*.

I cannot insist too strongly that the raw data of the psychoanalytic hour are speech, gesture, and posture *and*—unless the analyst claims to have telepathic powers—*nothing else*. If my psychodynamic theory allows me to say that I *"observed* the patient's hostility" when the existence of the patient's (latent, nonreported, denied) hostility is what is under dispute between me and, say, a certain kind of behaviorist, then my observations are here theory-infected to an undesirable extent, and I am scientifically obligated to move down in the epistemic hierarchy closer to the behavior flux itself (MacCorquodale and Meehl 1954, pp. 220 ff). If I refuse to do this, or at least to concede the epistemic necessity to do it (given your denial and request for my evidence), then I may be a perceptive analyst and a skillful healer, but I have opted out of the game "science." We have now reached little-boy impasse, "My Dad can lick your Dad," "Can not!", "Can so!" "Can not!," etc., irresolubly ad nauseam.

Now the rock-bottom data in the clinical example below I shall confine to the patient's speech and, in fact, to features of the speech that can be fairly adequately represented in the transcript without a tape, although everyone knows "something is lost" thereby. Thus pauses, gross fluctuations in rate of speech, and volume level can be measured objectively from the tape and indicated by a suitable notation in the transcript. We are, however, going to rely almost wholly upon the content, because that's what my proposed theme-tracing procedure deals with. (One cannot deal with everything at once!) In a way, our problem of tradeoff between errors of omission (insensitive, untrained listener not perceiving) and errors of commission (theoretically biased listener projecting) can be stated quite simply in the light of what was said above. Stating it thus leads fairly directly to my theme-tracing proposal. The problem is this: Restricted segments of the verbal output (blocks, hereafter) can be subsumed under a variety of thematic rubrics, and which rubric it is subsumed under by a psychoanalytically sophisticated listener or protocol reader will, of course, be affected by his tentative subsumption of the other blocks, all guided by his acceptance of the minimal Freudian theory. Now this interblock influence is what we want to get away from if possible, because we are sensitive to the skeptic's objection that "human ingenuity can fit almost anything together if you're not too fussy," based upon the several lines of criticism set out above. We want the clinician to use his psychodynamically sophisticated mind to achieve the subsumption of a verbal block. Yet we would like the conclusion of his subsuming activity to have as a work product something that cannot be easily defeated by the skeptic's "Well, so what? It's all very clever, you do make it sound as if it hangs together; but I am not impressed."

It is not that we think that a proper method of protocol analysis should be *coercive* with respect to a sufficiently determined skeptic, which it is foolish to require. We do not require that "good arguments" for the roundess of the earth be capable of convincing the (still surviving!) Flat Earth Society members. No, the problem is not that we foolishly seek an automated truth-finding machine, some kind of algorithm for particularistic inductive inference in this kind of material, which very likely cannot be come by and is not anticipated even for the utopian phase of psycholinguistics or psychodynamics. Rather the problem is that we ourselves, "pro-Freud leaners" with scientific training and epistemological sophistication, are fully aware of the merits of the skeptic's complaint. That being so, our aim is to use the clinically sensitive mind of the psychoanalytically trained listener or reader to discern the red thread running through the discourse, to carry out the theme-tracing that we are reasonably sure could not be done by a competent clerk or a non-Freudian psycholinguist (Meehl 1954,

Chapter 6, 7 and passim). But we would like those red thread discernings to be at what might be called an intermediate level of complexity and persuasiveness, so that given the thematic subsumption of blocks of the discourse, the preponderance or strength of allusions to the "segments of red thread" recurring in the various blocks will speak for themselves when we add them up across blocks. That is what the proposed method of theme-tracing tries to do.

So I present this somewhat simple-minded approach, which involves no fancy mathematics but which attempts to do partial justice to Freud's notion of the "red thread of allusions" running through the patient's material. The approach appeals to me because it combines the advantages of the skilled, empathic, thematically perceptive clinical brain with at least partial safeguards against contamination effects. Let's put the question thus: "How can we make use of the classificatory powers of the skilled mind without that mind's classifications being contaminated by theory, hence rendering the whole process especially vulnerable to the Achensee Question?" We want to use the clinician's brain to do something more complicated than associate tables with chairs or plug in *mother figure* for "empress," or, idiographically, to search his memory bank for previous references by this patient to green Tyrolean hats to find that they always relate to the uncle. But if we start using the skilled clinical mind to do something more interesting and complicated than these jobs, then comes the critic with the Achensee objection, telling us that we are brain-washed Freudians and therefore we naturally read our thoughts into the patients and "see" meanings, metaphors, and subtle allusions when they are not there.

To sharpen somewhat the distinction between the first method (doing a simple chi-square on a four-fold table showing the association between an objectively, i.e., clerically scorable dream content on the one hand and a subjective, impressionistic, skilled clinician's uncontaminated discernment of the theme in the subsequent associations on the other), and the fifth method, which I shall christen "Topic Block Theme Tracing," I choose a short clinical example that is in most respects as straightforward as the fire/ambition example, but *not quite*—and the "not quite" introduces terrible complexities for objective scoring, even by such a clever device as the computerized General Inquirer. A patient begins the session by reporting: *I dreamed there was a peculiar water pipe sticking into my kitchen*. My Radovian training suggests a minor intervention here, for clarification only, so I ask, "Peculiar?," to which the patient responds, "Yes, it was a peculiar water pipe because it seemed to have some kind of a cap on it, I couldn't understand how it could work." The standard symbology here [*waterpipe* = penis, *kitchen* = female genitals] is familiar to undergraduates, but knowing it would only permit the trained clerk or the supertrained clerk (e.g., the General Inquirer) to infer a heterosexual wish. What makes it interesting is the "peculiar cap," juxtaposed with the word "work" [= coitus, at least semi-standard]. Here an idiographic low-frequency consideration enters our minds, mediated by the fact that the patient is Jewish and I am gentile. I conjecture that the capped pipe is an uncircumcised (i.e., gentile) penis, and that the dream expresses an erotic positive transference impulse. I further conjecture (more tentatively) that the current manifestation of these transference feelings involves negative feelings towards her husband, unfavorable fantasied comparisons of me with him, and that the focus of these invidious comparisons will be something in the Jewish/gentile domain. Except for the one word, "Peculiar?," I remain silent until the last five minutes of the session. Everything the patient talked about during that period alluded directly, or almost directly, to the conjectured theme. Space does not permit me to present all of the associations, but to give you an example: She recounted a recent episode in which she and her husband visited a drugstore with whose proprietor the husband had formerly done business, and the patient was irritated with her husband because he slapped the counter and put his hand on the druggist's shoulder and asked in a loud voice how his profits were going. The patient noted the presence in the store of a slightly familiar neighbor woman named Stenquist, who the patient mentions is a Norwegian Lutheran. (She knew from the newspapers and other sources that I was a Lutheran and of Norse origins.) She had the conscious thought in the drugstore that her husband was "carrying on in exactly the way anti-Semites have the stereotype of the way Jewish people talk and act in public." She then talked about a non-Jewish boy she had gone with briefly in high school but quit because her parents disappoved, emphasizing that he was "quiet" and "somewhat shy" and had "very nice manners." She went on to say she liked men who were gentle (note further phonetic link between "gentle" and "gentile"), and after a bit of silence said that she realized it was my business to be gentle in my treatment of her but that she imagined I was the same way in real life. Some more hesitation, then a complaint that sometimes her husband was not gentle in bed; and then finally a reluctant expression of the thought that I would no doubt be gentle in bed.

Now this example presents only minor difficulties for an objective classification of the associative material, since I have picked out material that illustrates the theme. But in the "figure-ground" situation they are less clear, and one *could* miss the point without having his switches properly set by the symbolized uncircumcised penis in the manifest content of the dream. And although it is dangerous to impose limits on what the souped-up computer programs of the future will achieve, it will take some doing on the part of the content analyst who writes the General Inquirer's "dictionary" to deal with this case. That one (low-frequency) idiographic element, a manifest dream element that I have never heard before or since in thousands of hours of therapeutic listening, and that no one in any psychotherapist audience I have asked has had—that would mean a minimum of 100,000 hours of our collective experience—makes all the difference. It is for this reason that we need the psychoanalytic retrieval machinery of our own minds, i.e., that we are still the superprogrammed General Inquirers.

From this example it is only one step to cases in which the manifest content contains little of the received symbology, but is itself idiographic to such a degree that its meaning becomes evident only as we listen to the associations in the ensuing session. So we have a situation in which one cannot see a particular "fact" as bearing upon a particular conjecture except in the light of the conjecture itself. Crudely stated, we don't know that a fact is of a certain kind without knowing what it means, and in the work of the mind we don't know what it means without having purposes and intentions available as potential "explainers." In psychoanalytic listening we impose the relevant dimensions of classification on what, for the behaviorist, are just noises (of course he doesn't really proceed that way, but he pretends to, and tries to impose this impossible methodology upon us!) Our "imposing of relevant dimensions"—the important truth in the Kuhn-Feyerabend line as it applies to psychoanalysis—is precisely what makes us vulnerable to the Achensee Question.

It is almost impossible, as Freud points out in his introduction to the study of Dora, to illustrate the more complex (and, alas, commoner!) kinds of theme-tracing without making multiple reference to previous sessions and to the whole structure of the patient's life. I shall present you with a short example that involves a specific postdiction, namely, what the patient was reading the night before, and whose photograph looked like the person dreamed of. I use it as a kind of litmus test on my Minnesota colleagues to diagnose who is totally closed-minded against psychoanalytic thinking. If this sort of instance doesn't at least mobilize a rat psychologist's or psychometrician's intellectual curiosity to look into the matter further, he is a case of what the Roman Church would call "Invincible Ignorance."

Businessman, late thirties, wife had been a patient of mine; he refused to pay for her psychotherapy because "didn't believe in it, nothing to it." (She cashed an old insurance policy for payment.) Wife benefited greatly, to the point he became interested. Bright, verbal man of lower-class background (father was a junk-dealer). Patient went to University of Minnesota for almost two years, premedical course. Quit ostensibly because family needed money (in depression years), but he was flunking physics at the time and his overall grades would not get him into Medical School. Older brother Herman used to reprimand him for laziness. Brother got top grades, Ph.D. in chemistry. For a while patient worked in brother's drugstore. Brother was a pharmacist, who finished graduate school while continuing in drugstore business to make a living. I knew (from wife's therapy) that the brother had mysteriously died during surgery in what was thought to be a fairly routine operation for stomach ulcer, the family having been greatly impressed by the fact that on the night before the surgery, Herman expressed with great dread an unshakeable conviction that he would die on the operating table. (Some physician colleagues have told me that—whatever the explanation of this phenomenon, which has been reported in medical folklore before—they would have considered it undesirable to operate on the patient under those circumstances.) During the first interview the patient related to me in a mixed fashion, including: a) nonchalance, unconcern, "minor problems"; b) jocularity; c) deference; d) competitiveness. (I won't fully document these impressions as noted after the first hour, but an example: He fluctuates between addressing me as "Dr." and "Mr.," and once corrects himself twice in row—"Dr.,—uh, Mr.,—uh, Dr. Meehl.")

In the fourth session he had reported a dream that was quite transparent, even prior to associations, about a waiter who provided poor service, making the patient "wait a long time before providing anything. Also, I couldn't say much for the meal [= Meehl; I have learned, as have most analytic therapists whose names readily lend themselves to punning, that dreams about food, meals, dinners, lunches, etc. frequently are dreams of transference nature—as in my own analysis, I learned that, like some of my analyst's previous patients, if I dreamed about "buses," it likely referred to him because 'Bus' was his nickname]; and the bill was exorbitant, so I refused to pay it." He went off on discussion of ways his wife had and had not improved, alluded to the parking inconvenience on campus, time taken coming over here, expressed "hope we can get this thing over with fairly rapidly." At end of hour asked what bill was [$10—my modest fee back in the ancient days before inflation!] and "tell me something about the waiter" (blond, blue-eyed, crew-cut, mustached—an exact description of me in those days). I asked directly what thoughts he had about our sessions. He said he had wondered why I wasn't saying anything much. He contrasted our sessions disappointedly with some sessions his wife had told him about, in which she had been fascinated by interpretations of her dream material. I then interpreted the dream and summarized the corroborating associations. He asked "When did you figure out what I had on my mind?" I said I guessed it at the beginning, as soon as he reported the dream. (Debated matter of technique: Because of my Radovian analyst and supervisor, I frequently depart from the classical technique, following the practice of Freud himself and some of his early colleagues that part of the "educative" phase early on is to gain leverage by establishing a conviction in the patient that the process has a meaning, even if that involves saying something about the therapist's inferential processes and when they occurred. The dangers of this are obvious, but doing it carefully to avoid gross one-upmanship is part of what I believe to be involved in intellectually mobilizing suitable patients by engaging their cognitive needs, their need to understand themselves, and the sheer element of intellectual interest that is part of what aligns the observing ego to enter into the therapeutic alliance. Part of what one does early on is to engage the patient's reality-based, mature cognitive interest in the psyche and its machinery. Persons who cannot think psychologically and cannot distance themselves from their own puzzling experiences are unsuitable for psychoanalytic therapy; even patients who are able to think psychologically about themselves and others usually lack a firm, concrete, gut-level conviction about the unconscious. Any professional in psychiatry or clinical psychology who has had analysis can report his surprise during early sessions at the fact that "all this stuff is really true, even for me!") My patient seemed to be very impressed by this, chuckling and repeating, "By God, that's fascinating," and, "To think I didn't know what I was talking about, and you did!"

In the fifth session, following the session about the waiter dream, patient enters smiling, remarks before reclining, "Sure was interesting last time, *you knew* [emphasis!] all along what was going on and I didn't." First dream: *An Oriental, some kind of big shot, Chinese ambassador or prime minister or something—he's hurt—he has a big bloody gash in his abdomen.* Second dream: *I am measuring out some pills from a bottle.*

His associations continued as follows: Drugstore with brother—patient was partner but not registered pharmacist—at times he put up prescriptions when brother not in—nothing to some of them—silly to take the task so seriously—brother's Ph.D. degree—"he studied—I was lazy—lacked ambition—still do—make more money if I had more ambition—brother also brighter to start with—how bright am I, really?—always caught on to things quickly—poor work habits—wanted to be a physician, but not hard enough—I don't like doctors much—haven't seen one in years—some are pretty dumb"—(long discussion of wife's gynecologist who missed her diagnosis when patient got it right)—in drugstore patient used to advise customers about medication—often felt confident he had diagnosed a condition their doctor wasn't treating them for—(rambles on about various incompetent physicians he has known, detailed anecdotes with use of medical terminology, including narrative of brother Herman's unexplained surgical death, insensitivity of surgeon in pooh-poohing Herman's fear, then back reference to Herman's getting the Ph.D. by his brains and hard work). Comments on "experts who don't necessarily know much more than an intelligent amateur." But doesn't want to "overgeneralize that." (Pause—the first even short pause in the stream of the associations. Here one asks the tactical question whether to wait the silence out, which is sometimes appropriate, but in my experience, frequently not. I believe the tendency to wait it out regularly, as developed in this country during the twenties and thirties, is one reason for interminable analyses of people who are actually good prospects for help. We want to know what the patient is thinking that makes him pause, and I know of no really persuasive technical reason not to ask.) (Q Thinking?) "Still can't get over our last session—that *you knew* what was on my mind when I didn't—not just that I didn't—that's to be expected—but that *you knew*, that really gets me!"

Here, after the third over-stressed phrase "You knew," I had an association. The previous night I had been reading TIME, and saw a photo of the Burmese prime minister U Nu. So one hypothesizes a pun, mispronouncing the name, hence there is a linkage to me via this pun: Meehl = "You knew" = U Nu = Oriental in dream.

So I asked him for further thoughts about last session. "Surprised" [Pause] (Q Go on) "Impressed—what else can I say? [Pause] (Q Just keep talking) "Taken aback, sort of—why didn't *I* recognize it, if you did?— pretty obvious—then the $10 bill and all that stuff—shouldn't take an expert to see that!" [Laughs] (Q Any negative feelings at all?) "No—no negative feelings—irked with myself." (No resentment at all, toward me?) [At such moments one must be persistent] "No—or if so, very faint—I'd hardly call it resentment even." (Q But a feeling as if I had sort of won a round, or had one up on you, perhaps?) [Laughs] "A bit of that, sure—it's kind of humiliating to go yackety-yacketing along and then find out *you knew* all along—so I suppose you could say there was a little element of resentment there, yes."

I then asked if he was reading last night ("Yes, TIME"). Pressed for recall—"business, foreign affairs" (Q Picture in foreign affairs?) "Hey! By God—that oriental in the dream was a photo in TIME". Patient can't recall name— I tell him "U Nu" and point out that again, as last session, a play on words is involved. The dream shows how strong this reluctantly reported and allegedly faint resentment is, in that he has me wounded (killed, castrated, made into a woman?), linked perhaps by the associations to his wife's vaginal problem and the professional's misdiagnosis; then there is the obvious connection with the abdominal wound that surgically killed his competitor, harder-working Ph.D. brother Herman. The interpretation of this material led to some further fruitful associations regarding his competitive feelings toward his business partner, whom he had originally described to me as "entirely compatible" and "a sympathetic person." I had external evidence, not discussed with him before this session, from his wife that in fact the business partner tended constantly to put down the patient, to underestimate his abilities, to pontificate to him about cultural matters in which the patient was as well informed as he (the business partner, like brother Herman, had completed his college education with high grades). As a result of this interpretation, some of that ambivalence toward the business partner came out, and several subsequent sessions were especially fruitful in this regard.

I should be surprised if any psychoanalytically experienced readers disagreed about the essentials of this dream's meaning. (As stated earlier, I bypass here questions of optimal technique, the "output" side of therapist interpretation, except to remind ourselves how *avoiding* intervention helps avoid theory-contamination of the patient's associations). I have found in every audience of nonanalysts several listeners whose sudden facial "Aha!"-expressions showed the moment they "got it," sometimes with the irresistible bubbling up of laughter that so often accompanies a good interpretation (thus fitting Freud's theory of wit). It is equally clear that the lay audience (that includes some clinical psychologists in this context) displays wide individual differences in how soon various members begin to "catch on." And, of course, some tough-minded behaviorists or psychometrists (while smiling willy-nilly) shake their heads at the gullibility of Meehl and the other audience members. So Achensee—justifiably —is with us yet! The two diagrams say most of what needs saying by way of reconstruction. In Figure 1, the session's associative material is presented running sequentially (as it occurred) along the left and bottom. My first intervention (neglecting any unintended signals of changed breathing, throat-clearings, or chair squeakings) occurs after the first short pause by

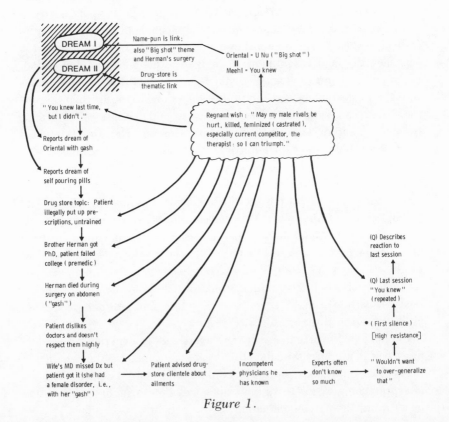

Figure 1.

the patient ("editing," in violation of the Fundamental Rule), after his mentioning not-so-knowledgeable experts he "wouldn't want to overgeneralize about." The postulated guiding theme ("unconscious wish-fulfilling fantasy," if you will) is shown at center. One sees that associations viewed as "topics" are all loosely connected with the second dream's manifest content, and with each other. In Figure 2, I have avoided the "causal arrow" in favor of nondirectional lines (without arrowheads), as here we merely conjecture associative linkages that perhaps strengthen some of the final verbal operants; but we do not say which way the causal influence runs, nor assign any time-order. The strengthening of associations here is loosely "contextual," and some connections are obviously more speculative than others.

The first dream finds no plausible place in this network, except via the (hence crucial) "U Nu = You knew" word-play (and, of course, the dream-day event involving TIME). We also, bootstrapping (Glymour 1980), invoke Freud's rule of thumb (Freud 1900/1953, pp. 315-316, 333-335, 441-444) that two dreams of the same night deal with the same thing and often express a cause-effect relation, the first dream usually (not always) being the antecedent of the causal "if. . . then." Read here: "*If* Meehl [= U

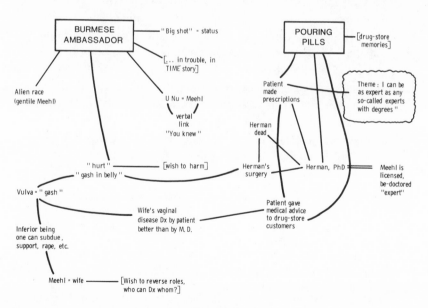

Figure 2.

Nu = "you knew"] is killed, like my earlier sibling competitor Herman, *then* I will be the triumphant, learned, be-doctored expert who is perfectly capable of prescribing pills, etc." This inferred latent structure I do not here pretend to "quantify," and I am not convinced it needs to be quantified. All that the theme-tracing method to be proposed would perhaps do for us is to reduce somewhat the "subjective ad hockery" component in the skilled clinician's discerning the "red thread" allegedly woven into the associations. (For a similar approach to a non-psychoanalytic interpretative problem of psychology see Meehl, Lykken, et. al., 1971).

What is nomothetic and, in principle, "computerizable" contributes to our understanding, but is rather feeble here unless combined with the idiographic components. A male figure with an abdominal wound would presumably occur in a psychoanalytic content analysis dictionary tagged with "castration" and "aggression" themes. But we don't have a place for pouring out pills, we don't have a place for Orientals, and we certainly couldn't get to the postdiction about TIME via the pun on U Nu. The pincers that "close together" to make the Achensee Question hurt do so, in this kind of situation, because the complex ontology (one pincher) requires a complex imposition of thematic content by the analytic listener and hence the other pincer (subtle epistemology) closes simultaneously. Detecting the "red thread" of allusions in the associative material, performing our psychoanalytic Vigotsky on blocks varying in many ways other than shape, size, and color, invalidates the jigsaw analogy, at least in the eyes of the skeptic. We have to discern what is common in the blocks of verbal output, but "what is common" resists any simplistic semantic or syntactic categorization. At the risk of overstating my case, I repeat, one must begin to formulate his conjectures before he can discern that a certain speech sequence tends to confirm them. To quote a previous paper of mine on this subject:

> Skinner points out that what makes the science of behavior difficult is not—contrary to the usual view in psychoanalytic writing—problems of observation, because (compared with the phenomena of most other sciences) behavior is relatively macroscopic and slow. The difficult problems arise in slicing the pie, that is, in classifying intervals of the behavior flux and in subjecting them to powerful conceptual analysis and appropriate statistical treatment. Whatever one may think of Popper's view that theory subtly infects even so-called observation statements in physics, this is pretty obviously true

in psychology because of the trivial fact that an interval of the behavior flux can be sliced up or categorized in different ways. Even in the animal case the problems of response class and stimulus equivalence arise, although less acutely. A patient in an analytic session says, "I suppose you are thinking that this is really about my father, but you're mistaken, because it's not." We can readily conceive of a variety of rubrics under which this chunk of verbal behavior could be plausibly subsumed. We might classify it syntactically, as a complex-compound sentence, or as a negative sentence; or as resistance, since it rejects a possible interpretation; or as negative transference, because it is an attribution of error to the analyst; or, in case the analyst hasn't been having any such associations as he listens, we can classify it as an instance of projection; or as an instance of "father theme"; or we might classify it as self-referential, because its subject is the patient's thoughts rather than the thoughts or actions of some third party; and so on and on. The problem here is not mainly one of "reliability" in categorizing, although goodness knows that's a tough one too. Thorough training to achieve perfect interjudge scoring agreement per rubric would still leave us with the problem I am raising. (Meehl 1970, p. 406)

I say again, we require the subsuming powers of the clinical brain, but we need a reply to the skeptic who says that there is so much play in the system that we can subsume arbitrarily, any way we want, by some mixture of general theoretical preconceptions and the prematurely frozen conjectures that we arrived at from listening to the dream and first association. My fifth proposal for making a dent in this problem is not very elegant, and I have not worked out any fancy statistics for doing it, partly because I think that they will not be necessary. We first have a clinically naive but intelligent reader break the patient's discourse into consecutive blocks, which I shall label "topics." This initial crude categorizing is done without reference to inferred motives by someone ignorant of such things as defense mechanisms, symbols, and the like, essentially in the way a high-school English teacher instructs students to paragraph a theme by topics. Passing intrusions from the manifest content of some other block are simply ignored (e.g., a one-sentence allusion "as I said, Jones was the sergeant" does not fractionate a block of discourse dealing with a single "non-Jones" episode of barracks gambling). In Table 1 I have done this by three crude topic designations running along the top of the table. The purpose of this breaking up by crude manifest topics is essentially to provide separable chunks of material sufficiently large for a clinician to discern possible themes, but sufficiently small to prevent his contaminating himself by

Table 1. Themes Discerned by Five Analysts Independently Reading Discourse Block I (Crude Topic: Brother Herman). Hypothetical Data.

	Theme Rubrics
Analyst A:	"Competition with males"
	"Sibling rivalry"
	"Object-loss" [Herman's death]
Analyst B:	"Object-loss"
	"Competition with another male"
	"Childhood family"
Analyst C:	"Sibling rivalry"
	"Educational failure"
	"Intellectual snobbery"
Analyst D:	"Competition with males"
	"Object-loss"
	"Self-sabotage" [didn't study]
Analyst E:	"Inadequacy-feelings"
	"Childhood family"
	"Sibling rivalry"

themes discerned when looking at other blocks. So each crude topic block of discourse is submitted to *different* psychoanalytic therapists with instructions to write down whatever theme occurs to them as "present in it," "underlying it," "alluded to by it." This requires several teams of readers who do not have access to any of the crude topic blocks that the other teams are reading. We then type up (on 3 x 5 cards) the set of conjectured themes that have been generated in our analytic readers by a particular block. These cards are given either to another team (or, in this phase, I see no harm in the same team doing it) and we ask them to rank (or rate, or Q-sort—I think the psychometric format unimportant here) each theme as to its likelihood (or strength?) as a thematic contributor to that block. Writing the instructions for this third phase will be tricky, because there is a certain opposition between base rates on the one side and low probabilities (as being stronger evidence) on the other side, which is one of the reasons we need clinicians as judges. We employ the 2-phase rating scheme because we believe that a clinician especially skilled (or hyper-responsive to a particular theme) may *sometimes* discern something that the other clinicians will quickly see as a good bet even if they didn't come up with it themselves.

When these batches of rated themes are colligated in a single table, one now reads horizontally instead of vertically, to see whether the thematic

Table 2. Summed (Weighted) Ratings of Themes Discerned within Blocks by Three Independent Sets of Analysts. Hypothetical Data.

Block I: Brother Herman		Block II: Drug-Store		Block III: Wife's Physician	
Theme	Σw_i	Theme	Σw_i	Theme	Σw_i
Competition with males	8	Self-aggrandizement	7	Intellectual snobbery	6
Sibling rivalry	6	Economic insecurity	5	Self-aggrandizement	6
Object-loss	6	Negative transfer	5	Competition with males	6
Childhood family	3	Hostility to experts	4	Negative transfer	5
Inadequacy feelings	3	Super-ego defiance	3	Dislike of doctors	4
Educational failure	2	Sibling rivalry	3	Object-loss	3
Self-sabotage	1	Competition with males	3		
Intellectual snobbery	1				

"red thread" is apparent. In Table 2 I have illustrated this with fictional ratings. The summed (weighted) ratings for "Competition with males" being the largest ($\Sigma w_i = 17$) the red thread would be crudely quantified by these imaginary results. I do not have an appropriate significance test for evaluating the end result of this process, but I am not much interested in statistical significance testing anyway. A couple of obvious possibilities are to establish a crude baseline for "chance congruency" by slipping in blocks that belong to a different session or even to a different patient. One interesting question is how often we can "make sense" of the associations given to a dream even if the manifest content was not dreamed by the associater—a claim made against psychoanalysis forty years ago by Rudolf Allers in his book *The Successful Error* (Allers 1940) and never, to my knowledge, answered.

Space does not permit an adequate treatment of such a method's limitations, but there is one major defect that must be mentioned. Sometimes the allusions are few in number, perhaps *only* two or three, buried in high-resistance "sawdust," but are given evidentiary weight because of some delicate mix of very low nomothetic base-rate ("expected-ness" in the denominator of Bayes' Formula) with very high idiographic linkage ("conditional probability" in the numerator of Bayes' Formula). In such sessions, the Topic Block Theme Tracing method would fail utterly; and, I fear, so would all the others.

Summary

Summarizing this essay is rather like pulling together the material from a murky psychoanalytic hour, which is perhaps diagnostic of my cognitive condition. I *do* have a theme of sorts, but it's hard to verbalize briefly. In a word, I am ambivalently saying that Fliess's Achensee question deserves a better answer than it has yet received. Granted that there are respectable documentary disciplines (like history) that rely mainly upon qualitative evidence, a mind's discernment of intentionality, and the argument from convergence—disciplines that are neither experimental nor statistical in method; and granted that the "experimental/quantitative" (often called the "scientific") is not coextensive with the empirical, nor with the reasonably believable; and granted that the usual behaviorist and psychometric objections to the *concepts* of psychoanalysis (e.g., not "operationally specified") are simplistic and philosophically uninformed; granted all this, it remains problematic just what *is* the state of our evidence from the best source, the analytic session. I have suggested five directions we might take in an effort to ascertain how much of what the "thought reader" reads— admittedly using his own mind—is also objectively there, in the mind of the other.

References

Allers, Rudolf. 1940. *The Successful Error*. New York: Sheed & Ward.

Campbell, N.R. 1920. *Physics: The Elements*. Reprinted as *Foundations of Science*. New York: Dover Publications, 1957.

Carnap, R. 1950. 2d ed 1962. *Logical Foundations of Probability*. Chicago: University of Chicago Press.

Freud, S. 1954. *The Origins of Psychoanalysis*. Ed. Marie Bonaparte, Anna Freud, Ernst Kris. London: Imago Publishing Co., Ltd.

Freud, S. 1900. *The Interpretation of Dreams*. In *Standard Edition of the Complete Psychological Works of Sigmund Freud*, ed. J. Strachey, Vols. 4-5, London: Hogarth Press, 1953.

Freud, S. 1974. *Standard Edition of the Complete Psychological Works of Sigmund Freud*. Ed. James Strachey, Vol. 24 (Index). London: Hogarth Press, 1974.

Glover, Edward. 1940. *An Investigation of the Technique of Psychoanalysis*. Baltimore: Williams & Wilkens.

Glymour, C. 1980. *Theory and Evidence*. Princeton: Princeton University Press.

Harman, H.H. 1976. *Modern Factor Analysis*. (3rd Ed.) Chicago: University of Chicago Press.

Kemeny, J.G., Snell, J.L., and Thompson, G.L. 1957. *Introduction to Finite Mathematics*. Englewood Cliffs, N.J.: Prentice-Hall.

Kordig, C.R. 1971a. The Theory-Ladenness of Observation. *Review of Metaphysics* 24: 448-484.

Kordig, C.R. 1971b. The Comparability of Scientific Theories. *Philosophy of Science* 38:467-485.

Kordig, C.R. 1971c. *The Justification of Scientific Change*. Dordrecht: D. Reidel.

Kordig, C.R. 1973. Discussion: Observational Invariance. *Philosophy of Science* 40:558-569.

Luborsky, L. 1953. Intraindividual Repetitive Measurements (P technique) in Understanding Psychotherapeutic Change. *Psychotherapy: Theory and Research*, ed. O.H. Mowrer, chapter 15, pp. 389-413. New York: Ronald Press.

Lykken, D.T. 1968. Statistical Significance in Psychological Research. *Psychological Bulletin* 70:151-159. Reprinted in *The Significance Test Controversy*, ed. D.E. Morrison and R. Henkel. Chicago: Aldine, 1970.

MacCorquodale, K. and Meehl, P.E. 1954., E.C. Tolman. In *Modern Learning Theory*, ed. W.K. Estes, S. Koch, K. MacCorquodale, P.E. Meehl, C.G. Mueller, W.N. Schoenfeld, and W.S. Verplanck. New York: Appleton-Century-Crofts, pp. 177-266.

Marmor, Judd. 1968. *Modern Psychoanalysis*. New York: Basic Books.

Meehl, P.E. 1954. *Clinical versus Statistical Prediction: A Theoretical Analysis and a Review of the Evidence*. Minneapolis: University of Minnesota Press.

Meehl, P.E. 1956. Wanted—A Good Cookbook. *American Psychologist* 11:263-272.

Meehl, P.E. 1957. When Shall We Use Our Heads Instead of the Formula? *Journal of Counseling Psychology* 4:268-273.

Meehl, P.E. 1970. Psychological Determinism and Human Rationality: A Psychologist's Reactions to Professor Karl Popper's "Of Clouds and Clocks." In Analyses of Theories and Methods of Physics and Psychology, *Minnesota Studies in the Philosophy of Science*, ed. M. Radner and S. Winokur, Volume IV. Minneapolis: University of Minnesota Press, pp. 310-372.

Meehl, P.E. 1970. Some Methodological Reflections on the Difficulties of Psychoanalytic Research. In *Minnesota Studies in the Philosophy of Science*, Volume IV. Minneapolis: University of Minnesota Press, pp. 403-416. Reprinted *Psychological Issues*. 1973, 8, 104-115.

Meehl, P.E. 1970. Theory-Testing in Psychology and Physics: A Methodological Paradox. *Philosophy of Science* 34:103-115. Reprinted in *The Significance Test Controversy*, ed. D.E. Morrison and R. Henkel. Chicago: Aldine, 1970.

Meehl, P.E. 1971. Law and the Fireside Inductions: Some Reflections of a Clinical Psychologist. *Journal of Social Issues* 27:65-100.

Meehl, P.E. 1972. A Critical Afterward. In I.I. Gottesman and J. Shields. *Schizophrenia and Genetics: A Twin Study Vantage Point*. New York: Academic Press, pp. 367-416.

Meehl, P.E. 1978. Theoretical Risks and Tabular Asterisks: Sir Karl, Sir Ronald, and the Slow Progress of Soft Psychology. *Journal of Consulting and Clinical Psychology* 46:806-834.

Meehl, P.E. and Feigl, H. 1974. The Determinism-Freedom and Body-Mind Problems. *The Philosophy of Karl Popper*, ed. P.A. Schilpp. LaSalle, Illinois: Open Court Publishing Co.

Meehl, P.E. and Golden, R.R. 1982. Taxometric Methods. In *Handbook of Research Methods in Clinical Psychology*, ed. P.C. Kendall and J.N. Butcher. New York: Wiley, 1982, pp. 127-181.

Meehl, P.E., Lykken, D.T., Schofield, W., and Tellegen, A. 1971. Recaptured-Item Technique (RIT): A Method for Reducing Somewhat the Subjective Element in Factor-Naming. *Journal of Experimental Research in Personality* 5:171-190.

Meehl, P.E. 1983. Consistency Tests in Estimating the Completeness of the Fossil Record: A Neo-Popperian Approach to Statistical Paleontology. (this volume)

Murray, H.A. 1938. *Explorations in Personality*. New York: Oxford University Press.

Neurath, O. 1959. Protocol Sentences. In *Logical Positivism*, ed. F.J. Ayer. New York: Free Press, pp. 199-208.

Read, R.C. 1972. *A Mathematical Background for Economists and Social Scientists*. Englewood Cliffs: Prentice-Hall, pp. 914-958.

Reik, T. 1948. *Listening with the Third Ear: The Inner Experience of a Psychoanalyst*. New York: Farrar, Strauss and Co.

Sears, R.R. 1943. Survey of Objective Studies of Psychoanalytic Concepts. New York: *Social Research Council Bulletin* No. 51.

Silverman, L.H. 1976. Psychoanalytic Theory: "The Report of My Death Is Greatly Exaggerated." *American Psychologist* 31:621-637.

Stone, P.J. Dumphy, B., Smith, M., and Ogilvie, B. 1966. *The General Inquirer: A Computer Approach to Content Analysis*. Cambridge, Mass: MIT Press.

Consistency Tests in Estimating the Completeness of the Fossil Record: A Neo-Popperian Approach to Statistical Paleontology

Most educated persons today, who have been repeatedly exposed since childhood to pictures of the famous "horse series," are likely to think of it, or of the fossil dinosaurs they have seen in museums, rather than other lines of evidence for the theory of evolution. They differ in this respect from Charles Darwin, who did not consider the fossil data to be supportive of his theory. On the contrary, he viewed the paleontological findings available in his day as perhaps presenting "the gravest objection which can be urged against my theory" (Darwin 1859, p. 280), emphasizing the fossil record's failure to provide the "infinitely numerous transitional links" (p. 310) that would illustrate the "slow and steady operation of natural selection". (Eldredge and Gould 1972, p. 87) Because he thought of the fossil record as adverse rather than supportive, his chapter on the record is appropriately titled by reference to its *imperfection,* and appears toward the end of the book, following the general chapter on "difficulties of the theory."

Philosophers of science, and psychologists of certain persuasions (e.g., Freudians who wish to defend the "scientific" status of psychoanalytic doctrine against super-operational critics), have sometimes invoked the theory of organic evolution as a paradigm case of a theory that is generally admitted to be empirical and scientific but that mostly offers after-the-fact explanations of data without being able to make predictions of those data. One thinks of evolutionary theory as explaining that there is such a creature as the rhinoceros, but nobody has ever claimed, given the general postulates of evolutionary theory, to *derive* the empirical statement that an animal of such-and-such rhinoceroid properties would evolve. Sir Karl Popper (1974, pp. 133-143), with his emphasis on falsifiability as a criterion of scientific theories, has gone so far as to suggest that we should look upon

Darwinism (he specifically includes neo-Darwinism) not as a scientific theory at all, in the strong sense he uses the term "theory," but as a metaphysical speculation that has been fruitful in science.[1]

Few of us find that characterization satisfactory. Aside from the philosophical puzzle of just how a metaphysical theory can be "fruitful" if it (literally) has no factual consequences, however vague or "weakly implied," most biological scientists could not rest content with Popper's diagnosis of the theory of organic evolution as not a scientific hypothesis but merely a fruitful metaphysical speculation. We must parse the issues, separating the well-known disputes about whether the notion of adaptation can be defined adequately without a vicious circularity (not all circularities are vicious!), and whether it is possible to explicate a logically adequate notion of after-the-fact explanation in a theoretical structure that would not have permitted before-the-fact prediction; and setting aside what factors other than slight selection pressures may play a critical role (e.g., founder effects, genetic drift, geographical isolation). Then we might hold that there is a minimal content of evolutionary theory that is falsifiable in principle, and hence would escape Popper's classification of the theory as metaphysical, to wit: There must have existed a large number of intermediate transitional forms, even under the modified hypotheses of neo-Darwinism. The probandum should, at least in principle, be falsifiable by the fossil record, *if suitable bounds can be put on find-probabilities*.

Even during the ascendancy of the somewhat dogmatic and overly restrictive Logical Positivism (Vienna Circle of the 1920s), everyone recognized the necessity to distinguish between the technical feasibility of confirming a scientific theory and the abstract possibility of doing so "in principle." Early on, there were controversies about how to construe "in principle." For example, attackers of the verifiability criterion of meaning made the telling point that unless the phrase "in principle" were construed so narrowly as to be nearly the same as "technically" (with present knowledge and instruments), one had to allow for the future possibility of novel auxiliary hypotheses, perhaps taken together with to-be-developed scientific instruments or analytic procedures. But allowing for such future possibilities meant that to deny the confirmability of the theory "in principle" required one to deny, in advance and a priori, the possibility of somebody's being clever enough to cook up the appropriate theoretical auxiliaries and associated technological instruments. This advance general proof of a negative would seem to be, on the face of it, an impossible intellectual feat.

The relevance of this old controversy to the present problem is obvious: One way to escape the charge of nonfalsifiability against evolutionary theory is to sketch out, in as much detail as the present state of knowledge permits, how one would go about estimating the numerical adequacy of the fossil record. This estimation is a necessary stage in evaluating the numerous intermediate probanda of the theory (e.g., the hypothesis that between the extant species of whales and their presumed land mammal ancestor, there existed numerous protowhale transitional forms.) It is not necessary, in order to escape the charge of untestability, for us to have *presently* available all of the kinds of data that would generate good estimates of the quantities that we conjecture will converge numerically. The philosophical point is that there is nothing unfeasible in principle about gathering such data, and in fact some of them, such as the dates at which the first finds of a given extinct species were made, are part of standard paleontological practice. I am prepared to argue that to the extent the completeness of the record remains an interesting live question (or, in some rare quarters, a persisting basis for doubt as to the macro-evolutionary hypothesis itself), then if the mathematics I suggest are essentially sound, this provides motivation for adopting whatever practices of recording, cataloging, and pooling information are necessary in order for the first step of raw data classification required by my proposed methods to occur. But even if that motivation were inadequate, I should think I had provided an answer to the Popperian complaint of empirical nonfalsifiability.

With the passage of time and the subsidence of theological and philosophical controversies, contemporary biologists and geologists can afford to take a detached and relaxed view, now that the overall concept of major evolution has become part of the mental furniture of almost all educated persons. The imperfection of the geological record is taken for granted, and we have causal and statistical considerations explaining why the record should be expected to be incomplete. Nevertheless the dearth of really good fossil sequences has remained a mild nagging source of intellectual discomfort, and from time to time some convinced evolutionist finds himself in hot water with his colleagues in dealing with it. Thus Goldschmidt (1940) was so impressed by this absence of transitional forms that when he combined it with skepticism about the accumulated micro-mutation *mechanism* of neo-Darwinism, he was led to propound an alternative idea of extensive chromosomal changes. This idea was criticized by Dobzhansky (1940, p. 358) as involving as much of a "miracle" as out-

and-out mystical extra-scientific hypotheses that Dobzhansky knew Goldschmidt would not have countenanced.

Quotations abound, and since universal knowledge of the epistemic role assigned to the record's imperfection may be presumed, I shall not bore the reader by piling up such quotations. I confine myself to mentioning a recent alternative theory, "Punctuated Equilibria: An Alternative to Phyletic Gradualism," set forth by Miles Eldredge and Stephen Jay Gould, largely motivated by the missing transitional forms problem. (Eldredge and Gould 1972) They suggest that many breaks in the fossil record are "real" (p. 84), that the interpretation of the incompleteness of the record is unduly colored by the paleontologists' imposition of the theory of phyletic gradualism upon otherwise recalcitrant facts, and that the rarity of transitional series "remains as our persistent *bugbear*." (p. 90) They refer to the reputable claims of Cuvier or Agassiz, as well as the gibes of modern cranks and fundamentalists, for whom the rarity of transitional series has stood as the bulwark of anti-evolutionists' arguments. (p. 90) They also say that "we suspect that this record is much better (or at least much richer in optimal cases) than tradition dictates." (p. 97)

The theory-demanded assertion of extreme record incompleteness, despite the fact that it no longer constitutes a serious objection to the essential truth of the theory of evolution in the minds of most scientists, does, as these contemporary authors indicate, still persist as a theoretical irritant, so that one would prefer, if possible, to have something more detailed and definite to say. It could also be argued, without tendentiousness, that for a science that is essentially "historical," having as its aim to reconstruct the unobserved past through a study of its material residues available to us today, the *quantitative degree of incompleteness* of the fossil record must surely be rated a "Big Parameter." Such a Big Parameter is so important theoretically that even very rough, crude estimates, putting bounds on orders of magnitude, are worth having if such could be achieved.

How poor is the fossil record? Unfortunately it is easy to present plausible arguments on both sides, and such counterposed armchair arguments (I do not use the word "armchair" invidiously, since, as Bertrand Russell once said against the American behaviorists, an armchair is a good place to think!) can lead us to expect that the record would be fairly complete, at least complete enough to help answer some of the big questions, such as whether evolution took place gradually or by saltation;

or the arguments can persuade us that the record is surely far too scant, *and must remain so*, for any such inferences. During the time that the substantial correctness of the theory of evolution itself was still a matter of dispute among educated persons, it was easy to find these kinds of considerations and counterconsiderations in the polemical literature. Nor have these old arguments on both sides been shown to be qualitatively invalid in the meantime. It is merely that the acceptance of the evolutionary paradigm has greatly reduced their importance, and therefore we do not find it easy to work up much interest in them any longer.

To summarize briefly the countervailing arguments: On one side, the evolutionists pointed out the unlikelihood of an animal specimen, even one with hard parts, dying in a situation in which the ravages of wind and rain and scavenger animals would leave it intact; then the unlikelihood, even if the animal escaped such quick destruction, that it would form a fossil; then the improbability that it would remain undisturbed by geological faulting, extreme temperatures, water erosion, and so forth; and finally the improbability that even an army of industrious geologists and paleontologists would happen to dig at just the right place. Although these arguments are not in quantitative form and were originally offered as ad hoc explanations of the scanty phylogenies (compared with what the theory might at first have led us to expect), we can rely on a quote from a recent historian of science that warns, "do not make a mockery of honest ad hockery."

Anti-evolutionists, admitting these considerations, were in the habit of rebutting them by saying that, although we don't expect any individual saber-tooth tiger to have been preserved from destruction and fossilized and dug up—a sequence admittedly of very low probability—nevertheless the saber-tooth tiger species was extant for many millions of years; at any one time, hundreds of thousands of tigers roamed the earth; so the net expected number of fossil finds is still "good enough," because we are multiplying a very big number by a very small number.

Of course the trouble with these arguments, rebuttals, and rejoinders is that although they are all intrinsically plausible as *qualitative* (or weakly "semi-quantitative") arguments, they do not lead to even a rough *numerical* value for anything. Consequently the role of the dominant theoretical paradigm is even more crucial in determining whether one finds them persuasive in his subjective, personalist probabilities than in other branches of scientific controversy (cf. Eldredge and Gould, pp. 83-86). In this essay I offer some tentative and rather general quantifying suggestions, which, despite their generality and the idealizations made, do advance us beyond the purely qualitative considerations found in the controversy's history. Formulas are derived that, if the methods are sufficiently robust under real departures from the idealization, provide *numerical estimates that we may hope will converge*.

Because of the novelty of my approach and the weight of received doctrine that speaks prima facie against it, I think it necessary to introduce certain general methodological considerations (stemming from a neo-Popperian philosophy of science) as a framework within which the more specific statistical suggestions are set forth. I do not attempt to "establish" or "justify" these methodological guidelines, but merely set them out so that the reader will understand the metatheoretical framework in which my statistical proposals find their place.

1. When qualitatively persuasive arguments of the kind quoted above are in such collision, and when rebuttals to them on the respective sides are not dispositive, so that subjective or personalistic probabilities (especially so-called subjective priors) are the basis of a scientist's choice, it is desirable, whenever possible, to reformulate such arguments in a quantitative rather than a qualitative form. If the state of theory and factual knowledge is not strong enough to permit that, it may be possible to express the expected *consequences* of these qualitative arguments in a quantitative form. That is the step attempted in this paper.

2. Quantitative treatment in all sciences, but especially the life sciences, unavoidably relies upon substantive idealizations in the embedding theoretical text and, either springing from these or added to them, mathematical idealizations in the formalism.

3. In the early stages of efforts to quantify a hitherto qualitative enterprise, roughness in numerical values arising from measurement and sampling errors, as well as from the idealizations in the text and formalism mentioned above, will usually be unavoidable. One hopes that these crude values can be refined later; but even as a first step, before refinement, such numerical estimates are scientifically preferable to a purely impressionistic subjective evaluation of nonquantitative arguments, however qualitatively meritorious.

4. Reasonably approximate numerical point values and reasonable tolerances of numerical error are almost always preferable scientifically to so-called exact significance tests or even exact confidence intervals, neither

of these even being really exact anyhow, for reasons well known to statisticians. (Meehl 1978, Morrison and Henkel 1972)

5. Numerical agreement, within reasonable tolerances, of two or more nonredundant modes of estimating a theoretical quantity is almost always scientifically preferable to single values estimated by only one epistemic route, even if the latter are accompanied by so-called exact standard errors.

6. The evidentiary weight with regard to a theoretically important quantity provided by several nonredundant convergent estimates increases with their diversity far more than replicated estimates relying upon the same evidentiary path (instrument or procedure of estimation).

7. Although we commonly speak of "hypotheses" in contrast to "required (auxiliary) assumptions," there is no logical or epistemological difference between the substantive theory of interest and the auxiliary assumptions, when both play the same role in a conjunction of assertions to explain or predict observational data. The only difference between what are usually called "hypotheses" (referring to the substantive theory of interest) and "assumptions" (referring to those auxiliary conjectures employed in the formalism or embedding text when testing the substantive theory) lies in the focus of the investigator's current interest. He may be treating only one of them as problematic, although if pressed, he would usually be willing to concede that both are somewhat so. In their logical, mathematical, and evidentiary status, the hypotheses and the auxiliary assumptions are often interchangeable.

8. A difference, however, sometimes exists when the auxiliary assumption has itself already been well corroborated by a variety of evidentiary avenues that do not themselves include reliance upon the substantive theory of interest, in which case we properly think of ourselves as mainly testing the substantive theory "in reliance upon" the previously corroborated auxiliary. But there is no necessity for this state of affairs, and even in the physical sciences, both the substantive theory and the auxiliary assumptions may have the same status at a given time with regard to the prediction of a particular observational result. Thus, for example, a conjecture concerning the structure of crystals and a conjecture concerning X-rays were both problematic at the time the epoch-making series of experiments on the use of crystals in X-ray diffraction was initiated. It is well known in the history of physics that these experiments *simultaneously* corroborated conjectures concerning the inner structure and molecular distances in crystals as well as the wave nature of X-rays and the order of

magnitude of X-ray wavelengths. It would have been impossible to answer the (seemingly reasonable) question, "Which are you presupposing in testing—the X-ray theory or the crystal theory?" The answer to that question, which physicists were too sophisticated to ask, would be "*Both, conjointly and simultaneously.*"

An extremely simple, clear, unproblematic example from the physical sciences, of estimating an unobserved true value via two fallible measures in the absence of a "direct, independent" avenue to either's accuracy, is Rutherford and Geiger's calculation of total (observed *and unobserved*) alpha-particle scintillation events via the numerical relations among the one-observer and both-observer tallies. High-school algebra cancels out the two nuisance parameters of "observer efficiency." (Segré 1980 pp. 103-104)

9. When a substantive theory and the auxiliary assumptions are both somewhat problematic at a given stage of our knowledge, a misprediction of observational fact refutes the conjunction of the two, and does not ordinarily give us much help in deciding which of them—perhaps both—must be modified or abandoned. It is of great importance, however, to recognize the other side of this coin, to wit, that successful numerical prediction, *including convergent numerical predictions by different epistemic avenues,* mediated by the conjunction of the two conjectures, tends to corroborate their conjunction—both of them receiving evidentiary support thereby.

I have spent some time on this matter of assumptions and conjectures because in developing the estimation methods, I shall rely on what would ordinarily be called "assumptions" that readers may find doubtfully plausible, as I do myself. If these "assumptions" are perceived (wrongly) as *necessary premises* for the use of the convergent methods I propose, the latter will of course appear to lack merit. If on the other hand these "assumptions" are seen as part of the system of interconnected conjectures or hypotheses, which network of interrelated hypotheses gives rise to a prediction that certain experimentally independent (instrumentally nonredundant) estimates of an unknown, unobserved quantity will agree tolerably well, then a different evidentiary situation is presented to the critic. Instead of being told, "If you are willing to assume A, then method M_1 will estimate something, and so will method M_2, and so will method M_3 . . . ," we say, "If we conjecture the conjoined auxiliary and theory (A.T.), then nonredundant methods M_1, M_2, M_3, M_4 should lead to consistent numerical results." If those concordant numerical results would be, without the conjunctive conjecture, "a strange coincidence," so that the predicted agreement of the methods takes a high risk, goes out on a limb, subjects itself to grave danger of refutation—then, as Popper says, surmounting that risk functions as an empirical corroborator. The usual use of "assumption" means something that is not itself testable. Whereas in my "assumptions," made to get the theoretical statistics going, I intend that the consequences of those assumptions are to be tested by the agreement of the empirically independent statistical estimates. So I shall deliberately avoid the use of the word "assumption" in favor of the words "hypothesis" and "conjecture," with the adjective "auxiliary" when appropriate.

Another way of viewing this matter is via the concept of *consistency tests*. (See Meehl 1965, 1968, 1978, 1979; Golden and Meehl 1978; and Meehl and Golden 1982.) If the parameter we desire to estimate is not directly measurable, either because its physical realization occurred in the past or because it is an inherently unobservable theoretical entity, we cannot subject the instrument of measurement, whether it is a physical device or a statistical procedure for manipulating observational records so as to obtain a theoretical number, to direct validation—the ideal procedure when available. Thus, for example, in assigning evidentiary weight to certain signs and symptoms of organic disease in medicine, we now usually have available fairly definitive results, reported by the pathologist and the bacteriologist, concerning the tissue condition and its etiology that under-lay the presented symptoms and course of the disease. But before knowing the pathology and etiology of the disease, the only way the physician could draw conclusions as to the differential diagnostic weight of the various symptoms and complaints, course of illness, reaction to treatments, and the like was by some kind of coherency, the clinical "going-togetherness" of the findings. Most of the theoretical entities inferred in the behavioral sciences, such as human traits and factors, or even temperamental variables of genetic origin, have to be "bootstrapped" in this way. (Cronbach and Meehl 1955; Golden and Meehl 1978; Golden and Meehl 1980; Meehl and Golden 1982)

There is nothing inherently objectionable about such procedures of bootstrapping by statistical manipulation of correlations between what are presumed to be fallible observable indicators of the desired unobserved latent state of affairs; but it is also known that many, perhaps most, methods of factoring and clustering such fallible phenotypic indicators of an alleged latent causal-theoretical entity are subject to an undue amount of ad hockery, to such an extent that some cynics in the social and biological sciences reject all such methods of factor and cluster analysis on the grounds that they will lead us to find factors, clusters, types, and the like even when there aren't any there. (Cf. Meehl 1979) It has therefore become an important task of the statistician and methodologist to derive procedures that will constitute internal checks on the validity of the attempted bootstrapping. One cannot say in advance that he will always succeed in finding a factor or a cluster if it's there, and that the method will surely give the right diagnostic weights to the fallible observable indicators; but some assurance should be provided. Similarly, an investigator should be reasonably confident that he will not wrongly find a factor, cluster, entity, or syndrome when it's not really there. We don't merely want statistical gimmicks that will "bring order out of disorder." We want procedures that will discern order only if it is latently present. We want, as Plato said, to carve nature at its joints.

In some of my own work developing new taxometric statistics for studying the genetics of mental diseases like schizophrenia, I have devoted attention to deriving multiple consistency tests by means of which the validity of a given inference made from a pattern of fallible observables to, say, the presence of the dominant schizogene would be possible. My reasoning again involves rejecting the usual "assumption-versus-hypothesis" model of inference in favor of a joint conjecture model. We avoid writing the conventional A: $(f_1 \cdot f_2 \cdot f_3) \to T$, which reads, "*If* you are willing to *assume* A, then the three facts $f_1 f_2 f_3$ *prove* the theory T." (This formulation is objectionable anyway because, as the logicians point out to us, facts in science never, strictly speaking, "prove" theories but only test them.) Instead we write the neo-Popperian $(A.T) \to (f_1 \cdot f_2 \supset f_3)$, which we read as, "Conjecturing the conjoined auxiliary and substantive theories A and T, we see this conjunction entails that facts f_1 and f_2 imply fact f_3." Further, since fact f_3 has a sufficiently low antecedent probability on $(f_1 \cdot f_2)$ lacking the theory, such a prediction takes a high risk, that is, exposes itself to grave danger of being falsified. If it escapes falsification, the conjoint conjecture (A.T) is corroborated.

The normal way, and the way that will be taken in this paper, of conceptualizing such low prior probability facts that are capable of subjecting the theory to a high risk of refutation is that the predicted facts are numerical point values or relatively narrow ranges. So we don't really rely on the "assumption" A in justifying our numerical methods, the traditional way of putting it which I am here strenuously opposing. Rather, we rely on the unlikely coincidence of converging numerical values in order to corroborate the "assumption" together with the main theory of interest that combines with it to generate an otherwise improbable numerical consequence.

Classic examples of this kind of thing abound in the exact sciences such as physics, chemistry, and astronomy. For instance, there are over a dozen independent ways of estimating Avogadro's number. In the early stages of the development of physical chemistry, each of these ways could have been criticized as relying on unproved "assumptions," and some of them were only doubtfully plausible at the time. But whatever the initial plausibility of the "assumptions relied on" (in one way of looking at each method singly), these are nonredundant independent methods and they converge at the same estimate of the number of molecules N in a mole of a substance. The fact of their numerical convergence tends to corroborate the correctness or near correctness of the assumptions (and, in some cases, the robustness of the method under departures from the idealization) *and*, be it noted, *at the same time gives us a high degree of scientific confidence in the theoretical number, Avogadro's constant, thereby inferred*. (Cf. Perrin 1910) The inference is made by multiple paths, and even though each path "relies on" what taken by itself might be only a moderately plausible assumption (and from now on I shall, I repeat, avoid this term in favor of "auxiliary hypothesis" or "conjecture"), once the convergence of these paths is found as an empirical fact, we view the a priori shakiness of the "assumptions" in a different light.

I do not, of course, intend to browbeat the reader by arguments from the philosophy of science into accepting the merits of my suggested procedures. I merely use them as motives of credibility in anticipation of the first criticism that springs to mind as soon as one begins to examine each of these methods, namely, that it "relies on assumptions" that are, to say the least, problematic and that some critics would think very doubtful.

Before commencing exposition of the methods, I remind the reader of our general strategy at this point, so as not to bore him with irksome repetitions of the preceding methodological guidelines. Each method will involve mathematical idealizations "based on" strong (and, hence, antecedently improbable) "assumptions" about the paleontological situation —both the *objective situation* (the "state of nature") and the *investigative situation* (the process of fossil discovery). I shall simply state these strong assumptions, emphasizing that they are intended as auxiliary conjectures whose verisimilitude (Popper 1962, pp. 228-237; Popper 1972, pp. 52-60) can only be guessed at from the armchair in our present state of knowledge, and, therefore should mainly be tested by their fruits. The fruits, as indicated above, consist of the agreement or disagreement among the four essentially independent (nonredundant) methods of estimation.

The quantity we are interested in may be called the *completeness coefficient*. For a given form, like a species, that coefficient, of course, takes on only the two values 1 or 0; a fossil of the species is known to us, or it is not. A certain kind of animal lived on the earth, but is now extinct, and we may have no knowledge of it because we have not found even a single recognizable fossil. For larger groups, such as families, orders, classes, or phyla, the completeness coefficient is not confined to the two-valued "found at least once (= 1)" or "never found, even once (= 0)," but may refer to the proportion of different species belonging to the larger group that have been found at least once. Example: If there were in the history of organic life on the earth 200 species of ungulates that are now extinct, and at the time of computing the completeness index, paleontological research had discovered 70 of these species, the ungulate completeness index would be .35. Then in the eyes of Omniscient Jones, the paleontologist knows of the existence of only about one in three of the species of ungulates that have existed but are now extinct. (The exclusion of extant species from the reference class that defines the denominator of the completeness index will be explained later, when I present Dewar's Method of Extant Forms.) The basic unit tallied in forming this index I shall take as a species, although arguably, because of the perennial disagreements between "lumpers" and "splitters" in taxonomy, it might be preferable to take as basic unit the genus, as was done by Dewar and Levett-Yeats. (Cited in Dewar and Shelton 1947 pp. 61-63, 71.) I do not see that anything except reliability hinges upon this choice, so I shall refer to species. The question of how high up in the hierarchical taxonomic tree the larger reference group should be chosen is also arbitrary to some extent, but not completely so because we have two countervailing methodological considerations from the standpoint of computing good statistics. Since the coherency of the four methods is, in the last stage of my procedure, to be examined by correlating estimated total numbers of extinct species within a given taxon, it is desirable to keep the larger classification sufficiently small so that there can be a goodly number of taxon-numbers to correlate. Thus we would not wish to take as denominator the number of phyla because then the correlation coefficients computed for estimates based on the four methods would be based on an N of only 20-30. On the other hand, for reasons of stability of the individual completeness coefficients computed by each method, we do not want to be dealing with very small numbers of species within each larger category. In the case of vertebrates, for instance, something like number of carnivora, classified at the level of the order, might be a reasonable value. As in the species/genus decision, nothing hinges critically upon this one. It's a matter of convenience and of statistical stability. So, in each of the methods, I shall refer hereafter to *species* as the basic taxon unit tallied, and have arbitrarily selected *order* for the larger group whose species tally generates a proportion.

Method I: Discovery Asymptote

The first completeness estimator I call the *Discovery Asymptote Method*. It relies on the simple intuitive notion, widely used in the life sciences, that if an ongoing process of producing or destroying some quantity is known to have a physical limit, and if the process goes on in a more or less orderly fashion, one can fit a curve to data representing the stage reached at successive points in time; and then, from that fitted curve one can arrive at an estimate of the total quantity that would be produced (or destroyed) in the limit, i.e., in infinite time. In our fossil-finding situation, the total (unknown) quantity is the number of species of a given extinct order that would ever be found if paleontological search continued forever, and the ordinate of such a graph at year 1983 represents the number of species of the order that are known to us today. For such a method to have any accuracy for the purpose of estimating the total number of extinct species of an order, the asymptote of total cumulative finds at infinite time must be approximately identifiable with the total that have lived. So the first idealization "assumed" (better, *conjectured*) is that every species in the order has been fossilized at least once and that fossil preserved somewhere in the earth's crust. The qualitative, common-sense considerations for and against that idealization have been set out above. At this stage we conjecture that the huge number of individual organisms of a species suffices to counteract the very low probability of an individual specimen being fossilized and preserved. We do *not* treat the exhaustion of the earth's crust by paleontological digging as an idealization (as we did with the previous conjecture), because that idealization is mathematically represented by the asymptote, which the mathematics itself tells us we shall never reach. But the abstract notion is, counterfactually, that paleontologists continue to explore the earth "forever," or at least until the sun burns out, and that no cubic yard of earth, even that under the Empire State Building or at the bottom of the sea, remains permanently unexamined. That this is of course a technological absurdity does not distress us, because we do not require to reach the asymptote physically but *only to estimate where it is*.

As to the choice of the function to be fitted, there is, as in all such blind, curve-fitting problems not mediated by a strong theory, a certain arbitrariness. But familiar considerations about growth processes can help us out. We know we do not want to fit a function that increases without limit, however slowly, as we get far out on the time scale, because we know that there is an upper bound, the number of different species of animals that have lived on the earth being, however large, surely finite. A choice among curves would be based upon trying them out, fitting the parameters optimally to each candidate function, and selecting the "best fit" by examining the residual sums of squares when each candidate function has been optimally fitted by least squares or other appropriate method. The most plausible function is what is sometimes called in biology and psychology a "simple positive growth function." Let N be the total number of species of, say, the order Carnivora that have ever lived. On the idealized assumption that the record in the earth (not, of course, the record in our museums or catalogues) is quasi-complete—that is, that every species of Carnivora is represented by at least one fossil somewhere in the earth's crust—the number of species known to us at time t, say in the year 1983, is the ordinate y in the known record of Carnivora. Then the residual $(N - y)$ is the number of species yet remaining to be found. Adopting the notion of a constant search pressure, say S (which, of course, is a gross oversimplification over the whole history of paleontological search, but I shall correct that in a moment), we postulate that the instantaneous rate of change of the ordinate is proportional to the number of species yet remaining to be discovered, the constant of proportionality being the search pressure S. This gives us the familiar differential equation for simple growth processes:

[1] $$\frac{dy}{dt} = S(N - y).$$

Integrating this differential equation, and determining the constant of integration by setting $y = 0$ at $t = 0$, we obtain the integrated form

[2] $y = N(1 - e^{-St})$,

which is a familiar expression to many biological and behavioral scientists, sometimes being called the "exponential growth function," sometimes "simple positive growth function" because it often approximates empirically the growth of organisms and their parts, as well as the growth of certain states and dispositions. For example, the psychologist Clark L. Hull had a predilection for fitting exponential growth functions in describing mammalian organisms' acquisition of habit strengths. (Hull 1943) The growth function attained visibility in the life sciences partly from the frequent use of it, or its autocatalytic form, by the Scotch-American-Australian biochemist T. Brailsford Robertson, who was highly regarded and widely known for his work on growth. Behind a trial-and-error attempt to fit empirical data, and successfully applied to a wide array of growth processes in the biological and social sciences, was the underlying concept from physical chemistry that the velocity of a monomolecular reaction is at any time t directly proportional to the amount of chemical change still remaining to occur. But aside from that historical scientific "rationale," the fundamental notion is our conjecture, in relation to any process of change, that the momentary velocity of the process is proportional to the amount of change still remaining to occur. So that for a fixed search intensity or find pressure, represented by the postulated constant S in the exponent of the growth function, the momentary rate of finding new species will, on an essentially random model, depend upon how many species there are left to find.

Experience in the life sciences has shown that the vast majority of data we get in the empirical world tend to follow a very few curve types. Among those most frequently encountered are the straight line, the parabola, the exponential curves of growth and decay, the logarithmic function, the Gompertz curve, the normal probability curve, and its integral, the normal ogive. Empirical curve fitting is still considered in applied mathematics to be partly a common-sense, intuitive, or theory-based business. Nevertheless we can, in principle, always select between two competing "admissible" curves on the basis of which fits better in the sense of least squares or another defensible criterion (e.g., Pearson's method of moments) when each competitor curve's parameters have been optimally fitted. In the case before us, the justification for giving special attention to the exponential growth function as a prime competitor is theoretical, for the reasons stated. Some functions, such as a straight line, a parabola, or an exponential function can be excluded confidently on theoretical grounds because they do not approach an asymptote. The total number of extinct species "to be found" (even on the fantastic assumption that paleontologists will dig up the whole earth's crust before the sun burns out) is finite, and that simple physical truth puts a constraint on our choice of functions.

But there is a pretty unrealistic idealization involved in the above text preceding our differential equation, namely, that the search intensity S remains the same from whatever arbitrary time dates (the year 1800 or 1700, say, or whatever one chooses) we begin to plot data for the science of paleontology at a search intensity sufficient to be significant in curve fitting. One must surely suppose that the search intensity has considerably increased from that time to the present, if for no other reason than the great increase in the number of professional paleontologists and the financial support provided for their work.

Thus the parameter S cannot have a fixed value over the last two centuries or so of paleontological searching, but must itself be increasing more or less steadily with time. As a better approximation, we could try, in the integrated form of the equation, to write instead of the parameter S a variable search intensity $s = a + bt$. This is, of course, still an approximation based upon the idealization that rate of finding is a linear function of amount of searching, and that amount of searching is a linear function of time. The first of these has considerable plausibility as probably a not-too-serious idealization, so that the mathematical expression of it would be robust under slight departures; but the second one, that the search pressure is a linear function of time, is presumably not a close approximation to the truth. Putting nonlinear functions into the exponent complicates the curve fitting process considerably, however, and it also leads to greater instability in assigning numerical values to the parameters. If one wanted to get some realistic notion of just how grossly distorted the linear growth of a search intensity exponent is, he had better go not to the historical record of fossil finds but to a more direct (social) measure of the search intensity. An alternate variable—like number of professionally active paleontologists in the world at a given time—would presumably serve as an adequate proxy; but we could also plot number of expeditions recorded, number of research grants given, or even number of dollars spent. The intent there would not be, of course, to determine the parameters, which cannot be done in this fashion, but to determine the exponent's likely function form so as to ascertain whether an approximation to s as a linear function of t is too grossly out of line with facts. These are crucial details that cannot be explored profitably here. If the linear model for variable search intensity exponent s as a function of time were provisionally accepted as good enough for the present coarse purpose, the integrated form of the growth function would then become

$$[3] \quad y = N(1 - e^{-(a + bt)t}) = N(1 - e^{-(at + bt^2)}).$$

So we have a more complicated growth function with the same parameter asymptote (total fossil species N to be discovered in the limit), but the growth constant is now replaced by a second-degree polynomial as a function of time.

If it is found that this modified exponential growth function with a variable growth rate graduates the empirical data for a given order, say, Carnivora, satisfactorily (which means in the practice of curve fitting that it graduates it about as well as the leading competitor functions that have some theoretical plausibility, and preferably somewhat better than these others), then the best fit by least squares determines the parameters N, b, and a. If the original cruder form were an adequate approximation to our paleontological discovery data, it determines the two parameters N and S. The parameter N is the asymptote that the search process is approaching in the limit of infinite time exhausting the earth's crust. On our idealization that the record held in the rocks—*although not our record at any time*—is quasi-complete, this asymptote is an estimator of the total number of species of the order we are investigating that ever lived and have become extinct. At the present moment in time (year 1983) the value of y in the fitted function is the number of species of Carnivora that have thus far been discovered in the fossil record. So the completeness index, the "percent adequacy" of the Carnivora fossil record, is simply y/N, that is, the proportion of all (known + unknown) extinct species of Carnivora that are known to us as of 1983. The same curve-fitting operation is independently carried out for each of the orders of animals with hard parts that we have decided to include in our study of the agreement of the four methods.

The ratio y/N is of considerable intrinsic interest, in that most paleontologists would be astonished if it were greater than 50 percent, indicating that we know today the majority of species of extinct Carnivora; so that the record would be much more complete than it is customary to assume in paleontological writings. But the number we want for our subsequent correlational purposes when investigating the "methods" consistency is not the percentage 100 (y/N) but the absolute asymptote N. The statistical reason for this will be explained later. The point is that we have fitted the discovery growth function for the order Carnivora and have recorded its asymptote N, the total estimated species of Carnivora, *both those known to us and those not presently known*, for our later use. We do the same thing for primates, rodents, and so forth, each time the curve-fitting process terminating in an estimation of the unknown parameter N, the total number of species of extinct members of that order, known and unknown.

The Discovery Asymptote Method just described relies on a process taking place through recorded historical time, and each extinct species (of a given order for which the cumulative discovery graph is plotted) appears only once as a datum contributing to the determination of that curve; i.e., it appears associated with the abscissa value of t at its first discovery. This is important because it means that the *number* (absolute frequency) of individual fossil specimens of the same extinct species plays no role whatever in determining the parameters of the function fitted in the Discovery Asymptote Method. The second method I have devised pays no attention to date of find, or otherwise to an ongoing time-dependent discovery process, but is wholly "cross-sectional." It ignores time but attends instead to the very thing that the first method systematically excludes from consideration, namely, how many individual fossil specimens of each known extinct species are available at the present time. This is crucial because in order for two or more methods of estimating an unknown quantity by means of some proxy measure (such as we must rely on in paleontology) to be mutually corroborative, the methods must not be redundant. That is, they must not be essentially numerically trivial reformulations of the same raw data, or of summary statistics of those raw data which are known algebraically or empirically, via dependence on some powerful "nuisance variable" (Meehl 1970), to be highly correlated.

Method II. Binomial Parameters

The second method, which I call the Method of Binominal Parameters, again takes off from a well-known intuitive idea in the application of statistics to the life sciences. This idea is that the distribution of the frequencies with which an improbable event is known to have occurred is a reflection of its improbability and can therefore, under suitable circumstances, be employed to estimate that underlying unknown probability. I shall deliberately sidestep the deep and still agitated question of the basic nature of the probability concept, which remains unsettled despite the best efforts of mathematicians, statisticians, and philosophers of science. The reader may translate "probability" either as a partially interpreted formal concept, the notion apparently preferred by most mathematicians and by many statisticians who are not primarily identified with a particular substantive science; or as some variant of the frequency theory, still very popular despite recent criticisms; or, finally, as a propensity—the view favored by Sir Karl Popper. I hope I am correct in thinking that these differences in the basic conceptual metaphysics of the probability concept, surely one of the most difficult notions to which the mind of man has ever addressed itself and, unfortunately, one of the most pervasively necessary in the sciences, do not prejudice the argument that follows.

We are forced to get our method off the ground by postulating a strong idealization that we are confident is literally false, and whose verisimilitude is unknown to us. Here again the verisimilitude is not estimable directly, and for that reason is only indirectly testable by the consistency approach, i.e., by its convergence with other nonredundant modes of estimating the record's completeness. Consider a particular order (say, Carnivora) and a particular extinct species (say, the saber-tooth tiger) belonging to that order. We represent the resultant at any time (e.g., 1983) of the paleontological discovery process as a collection of individual, physically localized excavations, each of which we arbitrarily designate as a "dig." Since this idealization of a dig is unavoidably imprecise and coarse, we consciously set aside difficult questions about its area, returns to excavate nearby, distinguishing between individual investigators or teams, and the like. We also set aside the question of whether multiple specimens of the same species found in the same dig are counted as one or many, a decision that should be made on the basis of the availability of the relevant data in the catalogued museum record. Choice among these first-level data reductions (instance classifications and counts) may, without vicious circularity, be made on the basis of their comparative conformity to the conjectured mathematical model. We ask: "Which is more orderly (= Poisson-like) —frequency of finds as tallied by method A or by method B?" If, for example, counting three saber-tooth skulls as one find because they were found by one team digging in a "small area" reveals operation of a cleaner

Poisson process than would counting them as three, then the former tallying rule is the rational choice.

Then we consider the abstract question of whether in a given dig, a fossil specimen of the particular species is or is not found. (We are not here tallying actual digs, of course—only conceptualizing them.) We then conceptualize the probability p of its being found. From the frequency interpretation of the probability concept, what this amounts to is that we conceptualize the huge class of digs in the hundreds of thousands (setting aside the complication "what the paleontologist is looking for") and consider what proportion of them contain a saber-tooth tiger fossil.

Let us designate as n the total number of digs on which this propensity (or relative frequnecy) of finding is based. Given the usual idealizing assumption of essential independence (not in this case too unrealistic), the probability of finding a saber-tooth in all n digs is p^n; the probability of finding a saber-tooth in all but one of the digs is $n\,p^{n-1}\,q$ (where $q = 1-p$ is the "not-find" probability); and so on with the familiar series of terms in the expansion of the binomial $(p + q)^n$. Then we have as the last term of this sequence of probabilities, each of which specifies the probability of finding precisely k saber-tooth fossils among n digs, the probability q^n, which is the probability of failing to find a single saber-tooth fossil among any of the n digs. This epistemologically regrettable event would correspond to our total ignorance that such an organism as the saber-tooth tiger ever existed. Consequently the last term of the binomial expansion is the one of interest for the present purpose, its complement $(1 - q^n)$ being the completeness coefficient when we apply it over the entire order Carnivora.

On the further idealizing assumption that the same find probability over digs applies to each species of the order being investigated, or on the weaker assumption that the mean value of these probabilities can be applied without too much distortion over the board (a matter of robustness), let the total number N of extinct Carnivores be multiplied by the expansion of $(p + q)^n$. If each probability term in the expansion were thus multiplied by the unknown species number N for the order Carnivora, this would generate the theoretically expected *absolute frequencies* with which the various species of the order are represented as fossils in the museum catalog. Of course, we do not know the number N, and the point of the procedure is to estimate it. What the theoretical model just developed tells us is that the observed frequencies of species represented by only a single fossil, by two fossils, by three fossils, and so on up to hundreds or thousands of fossils (as we have of the woolly mammoth or the trilobites), is latently generated by the process of multiplying the terms of the binomial by the unknown constant N.

The big fact about the *empirical* graph we obtain by plotting (starting at the left) the numbers of extinct species of the order Carnivora that are represented by only one fossil, by two fossils, by three . . . ,—a tallying process performable empirically without having recourse to the above theoretical model of how these frequencies were generated—is of course the fact that the frequency of species corresponding to the extreme left-hand value of the abscissa variable: "number of fossils representing the species" is an empty cell. It represents the number of extinct species of Carnivora for which the paleontologist has at present found not a single fossil. Of course that number is not known, because those are the species we do not at present know to have existed at all. The basic idea of the Binomial Parameters Method is that we can make an extrapolative estimate of the unknown number of species in that last column by studying the properties of the distribution of those in the remaining columns.

Before presenting the development of the procedure for this extrapolation, which unavoidably involves some alternatives, each of which must be tried on real data, I shall attempt to give the reader an intuitive appreciation of the way of thinking underlying this method by beginning with the very coarse-grained question, "Is the record at least half complete?" That is, taking the order Carnivora as our example, can we at least conclude, without attempting more precise parametric inferences about the underlying binomial properties, whether we presently have in the museums at least one fossil representative of over half of the extinct species of Carnivores?

Suppose we conjecture the contradictory, with all contemporary evolutionists, that the record is considerably *less* than 50 percent complete. That is, of all the species of Carnivora that ever walked the earth, paleontology in 1983 knows of fewer than half. For over half of them, not a single fossil has been found. This means that the latent quantity $q^n > 1/2$. Since on our idealized model the find probabilities are conjectured to be approximately equal and independent for different carnivores, it doesn't matter whether we consider the terms of the binomial here or the result of multiplying the expanded binomial by the unknown total Carnivora species number N, since these are proportional. We can therefore consider the observed distribution that will run from the extreme right, where we have hundreds or even thousands of fossils of certain carnivores, and move to the left with diminishing numbers. In the far left region we have a group of species, each of which species is represented by only two fossils; then to the left of that is abscissa value $= 1$, which designates those species each of which is known to us by virtue of only a single fossil specimen; and then finally the empty cell of interest, the abscissa value $= 0$, corresponding to those species of Carnivora that are unknown to us because no specimen has yet been found as a fossil even once. In the graph of the distribution of finds, this unknown, were it plotted, would be the highest ordinate. But if $q^n > 1/2$, the sum of all the other terms in the binomial, i.e., $(1 - q^n)$, must be $< 1/2$. Therefore each individual term except q^n must be less than $1/2$. In particular, the term adjacent to the unknown term at $x = 0$, i.e., the probability corresponding to the number of species that are represented only by a single fossil in the record known to us, must be less than the height of the unknown ordinate at $x = 0$. Since it is a theorem of algebra that the binomial—even if extremely skew and whatever its parameters—cannot have multiple local modes, it follows that there cannot be another mode anywhere to the right of the mode latently existing for $x = 0$. But if the coefficients ever increased as we moved to the right, there would be two local modes and the theorem would be violated. Hence the model requires that.there cannot be an increase as we move to the right, toward the species columns represented by large numbers of individual fossils. Since no two terms can be equal, it follows that as we move to the right toward species represented by very large numbers of fossils, the graph is everywhere decreasing. So we see that if the record is less than 50 percent complete, the observed distribution of numbers of species represented by various numbers of individual fossil finds must slope everywhere upward toward the left. This corresponds to a latent situation known in statistics as a "J-curve" found in studying low probability events, such as radioactive decay, behavioral studies of social conformity, and the like.

So it is interesting to note that before starting a more complicated mathematical treatment of the data by studying the statistical properties such as skewness and kurtosis in the higher moments of the distribution form, choosing between a binomial and a Poisson distribution for this purpose, etc., we can already see that a universally-held doctrine of paleontology, to the effect that the record is considerably less than half complete, can be subjected to rather direct test over a group of orders by simply plotting the "find replication" numbers for various species. The test prediction is that these graphs, which should be fairly stable statistically (the numbers involved being large provided that the unit tallied is genus or species and the supra ordinate taxon is broad enough, such as order or class), should reveal a monotone function rising to the left. Further, on the assumption that the completeness index is less than 50 percent and usually taken to be very considerably less than this (the vast majority of species in any extinct order being unknown to us), we must infer that $q^n >> 1/2$. Then, since the complementary fraction $(1 - q^n) << 1/2$ must be dispersed over a very large number of integral values on the abscissa (covering the range of species of fossil replication numbers up to those fossils represented in the hundreds or thousands), it follows that we expect a nearly flat and very low graph until we get to the extreme left end, where it is expected to undergo a steep rise shortly before we come to the empty interval at $x = 0$. More will be said below about this problem of shape in a discussion of the binomial and Poisson distribution question.

Since the physical situation of fossil discovery reflected in empirical statistics is surely one in which the individual event of interest ("a find"—of a single specimen) has extremely low probability, this being countervailed to some (unknown) degree by the sizable number of paleontological digs, a highly skew binomial distribution can be anticipated with confidence. The highly skew binomial distribution is customarily represented by the Poisson distribution, the terms of which involve only the one parameter $m = np$, where n is large and p is small, their product being a (proper or improper) fraction near to one. It is known from wide experience in a variety of scientific contexts (e.g., emission of alpha rays, occurrence of industrial accidents, overloading of telephone exchanges, rare cell counts in a haemocytometer) that the Poisson distribution provides an excellent fit to observational data arising from this combination of a small probability for an event with a large number of opportunities for it to occur. Mathematically the Poisson distribution is generated from the binomial by a limiting process in which, holding the product $np = m$ fixed, we imagine p going to 0 as n increases without limit, and the defined binomial moments are then examined. So the Poisson distribution is, strictly speaking, an approximation to the precise numerical values we would get by using the binomial, even for cases of extreme asymmetry in its complementary probabilities. (This explains why I have called Method II "Binomial Parameters," to designate the general case—covering the unlikely situation of only slight skewness—and the rough bound-setting process described supra. Mainly I rely in what follows on the special properties of the Poisson distribution for numerical estimates, so Method II could as well have been labelled "Poisson Parameter" instead.) One should, however, beware of assuming that we must take *very* extreme values (infinitesimally small p's and gigantic n's) to get a good Poisson approximation, since even moderate degrees of asymmetry are graduated quite well. (Cf. Kyburg 1969, pp. 146-151, noting that Table 6.1 has a misprint heading penultimate column; Lewis 1960, pp. 243-258; Feller 1957, pp. 135-154; Hays 1973, pp. 202-208; Cramér 1946, pp. 203-207; Yule and Kendall 1940, pp. 187-191.) Fitting a Poisson distribution with the single parameter m to data that would be well graduated by the binomial for asymmetrical ($p \ll q$) would give substantially the same answer, an answer that would differ numerically from the precise one by amounts less than the errors introduced by our raw data and the literal falsity of the conjectured discovery model (independence, crude approximations in defining what is "a dig" and "a find," and the like). The mathematical theory of statistics provides no criterion for when the Poisson distribution approximates the asymmetrical binomial closely enough, since this is not a mathematical but a "statistical tolerance" question having no precise mathematical meaning. Appendix I suggests a way of using the binomial proper instead of the Poisson approximation to it, but it has the disadvantage of estimating more latent parameters via expressions of unknown robustness under departures from the idealized model.

As explained above, the basic problem is an extrapolation problem (not soluble by any simple extrapolation procedure that we can trust), namely, to fill in the frequency of species occupying the extreme left-hand interval in the distribution of number of species represented by various numbers of single specimens from 0, 1, 2, 3, . . . to very large numbers of specimens. No observation is available for the number of unknown species whose find frequencies are 0, i.e., we don't know how many species there are of extinct Carnivora for which we have not found a single fossil. By definition, since we haven't found any of them, we don't know how many there are. The Poisson distribution conjecture, suggested by the plausible physical considerations supra (small find probability but numerous digs), engenders two submethods for estimating the missing ordinate at $k = 0$.

The basic idea of the first Poisson method is that if the distribution, including the unknown ordinate of the graph at $k = 0$, is generated as theoretically conjectured by a Poisson process, we can make use of a simple, well-known relation between the successive terms of a Poisson distribution to solve for the missing species frequency. In the graph, the observations plotted are the frequencies with which different species of Carnivora are represented in the record by 1, 2, 3, . . ., k individual fossil specimens. So each ordinate of the graph is $Np(k)$ where N is the fixed (unknown) number of species of Carnivora that ever lived. Since the N will cancel out in the first step, we neglect it for present purposes. Terms of the Poisson distribution giving probabilities of k and $(k + 1)$ fossil specimen finds are

[4] $\quad p(k) = e^{-m} \dfrac{m^k}{k!}$

[5] $\quad p(k+1) = e^{-m} \dfrac{m^{k+1}}{(k+1)!}$

so the ratio of two successive terms is

[6] $\quad \dfrac{p(k)}{p(k+1)} = \dfrac{m^k}{k!} \cdot \dfrac{(k+1)!}{m^{k+1}} = \dfrac{k+1}{m}$

Taking logarithms and their differences (the $\triangle$'s in standard interpolation theory) we obtain

[7] $\quad log \ \dfrac{p(k)}{p(k+1)} = log \ p(k) - log \ p(k+1) = \triangle^1$

and taking the second difference $\triangle^2 = \triangle_i^1 - \triangle_j^1$,

[8] $\quad log \left(\dfrac{k+1}{m} \right) - log \left(\dfrac{k+2}{m} \right) = log \dfrac{k+1}{k+2} = \triangle^2$

That is, the second order differences $\triangle^2$ of the log frequencies are a simple sequence of fractions, the unknown parameters having cancelled out. At the rare-find end of the distribution these fractions are small integers, their generative law being the addition of one [$=1$] to both numerator and denominator, thus:

[9] $\quad$ for $k = 4, \ \triangle^2 = 5/6$
$\qquad \quad \ k = 3, \ \triangle^2 = 4/5$
$\qquad \quad \ k = 2, \ \triangle^2 = 3/4$
$\qquad \quad \ k = 1, \ \triangle^2 = 2/3$
$\qquad \quad \ k = 0, \ \triangle^2 = 1/2$

Note that we can test the observed frequencies for their closeness to the Poisson before proceeding to our main step, which is extrapolating to the unknown frequency N_k at $k = 0$, "species for which no fossil has been found."

Taking the $\triangle$'s at this low-frequency end (k = 2, 1, 0) with the variable being *log raw observed frequency*, we have

[10] $[log\ N(2) - log\ N(1)] - [log\ N(1) - log\ N(0)] = log\ (\frac{1}{2})$,

so the unknown is obtained:

[11] $log\ N(0) = 2\ log\ N(1) - log\ N(2) + log\ (\frac{1}{2})$,

the desired unobserved frequency on our graph at $k = 0$.

A second submethod of employing the Poisson distribution properties to estimate the unknown missing species frequency relies upon the fortunate mathematical fact that the first and second moments of a Poisson distribution are equal (as is also the third moment, not used here but perhaps usable as a check on numerical consistency of the results). The frequencies of the Poisson distribution being probabilities for discrete values of k, the whole system is, of course, multiplied by the unknown total number of species of Carnivora N to generate the observed raw frequencies for numbers of fossil representatives per species. Let N^* be the total observed frequency for the known fossil species, that is,

[12] $N^* = N(1) + N(2) + \ldots + N(k) + \ldots$

and unstarred N is the true total number of species, known and unknown,

[13] $N = N^* + N(0)$,

the second term on the right being the missing point on our "finds" graph at $k = 0$. Take the mean of known points

[14] $\bar{x} = \frac{1}{N^*} \sum^{N^*} x$

The true unknown mean is

[15] $= \frac{1}{N} \sum^{N} x = \frac{1}{N} \left(\sum^{N^*} x + \sum^{N(0)} x_0 \right)$

$= \frac{1}{N} \sum^{N^*} x + 0$

i.e., the unknown cell, if known, would affect the N we divide by but not the sum of x's, since this sum is a count of "number of specimens found" and the missing cell at $k=0$ refers to those species *not* found. Then since

[16] $\quad \mu = \dfrac{1}{N} \displaystyle\sum^{N*} x$

$$\sum^{N*} x = N\mu$$

and

[17] $\quad \displaystyle\sum^{N*} x = N*\bar{x}$

[18] $\quad N\mu = N*\bar{x}$

hence, as $N = N* + N(0)$, $(N* + N(0))\,\mu = N*\bar{x}$

so

[19] $\quad \mu = \bar{x}\,\dfrac{N*}{N* + N(0)} = \bar{x}\,\dfrac{N*}{N}$

which is in two unknowns μ and N, so indeterminate as yet. But we now consider the second moment

$$\mu_2 = \dfrac{1}{N} \sum^{N} x^2 - \mu^2$$

[20] $\qquad = \dfrac{1}{N} \left(\displaystyle\sum^{N*} x^2 + \displaystyle\sum^{N(0)} x_0^2 \right) - \mu^2$

$\qquad = \dfrac{1}{N} \left(\displaystyle\sum^{N*} x^2 + 0 \right) - \mu^2$

Designate by $S*$ the observed second moment (about zero) for intervals $k = 1, 2, 3, \ldots$

[21] $\quad S* = \displaystyle\sum^{N*} x^2$

in the above. Then we have the true mean and true second moment

[22] $\quad \mu = \bar{x}\,\dfrac{N*}{N}$

[23] $\quad \mu_2 = \dfrac{1}{N}\,S* - \mu^2$

The "completeness index" for total species count is

[24] $C = \dfrac{N^*}{N}$

i.e., the proportion of all species, known and unknown, that are known to us.

If the underlying find-function is Poisson, we may infer that the first and second moments are equal, writing

[25] $\mu_1 = \mu_2$

which from the preceding amounts to saying that

[26] $\bar{x}\dfrac{N^*}{N} = \dfrac{1}{N} S^* - \mu^2$

$= \dfrac{1}{N} S^* - \bar{x}^2\left(\dfrac{N^*}{N}\right)$

and transposing

[27] $\left(\dfrac{N^*}{N}\right)^2 \bar{x}^2 + \left(\dfrac{N^*}{N}\right)\bar{x} - \dfrac{1}{N} S^* = 0$

which in "completeness index" notation reads

[28] $C^2 \bar{x}^2 + C\bar{x} - \dfrac{1}{N} S^* = 0$

and dividing by completeness index

[29] $C\bar{x}^2 + \bar{x} - \dfrac{1}{N^*} S^* = 0$

which transposed and dividing by $\bar{x}^2$ yields

[30] $C = -\dfrac{1}{\bar{x}} + \dfrac{1}{N^*\bar{x}^2} S^*$

$= \dfrac{S^* - N^*\bar{x}}{N^*\bar{x}^2}$

[31] $= \dfrac{\displaystyle\sum^{N^*} x^2 - N^*\bar{x}}{N^*\bar{x}^2}$

and all quantities on the right are observable from the fossil frequency distribution of knowns (k > 0), so we obtain the completeness index C. Dividing C into N* gives us our estimate of the total number of Carnivores, found and unfound.

I have no grounds for preferring one of these methods to the other, and the existence of two Poisson-based methods helps us to get a consistency check. It might be argued that since they both rely on the conjectured latent Poisson process, they *must* be equivalent; and hence their numerical agreement is tautologous and cannot bear on anything empirical. That is a mistake, because the algebra does not permit a direct derivation of the equality relied on in the second method from the equality relied on in the first method. (I invite the reader to try it if he thinks otherwise.) We have here a nice little point in philosophy of science (one sometimes forgotten by scientists in making claims of equivalence between *consequences* of theories) that two theorems may flow from the same postulates—and hence each provides a test of the postulates empirically and an estimate of quantities referred to in the postulates theoretically—but the two theorems do not follow *from one another*. It's really a trivial result of the difference between one- and two-way derivation chains in a formalism. There is a "deep" sense in which they're equivalent, in that they both spring from the same conjecture. But since they are not mutually derivable from the formalism alone, they tend to reinforce one another and corroborate the conjectured latent process that leads to each of them. The nice distinction to see here is that between a truth "*of* the formalism" (i.e., an algebraic identity) and a conjecture concerning the physical world formulable "*in* the formalism," which then has a consequence *derivable via the formalism*.

It is perhaps unnecessary to say that one can test the observed frequencies for their closeness of the Poisson distribution before proceeding to either of these main steps of extrapolating to the unknown value at k = 0, "species that have *no* fossils found as yet." Here again a significance test is not appropriate. We know it doesn't exactly fit. The question should be, "Is it a bad fit or a reasonably good one?"

If the two methods agree within a reasonable tolerance and if no systematic bias can be shown theoretically to obtain, nor is there other reason why one of them should be privileged, one might simply strike an average between the two estimates of the missing frequency in the class interval k = 0 for species not found. Then, as in the Discovery Asymptote

Method, one records the estimated total number of species $N^* + N(0) = N$ for the order Carnivora. The same process is carried out for each of the orders being considered, thus generating an array of estimates of total extinct species over the various orders. These are elements in a second column of estimates for the frequency of extinct species (known and unknown) for the various orders, corresponding to estimates previously obtained by the Discovery Asymptote Method.

Method III: The Sandwich Method

The third method, which I have christened "Sandwich" for lack of a less clumsy term and for reasons that will be obvious, is superficially the most straightforward of the three methods invented by me, and at first blush seems as simple as the fourth method (Douglas Dewar); but a little critical reflection shows the simplicity to be illusory. The method turns out to be the most complicated and treacherous of the four, and despite many hours of manipulating algebra and calculus, Monte Carlo trials, and consultation with four mathematically competent colleagues, I am not entirely satisfied with the results. Intuitions on this one can be quite misleading, but the core idea seems (to me and everyone to whom I have explained it) pretty clearly sound. Rather than wait upon publication of this paper for a more satisfactory estimating procedure, or for a simpler approach via the same basic idea (which my intuitions persist in telling me does exist, though I have not been able to hit upon it), I present it here for criticism and improvement.

The basic idea of the Sandwich Method is simple and compelling, despite the complications that appear when we seek an exact formula. If we know that the saber-tooth tiger existed at a certain point in geologic time, and we know that he was still extant at some subsequent point of geologic time, then we know that there must have been saber-tooth tigers at all times during the intervening period, since the process of evolution does not exactly duplicate itself by the formation of a species identical with one that has previously become extinct (Dollo's Law). On this assumption, an evolutionary truism not disputed by anybody, one approach to the question, "How complete is the fossil record for extinct species belonging to the order Carnivora?" runs as follows: We presuppose a high confidence determination of the geologic time for the earliest stratum in which a reliably identified saber-tooth tiger fossil has been found, and similarly the identification of other fossil(s) of the saber-tooth tiger in a reliably dated stratum different from the first one. (We pay no attention in the Sandwich Method to *when* it was found nor to *how many* individual fossil specimens are found—so the Sandwich Method is nonredundant with the first two methods.) We then consider strata dated anywhere in between these two anchor points in geological time, and inquire whether at least one saber-tooth fossil has been found in any of them. Since we know that the species must have existed for the entire intervening time period between the two extreme values anchored, the fossil record is incomplete, with regard to this extinct species, for the time interval between the anchoring fossil finds if we have not found it anywhere in that interval.

Using such an approach, we may consider every extinct species of Carnivora for which anchoring fossils have been reliably identified in strata accurately dated, and separated by some specified geologic time interval, as defining a "sandwich." The two strata anchoring the species in geologic time constitute, so to speak, the "bread slices" of the sandwich, outside of which no fossil specimen of this species has been found. The completeness index is then based upon tallying, for the various species sandwiches thus defined, what percentage of such sandwich-associated species are also represented, following the metaphor, in the "innards" of the sandwich. This tally would be made by inquiring, for each species defining a sandwich by bread slices, whether at least one fossil specimen is found in a stratum reliably dated as lying between the two bread slices. We then calculate from this tally of all extinct species that define a sandwich (so as to be available for such a calculation), what proportion of them are also represented in the innards of the sandwich (so to speak, in the Swiss cheese or ham as well as in the defining bread slices). The resulting number would be a crude measure of how complete the record for Carnivora is. One may conceive loosely of a "total antifind pressure," the combined factors opposing the [forming + preserving + discovering] of fossils, a quantitative property for extinct species collectively of a given broad taxon. We try to estimate the net impact of these counterfind pressures on the taxon Carnivora by tallying representations of species in that taxon for geologic periods during which each is known to have existed, and the numerical value of that impact is then used to infer how many "unfound" Carnivorous species there were.

As I say, the reasoning seems initially quite straightforward: If we know that a species existed during a certain time period, which we can know by knowing that it existed before that time period (in the bottom bread slice of the sandwich) and after that time period (because we find it in the top bread slice), then a failure to find even a single fossil specimen of it anywhere in any strata dated between the slices of the geological sandwich thus defined tallies one "incompleteness instance."

But the percentage completeness thus computed as a crude fraction of total sandwich-defining fossil species is not a valid estimate of true "completeness of the record," as we shall see shortly. I do not say that such a crude completeness index, obtainable by summing such tallies over all extinct species of Carnivora that define various sandwiches of different sizes (and different absolute locations in the time scale), is utterly without value. For example, it could set a safe lower bound that might be illuminating in our present weak state of knowledge about completeness. But if we desire to obtain a quantitative estimate of the incompleteness index that would yield a number for missing species that could be meaningfully correlated over orders with the (allegedly correspondent) numbers of missing species as estimated from the Discovery Asymptote Method and the Binomial Parameter Method, we need a Sandwich Method that will not suffer from a gross systematic bias.

It is easily seen that a crude count of within-sandwich representations over the set of sandwich-defining species suffers from an unknown but non-negligible amount of bias downward. That is, the completeness of the record is seriously underestimated by this crude, unweighted procedure. Two or three home-baked Monte Carlo trials (with differing find-probability assignments and modest species $N = 100$) can be done with a desk calculator, and the bias is so large as to persuade immediately. The first reason for this is that in calculating the percentage completeness in this crude way by simply summing the unweighted tallies of species that define sandwiches and that are also found at least once within the sandwich, we must, of course, ignore the two (or more) fossils definitive of the sandwich slices since, obviously, the proportion represented here $= 1$; otherwise we wouldn't know about the species in the first place, or have a sandwich available to make the count upon. For sandwiches in which the defining bread slices are far apart in geologic time, so that the sandwich is very "thick" in its innards (a geological "Dagwood"), we have many geologic time slices in between, where there exists an opportunity for a saber-tooth tiger fossil to be found. But consider species which in reality (unknown to us) had a very short geologic life span, i.e., in which the true species longevity L_3 amounts to only three time intervals, so that the fossil specimen found in the latest interval and the one in the earliest interval defining the two bread slices of the sandwich leave only one layer of Swiss cheese in between. If that inner sandwich does not contain a specimen, this species is tallied as a failure of completeness. But of course we have in fact found him twice, in two distinct time slices. We need not have an accurate algebraic expression for the size of this bias to realize that short longevity species will, for any small or even intermediate level of *within-sandwich probability of a find*, be contributing adversely to the completeness index. More generally, the completeness index is being unavoidably reduced by the neglect of the fossils found in the bread slices that define the sandwich for each species because, of course, if we haven't got the bread slices, we haven't found a species.

A second consideration is that it would be somewhat surprising, under usual evolutionary assumptions, if there were no correlation whatever between the within-sandwich-slice probability p_i and the true species longevity L_i. On the average, at least if we consider animals falling within a given broad taxon such as the order Carnivora, one would expect variables such as *total number of specimens alive at a given time* and their *range of geographical distribution* to be correlated with ultimate species longevity, and this should generate a correlation between longevity L_i and find-probability p_i within slices of the sandwiches.

I shall present two methods that conjecture (and test for) a relatively fixed p_i over the L_i's, then a third method that approximates variable p_i's. Finally, a complete, nonapproximative method, but one that involves a rather messy system of nonlinear equations, will be explained. Appendix II presents some interesting theorems concerning the latent generating system that are not used in the four methods but should help to fill out the picture and perhaps inspire readers more mathematically competent than I to derive other estimators or consistency tests motivated by the "sandwich" concept.

Our idealized body of observational data consists of geological sandwiches defined by fossils of the various extinct species of our exemplary Order Carnivora. Units of geologic time (rather than named strata corresponding to times) will be expressed for convenience as a tenth portion of the time corresponding to the species of greatest longevity as indicated by the thickest sandwich. Although some species will, under our circumstances of incomplete knowledge, have defined sandwiches of shorter duration than the species' true longevity L, we take the longest longevity as being safely estimated by the thickest sandwich found for any species of Carnivora. This is a reasonable assumption because, despite the fact that some species of great longevity may have been found in considerably smaller sandwiches, it is improbable that *none* of the species of the extremest longevity has defined a sandwich of that thickness. Since even one species defining a sandwich that matches its own longevity suffices to determine this upper value, for us to be in error in the downward direction would require that *every* species of the maximum true longevity fails to be found at its sandwich edges—a highly unlikely situation. (Should this happen, we have a somewhat smaller maximum longevity than the true—not a vitiating error for the method.) Dividing the longest species survival time into ten equal time units, I shall designate each subinterval as a *decichron*. Then the true unknown species longevities range from $L = 1$ through $L = 10$ decichrons, and it will be convenient to refer to species longevities with subscripts as $L_{10}, L_9, \ldots, L_1$. To avoid repetition of the language "true" or "actual but unknown" longevity, hereafter the term "longevity" will be taken to refer, as will the capital letter L, with or without subscripts, to the true historical time period through which the species persisted.

The sandwiches defined by identifying one or more fossils of a species in two distinct geological time intervals separated by another, this being the necessary condition to have a sandwich for which presence or absence of the species' fossils within the sandwich contents can be tallied, will be called a *total sandwich* when we include the two sandwich-defining outer layers (metaphorically, the "bread slices"). The phrase *inner sandwich* will refer to the layers between the two bread slices. So that if $s = $ the total sandwich including the bread slices, and $k = $ the number of layers within the sandwich of size s, we have for every sandwich size, $s = k + 2$. The term "sandwich," when unqualified, means one having innards $k > 0$, $s > 2$.

It will not be confusing that sometimes I use "degenerate sandwich" to designate cases $s = 2$ and $s = 1$, two adjacent bread slices (no innards), or a single bread slice. Although we are interested ultimately in species of longevities down to L_2 and L_1, they will not enter our initial calculations because such short-lived species cannot (except through errors of identification which we are here setting aside) generate a sandwich of the minimum possible sandwich size $s = 3 = k + 2$ where $k = 1$. Neglecting these short-lived species does not introduce a bias because the equations first employed refer only to those of longevity greater than $L = 2$.

For all species of a group that have defined sandwiches of any given size *s*, we could ask whether or not that species is represented in any given intra-sandwich slice. "Represented" here means found *at least once* as a fossil in that slice, as here we are not concerned with how many (as we were in the Binomial Parameter Method) nor with when found (as in the Discovery Asymptote Method). The tally of representations per slice is a crude initial index of "completeness" for a slice, but it is of course not the number we are ultimately interested in. I shall refer to that number as the *crude observed slice-p*, when it is averaged over the *k* within-sandwich slices for sandwiches of total thickness $(k + 2)$. These slice probabilities for sandwiches of a given size, as well as their dispersion over the slices, constitute another portion of our observational data. The relationship between the dispersion of the slice probabilities and the average slice probability of a sandwich of a given size should be related, on a substantially random model of within-sandwich finds, by a simple formula associated with the nineteenth-century satistician Lexis. (Uspensky 1937, pp. 212-216) I do not reject that approach, as I think it should be pursued further; but it is not the one I adopt presently. The slice-*p* value will instead be treated as an inferred "latent construct" variable. The latent state of affairs includes the distribution of longevities $p(L_{10})$, $p(L_9)$, . . . , $p(L_3)$; and to each of these longevities, which include, of course, species that we have not found at all (even in a single bread slice incapable of defining a sandwich, i.e., species not known to us), there is a corresponding conditional probability of generating a sandwich of a certain size. Also there is a probability, for a species that has generated a sandwich of that size given a true longevity L, of its being represented anywhere within a specified inner sandwich slice and therefore of its being represented in the sandwich at all. If *p* is the single-slice probability, its complement $(1 - p) = q$ is the nonrepresentation-slice probability. Then q^k is the probability of not being represented anywhere within the sandwich, so its complement $(1 - q^k)$ is the probability of its being represented in a sandwich *if that sandwich is defined*.

Designate species longevities by L_1, L_2, . . . , L_{10} and the proportions of carnivorous species having these longevities as $p(L_1)$, $p(L_2)$, . . ., $p(L_{10})$. These proportions would appear as Bayes's theorem "prior probabilities" if we could estimate them. The slice-probability that a species of longevity L_1 is found represented in any one time slice $\triangle T = (1/10)L$ (L = largest longevity = 10 decichrons) is designated by p_1, the subscript indexing the source longevity. The conditional probability that a species of longevity L_1 generates a sandwich of total ("outside") thickness s is designated $p(s/L_1)$, which is the same as the probability of a sandwich whose "innards" thickness (excluding the bread slices in which the sandwich-defining fossils were found), is $s - 2 = k$.

$$k = 2 \qquad\qquad\qquad L{-}k{-}2 = 6$$

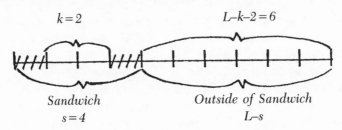

$$\text{\emph{Sandwich}} \qquad\qquad \text{\emph{Outside of Sandwich}}$$
$$s = 4 \qquad\qquad\qquad L\text{-}s$$

Figure 1

What is the probability that a species with longevity L generates a sandwich of inner size k? The diagram illustrates with $L = 10$, $k = 2$ how the slice-probability p_i determines this sandwich-generating conditional. The longevity L_{10} is divided into 10 equal decichrons. One way to get $k = 2$ from $L = 10$ is shown, the earliest bread slice (shaded interval) occupying the extreme left position. The slice-probability p gives p^2 as probability of the joint event "finding a fossil in these two (specified) bread slices." The sandwich occupies $s = k + 2 = 4$ decichrons, leaving $L - k - 2 = 6$ slices "outside the sandwich." All these slices must be empty, this joint event having the probability q^6 (no fossil representation in any of them, $q = 1 - p$ being the complement of the slice-probability p). Since there are 6 additional positions it could occupy, the composite event can occur in 7 ways, counting the one shown. The general expression for a sandwich of inner thickness k arising from longevity L is therefore

$$[32] \quad p(k/L) = (L - k - 1)\, p^2 q^{L - k - 2}$$

We shall employ this conditional probability of sandwich size on a generating longevity in deriving the second and third approximate methods now to be explained.

I present first an approximation that may seem rather crude, but that turns out to be remarkably close, as the reader may easily verify by trying a few selected numerical examples representing various combinations of parametric values. I myself am satisfied that it is sufficiently accurate, given our modest goal—to get rough estimates of the fossil record's completeness. Next I present two improvements on this approximation. The precisifying development I present finally will probably not be much better than these approximations, since the precision expressed in the formalism will not be physically realized. Departures from the idealization expressed in the more "exact" formalism will, I fear, wash out any theoretical improvement in exactitude over the rather crude methods I now explain.

If the slice-q values associated with each longevity were known to us, we could compute the true incompleteness index by multiplying the base-rate probability $p(L_i)$ for a species of longevity L_i by the probability that it is *not* found represented in any of the time slices from 1 to i. But this (assuming independence) is simply the slice-q improbability raised to the L_i power. The sum of these products over all longevities is then the true incompleteness for all species, all longevities, in the order Carnivora, thus:

$$[33] \quad I = p(L_{10})\, q^{10} + p(L_9)\, q^9 + \ldots + p(L_1)\, q^1.$$

We do not know the numerical values of any of these 20 quantities, but we are now going to approximate them crudely. We may first inquire whether there is evidence that the q's depend strongly on the L's. Since we do not know the L's, we take as proxy the associated k's, i.e., the observed sandwich thicknesses. There is a strong correlation between the L's and the k's because a short longevity cannot be associated with a sandwich larger than it, although a large L may generate a few very small sandwiches. Hence if there were an appreciable correlation between the longevities L_i and their associated slice-q values in the underlying stochastic generating situation, there would be a corresponding (although somewhat attenuated) correlation between observed sandwich thicknesses k and the q-values that generate the intra-sandwich completenesses.

Given a slice-q value q, the probability of a species not being found in any of the k slices of a sandwich of inner thickness k (on a random model of fossilization and finding) is simply q^k. If only a single value of q were the underlying generator of the observed incompleteness tally for sandwiches of size k, that generating q could be obtained simply by taking the kth root of the observed incompleteness index I_k for sandwiches of that size. So what we do first is compute the observed proportions of species not represented in the sandwiches of various sizes against the sizes $k = 1, 2,$ . . . , 8, to see whether there is evidence of a trend line. I do not think it appropriate to carry out formal significance tests; but if desired, that can be done. Mainly what we want to know is whether there is a moderate to strong relation between incompleteness as an estimator of the q values and the sandwich sizes taken as proxies for the true generating longevities. If there is not—if the graph is flat (or even if it bounces around somewhat but shows no clear trend)—we choose the kth root of the crude incompleteness as our guess at the underlying slice-q value for all longevities.

Then the grand true incompleteness index over all longevities would be obtained by multiplying the longevity proportions by the corresponding powers of q and summing as in the above equation [33]. This we cannot do, not knowing the longevity distribution, but we again make a crude approximation by ignoring the distribution of longevities—treating it in effect as if it were rectangular—and raising the kth root of the grand crude incompleteness index I_c to the power L/k. Here k is the mean observed sandwich size that we compute directly from the data. L is unknown to us, but assuming approximate symmetry of the distribution of longevities (whether they are normally, platykurtically, or rectangularly distributed), since they range from a low of $L = 1$ to a high of $L = 10$—the mean of these integers for a symmetrical case being 5.5—this value is taken as our approximation to the unknown L. In effect what we are doing is substituting an exponent that is a ratio of two estimated mean values of roots and powers, and applying this fractional exponent to a crude sum that is the unweighted, uncorrected incompleteness index over all sandwich sizes. It's essentially a matter of raising a sum to a certain power rather than summing the component values taken to differing powers initially before adding them. Oddly enough, one finds by taking various combinations (including the cases of equal or extremely variable q-values and the cases of rectangular or normally distributed longevities) that these coarse approximations result in a systematic error of only one or two points in the second decimal place in estimating the true underlying incompleteness index I. This works because of the sorts of numbers we are dealing with, to wit, fractional variables raised to powers that are ratios of small integers.

Perhaps a better approximation, although it is not obviously better given the probable distortions of the idealized model and the statistical instability of the components being added, is obtainable by solving for the distribution of longevities as an intermediate step. If the initial graphing of the quantities $(I_k)^{1/k}$ for various k's reassures us that the generating q-values do not correlate strongly with the longevities, we may use the conditional probability of a longevity generating a sandwich of a certain size (empty or occupied, not relevant) derived in the text above, equation [32]. Then the expected value of the proportion of sandwiches of a given size can be written as a sum of products of the unknown longevity probabilities by the appropriate conditionals for each longevity's generating a sandwich of each size k, including the two degenerate cases $s = 2$, $s = 1$. The sum of these products is the expected value of sandwiches of size s over all longevities, the entire system of such equations being soluble since the conditional probabilities can be computed from the quasi-constant q-value we had estimated over all sandwich sizes. The proportion of sandwiches of each size being found directly from the data, we can solve for the unknown latent longevity probabilities $p(L_{10})$, $p(L_9)$, . . . , $p(L_3)$, $p(L_2)$, $p(L_1)$. When these longevity probabilities are thus estimated, we then multiply each of them by the constant value q raised to its appropriate exponent L_i for that longevity, and sum to get the grand true incompleteness.

A somewhat better approximation makes use of the fact that a species defining a sandwich of total (outer) size s must have a longevity $L \geq s$. This being so, we can restrict the range of latent longevity sources L_i that underlie the observed incompleteness index for a sandwich of inner size k. This "averages" the unknown latent generating values $(L_i, L_j, . . .)$ over a smaller range, so that striking a plausible representative value (I use midpoint) involves less "smearing" than taking kth root of total crude incompleteness I_c over all sandwich sizes. Furthermore the procedure locates enough points so that if a straight line seems to be a reasonable graduation, we shall wash out some of the fluctuations by fitting that line. The procedure is as follows:

Consider the observed sandwiches of size $s = 7$ (innards size $k = 5$). These sandwiches can have arisen only from longevities $L = 7, 8, 9$, or 10. Although the latent slice-p values vary over these four longevities, the composite crude incompleteness index I_c (5) is employed as a "smeared" value in the approximation. We estimate a representative latent slice-q underlying the observed sandwiches of size s by taking the kth root of the crude observed incompleteness $I_c(k)$ for sandwiches of that size. We do not know the latent longevity distribution, and the relationship between that and the smearing effect is rather complicated. (Thus if extremes of longevity are somewhat rarer, the distribution not being strictly rectangular at least at the ends, this softening influence at the ends is modified by variation in the shape of the conditional probability graphs of $p(s_i/L_j)$ for various sandwich sizes on each longevity.) What we do is strike a middle value for the possible longevities capable of originating sandwiches $s = 7$, taking the midpoint of longevities L = 10, 9, 8, 7, that is, $L_m = 8.5$. (We are not bothered by the fractional longevity since, of course, the original time division was quite arbitrary, simply taking tenths of the maximum longevity for a given taxon of intermediate level.) We do the same for sandwiches of all eight "complete" sizes, i.e., those for which an incompleteness index is computable because it has innards $k \geq 1$. We fit a straight line to these eight points (notice that one of them, s = 10, does not involve any smearing approximation). The slice-q values for the two short longevities (L = 1 or 2) that can give rise only to degenerate sandwiches (s = 1 or 2) are then read off the extrapolated line.

Given these 10 estimates of the latent slice-q's, and taking each one's complement $p_i = (1 - q_i)$, we can then write directly the expected values of observed frequencies of sandwiches of all 10 sizes as functions of the conditional probabilities $p(s_i/L_j)$. We have a system of equations linear in the unknown latent longevity frequencies $N(L_i)$ thus:

$$[34] \quad \begin{aligned} N(s_{10}) &= N(L_{10})p(s_{10}/L_{10}) \\ N(s_9) &= N(L_{10})p(s_9/L_{10}) + N(L_9)p(s_9/L_9) \\ N(s_8) &= N(L_{10})p(s_8/L_{10}) + N(L_9)p(s_8/L_9) + N(L_8)p(s_8/L_8) \\ N(s_1) &= N(L_{10})p(s_1/L_{10}) + N(L_9)p(s_1/L_9) + \ldots + \\ &\quad N(L_1)p(s_1/L_1) \end{aligned}$$

Solving these for the latent longevity frequencies $N(L_1)$ by putting our *observed* sandwich frequencies $N(s_j)$ on the left, we add $N(L_1) + N(L_2) + \ldots + N(L_{10})$ to obtain the estimated total number of species of all longevities

$$[35] \quad N_t = \sum_{j=1}^{10} N(L_j).$$

Numerical trials and inspection of the graphs of conditional probabilities show that this approximation is remarkably good, despite the smearing of the latent values. It is likely to do about as well as the "exact" procedure described next, because the exact procedure involves the solution of a system of 16 equations of high degree in which both the slice-p values and the longevity distribution frequencies are unknowns, whereas in this third approximation, the conditional probabilities are first obtained by substituting the slice-q values in the formula so that the system of equations to be solved is linear.

The only "exact" solution I have managed to come up with makes use of the fact that we possess two items of information about all 8 of the nondegenerate sandwich sizes ($k \geq 1$, $s \geq 3$), namely, for each such sandwich size we know its empirical frequency N_k and its crude completeness index I_k. Since these observable quantities are expressible in terms of the latent longevity-frequencies and the latent slice-q values, we obtain a soluble system of 16 equations in 16 unknowns. The remaining four latent values associated with short longevities (N_2, N_1, q_2, q_1,) we get by extrapolating and check for consistency by predicting the two degenerate sandwich frequencies $N(s_2)$, $N(s_1)$. The "exact," method, using all the sandwich frequency and incompleteness data at once, proceeds as follows:

Let us confine attention first to the nondegenerate sandwiches ($s = k + 2 \geq 3$). It is convenient to employ raw species frequencies rather than probabilities, avoiding the normalization problem to $\Sigma P_i = 1$. Only species with longevities $L_i \geq 3$ can generate these sandwiches, so we have to consider eight longevities and eight (associated) slice-p's. It is convenient to treat the complementary slice-improbabilities q_i instead of the p_i's themselves. The latent situation is then

Longevities: L_{10} L_9 L_8 . . . . . . L_3
Base-frequencies: N_{10} N_9 N_8 . . . . . . N_3
Slice-q's: q_{10} q_9 q_8 . . . . . . q_3

All the N's and q's are unknowns, and there is no constraining equation on either of them (except, of course, that $\Sigma q_i < 8$, which is not helpful).

The observed sandwich frequencies are expressible in terms of these latent quantities (writing "observed" on the left, although strictly speaking these equations yield only the *expected values* of the several species N's). Thus:

$$N(s_{10}) = N_{10}\, p(s_{10}/L_{10})$$
$$N(s_9) = N_{10}\, p(s_9/L_{10}) + N_9\, p(s_9/L_9)$$

[36]

$$N(s_3) = N_{10}\, p(s_3/L_{10}) + N_9\, p(s_3/L_9) + \ldots + N_3\, p(s_3/L_3)$$

For each sandwich (inner) size k there is available an observed crude incompleteness index I_k. These observed indexes for each inner thickness k are compounded from latent incompleteness indexes arising from all sources with a sufficient longevity $(L \geq k + 2)$ to generate such a sandwich; but for each such latent source L_i the expected incompleteness among the subset of sandwiches *it* generates will be $q_i{}^k$. These latent component incompletenesses have expected contributions to the sandwiches in proportion to the proportional composition of each sandwich size's source species. Thus if $p(L_1 / k_j)$ is the posterior ("inverse," Bayes' Theorem) probability that a sandwich of size k_j has originated from a source of longevity L_i, we can write the expected values of observed sandwich incompletenesses as follows:

$$I(k_8) = p(L_{10} / k_8) \; q_{10}^8$$
$$I(k_7) = p(L_{10} / k_7) \; q_{10}^7 + p(L_9 / k_7) \; q_9^7$$

[37]

$$I(k_1) = p(L_{10} / k_1) \; q_{10}^1 + p(L_9 / k_1) \; q_9^1 + \ldots + p(L_3 / q_3) \; q_3^1$$

This system of 8 equations adds no new unknowns, because the inverse probabilities $p(L_i / k_j)$ are Bayes' Theorem functions of the priors and conditionals in the system [36]. Hence we have a total system of 16 equations in 16 unknowns that we can solve.

After solving these to obtain the 8 latent longevity-frequencies $N(L_{10})$, $N(L_9), \ldots, N(L_3)$ and their 8 associated slice-improbabilities q_{10}, q_9, $\ldots, q_3$, we still have to estimate the latent species frequencies $N(L_2)$, $N(L_1)$ for species too short-lived to produce a nondegenerate sandwich, and the slice-improbabilities q_2, q_1 associated with them. This we do by extrapolation, thus: Plot the $N(L_i)$ values against the longevities L_{10}, L_9, $\ldots, L_3$ and fit a curve—a straight line should be good enough, tested for linearity if desired—to the 8 points now known. The two N's at the low longevity end are then assigned by reading the ordinates off that fitted line. Similarly we obtain q_2, q_1 by reading off a line fitted to the eight pairs (L_i, q_i) for $L_i \geq 3$. A consistency check for the trustworthiness of these extrapolated values is available by writing the equations for the 10 "sandwich" frequencies (now including the degenerate cases $s = 2$, $s = 1$) in terms of the latent generating values.

We now possess numerical estimates of the values of all 20 latent variables conjectured to generate the observed distribution of sandwich sizes and the crude incompleteness indexes tallied for nondegenerate sandwiches. The longevity frequencies $N(L_{10})$, $N(L_9), \ldots, N(L_1)$ can be summed to obtain the total (found and unfound) carnivorous species number, and the completeness index is the ratio of the known species tally to this sum:

$$[38] \quad C = N^* \Big/ \sum_{i=1}^{10} N(L_i)$$

We can also obtain this estimate from Equation [33] supra (where $I = 1 - C$) and the two should agree. We record this species number in the third column of our master table of four estimates, in the row for the order Carnivora.

IV. Method of Extant Forms (Dewar's Method)

The fourth and final method of completeness estimation was not invented by me but by Douglas Dewar. I am especially indebted to his rarely mentioned work because it was my reading of the exchange of letters between Dewar and H.S. Shelton that first interested me (alas, over thirty years ago, when I set the problem aside) in the possibility of estimating the completeness of the fossil record, although I was informed by colleagues in zoology and geology that such a numerical estimate is obviously "impossible." It has been easy to minimize the importance of Dewar's contribution because despite having some professional distinction as an ornithologist and, as his writing clearly shows, an extraordinarily intelligent and resourceful mind, he was also a die-hard anti-evolutionist, whose espousal of creationism was apparently not associated with conservative Christian theology. A scientifically legitimate objection to the Method of Extant Forms is that it relies (in the traditional non-Popperian notion of "rely" to mean "justification") on questionable "assumptions" concerning some quantitative equivalences that are, to say the least, doubtful. In the eyes of an evolutionist who wishes to dispute even the possibility of a completeness index for the fossil record, these equivalences would not be acceptable.

The essential notion of Dewar's Method is extremely simple, and if the questionable assumption about certain equivalences were granted, it is the most straightforward and least problematic of the four methods here discussed. In all three of the methods I have devised, we are concerned, so to speak, in estimating the denominator of a fraction—the completeness index—from its numerator, given some kind of conjectured function relating the two. The numerator is the number of extinct species of a group known to us from paleontological research, and the desired denominator is the unknown total number of species of that group (such as all animals with hard parts, or all vertebrates, or all carnivores, or all ungulates) that have ever walked the earth. We are moving from the known to the unknown—not inherently a methodologically sinful procedure, as Popper points out—and the confidence we have in that epistemic movement hinges at least partly on the plausibility or independent testability of the alleged functional relationship between the known term in the numerator and the unknown term in the denominator. I say only "at least partly," because the essential independence or nonredundancy of the three paths for estimating an unknown number provides a strong Popperian test of the accuracy of the estimation procedures, as I said in my introductory remarks. The relative importance of (a) *the a priori plausibility* of the postulated functional relationship between the known and unknown quantities and (b) *the mutual corroboration provided by numerical agreement*, no one presently knows how to assess, particularly if the agreement permits large error tolerances. But as we desire here to estimate an unknown quantity of great scientific importance, we are willing to attempt it even if the percentage of error for each estimate is quite large. (I shall say more below about assessing the convergence of values.) But Dewar's method, the Method of Extant Forms (as I may call it to indicate its character, as the other three were labelled, although I think Douglas Dewar should receive posthumous recognition for his method), is unlike my three methods, because in Dewar's Method the desired denominator is known to a high accuracy for many taxa. That is due to the fact that the tally of species in the denominator consists of extant forms. The discovery rate for new extant species of the order Carnivora is now extremely small and has been for many years, so that zoologists can say that contemporary science "knows about" the existence of close to 100 percent of the species of carnivores living today. Or, moving up from order to even such a broad taxon as class, of "large mammals" only two new species have been discovered in the last 150 years (the giant panda and the okapi). So, to answer the first-level question, "What is the completeness of the fossil record for *extant* species of Carnivora?" we have a simple and straightforward computation, to wit, we divide one number known with high accuracy (how many species of extant carnivores have been found at least once as a fossil) by another number known with very high accuracy (how many extant species of Carnivora there are). No complicated algebra or doubtful auxiliary postulates about randomness, linearity, etc., are required for applying Dewar's

Method. For those groups of animals in which the current discovery rate of new species is considerable, even as adjudged by taxonomists of a "lumper" rather than a "splitter" bias (such as the hundreds of thousands of insect species that are still being added to at a respectable rate by entomologists), the method presents more serious problems, so those taxa might better be set aside for that reason. The vexed question of which phyla, classes, orders, and so forth should be candidates for the four completeness index estimation procedures here presented, I discuss below.

But corresponding to the delightful simplicity of Dewar's Method is, of course, the transition from a completeness index for extant forms to the one we are really interested in, i.e., the completeness index for extinct forms. We pay for the extreme methodological straightforwardness and algebraic simplicity of Dewar's index, the mathematics of which is so much simpler and the embedding justificatory text so much more plausible, with a slippery and dangerous assumption, that *the completeness index for extant and extinct forms is about the same size fraction*. With my neo-Popperian approach, i.e., auxiliary conjectures as part of the network to be tested by the fact of convergence rather than pure faith "assumptions" (which an optimist may be willing to rely on and a pessimist will reject), I don't face quite the problem that Dewar had with Shelton, when the latter objected to the use of extant forms as a basis for estimating the completeness of the fossil record for extinct forms. The reason is again that the numerical convergence of nonredundant estimation procedures is taken as a Popperian test of the validity of each, including its auxiliary conjectures, however plausible or unplausible these may seem when examined qualitatively.

Dewar's Method does not strike me as so implausible as Shelton seemed to think, though. One must remember that we are not comparing groups of very diverse kinds, where the number of the individual specimens, the geological period during which they flourished, of generations per decichron, the evolutionary rate, the geographical distribution, the ecological niche, and so forth, are often very different (as they would be if one were comparing the completeness index for extant bears with the completeness index for, say, extinct echinoderms). What we are comparing is the completeness index for animals like the polar bear with animals like the saber-tooth tiger, because the final stage of our procedure is to involve the correlation of estimated species numbers over various groups, and these groups are defined as animals of roughly the same sort, i.e., carnivores, ungulates, crustaceans, or whatever. So long as the conditions of the earth are such that a given order, say, carnivorous mammals, can exist in appreciable numbers at all (part of this, of course, is simply Lyell's uniformitarian assumption that evolutionary theory postulates instead of catastrophism), it is not obvious that the conditions operating statistically to determine the laying down of a fossil, its preservation, and contemporary searchers' luck in finding it are very markedly different between extant and extinct species *of the same order or family*.

The idealized conjecture relied upon in Dewar's Method is that the terrestrial conditions for preservation of the hard parts of an animal as a fossil; the distribution, intensity, and duration of fossil-destroying geologic occurrences such as erosion and faulting; and—what presumably will not have varied appreciably—the statistical odds of the paleontologists' search locating one and identifying it, have remained, in their aggregate influence, roughly comparable for extant carnivores and extinct carnivores. As suggested above, we conceptualize (without claiming to quantify causal-theoretically) a sort of "anti-find pressure," the net influence of these collective factors acting adversely to paleontological discovery. The incompleteness index I_k (= species not found/total species) is some monotone function of this antifind pressure, and the rationale of Dewar's Method is the conjecture that the antifind pressure is likely to be about the same for extant and extinct species of the same broad taxon, as carnivores, primates, ungulates, etc. The manner in which we employ the simple completeness index based on fossil finds of extant species is also simple. The fraction C* obtained on extant forms is taken as an estimator of the corresponding completeness index C for extinct forms of the same kind, e.g., order Carnivora. We therefore take the number of species N* (= number of extinct carnivorous species found at least once as a fossil), and, writing N* = NC, we solve for N, the desired total number of extinct species of Carnivora, known and unknown.

Reflection on Dewar's Method quickly suggests to the mind a different and powerful use of the fossil data on extant forms, namely, a direct test of the numerical accuracy of what I may call "Meehl's Methods." The main use of Dewar's Method is as an estimator of the completeness index for extinct forms, where as we have seen, it has advantages over my three methods in its conceptual and mathematical simplicity, but with the attendant disadvantage that it relies on the vague concept of "aggregate counter-find pressures," postulated to be approximately the same for extinct and extant forms of the same intermediate level taxon (e.g., order

Carnivora) to rationalize the extrapolative use of the index to the extinct groups. Assigning some degree of antecedent probability to that auxiliary conjecture, we test its approximate validity indirectly, by means of the empirical test of convergence among the four methods taken together, as explained in the next section below. Although this indirect test is perfectly good scientific procedure, whether we think in a neo-Popperian fashion or simply reflect on the history of the developed sciences in their use of auxiliary conjectures, it is nevertheless true that the famous Quine-Duhem problem presents itself, in that if convergence fails, all we can say with confidence is that *something* is wrong in the conjectured conjunction of main and auxiliary statements. It is well known to logicians and philosophers of science that no straightforward automatic touchstone procedure exists for identifying the culprit in such circumstances, although very few working scientists would accept the strong form of the Quine-Duhem thesis sometimes stated, that *since* the formal logic of a negated conjunction is a disjunction of negations, *therefore* a scientist's first-round choice of the likely culprit cannot be principled, "rational," or other than whimsical. Although from the standpoint of formal logic, *modus tollens* refutation of a theoretical conjunction amounts to a disjunction of negations of the conjuncts, when we distinguish considerations of formal logic from methodology of science, we know that in ongoing scientific research some of the conjuncts are considered less problematic than others. This is true either because they flow quasi-deductively from well-corroborated background knowledge, or because the investigator has meanwhile tested them separately from the others by a suitable choice of the experimental context. These complexities have to be accepted, and they will not be troublesome to us if the four methods converge well when examined as described in the next section of the paper. The disturbing case is the one in which they do not converge, and that would lead us to look at whether some agree fairly well with each other but one of them, as for instance Dewar's Method, is markedly out of line.

But the existence of a fossil record for extant forms provides a more direct path to assessing the methodological power of my three methods. Instead of inferring that the methods have some validity because they converge on the unknown latent value of the species number, we can apply my three methods to the population for which that species number N is known, to wit, extant forms.

By applying my three methods to a suitably chosen set of orders of extant forms (qualitatively diverse and with species numbers both sizable and variable), we can ask about each method how accurately it estimates the total number of extant species of carnivores, primates, crustaceans, insectivores, and so on. We can also get an idea of how well the composite of three weighted estimates performs as estimator. There are several interesting and important questions about the three indirect methods that can be studied in this way, of which I list the more obvious and crucial ones, but I daresay others will occur to the reader:

a. What is the average accuracy of each of the three indirect methods? ("Accuracy" in this list means numerical closeness to the true species number for each order estimated.)

b. What is the variation in accuracy of each method over the different orders?

c. Are the three indirect methods more or less random in their discrepancies, or can we discern a ranking in the species numbers estimated?

d. If there is such a ranking, can one strike a reasonable average bias, so that a corrected estimate for the clearly biased method can be applied to the extinct orders, the anticipation being that this will improve the consistency of the results on that group?

e. How do some of the auxiliary conjectures fare in this situation, where they can be checked directly against the true numbers rather than indirectly, i.e., by their success when used in a derivation chain eventuating in a numerical value testable for its consistency with other numerical values? For example, the conditional probability of sandwich sizes for a specified longevity and slice-probability generates a family of graphs of expected frequencies. How well do those theoretical graphs fit the empirically computed ones for the case in which the slice-q values are accurately determinable from the observations on extant forms, and only the lower end of the longevity is problematic?

A good convergence of the numerical values of the four methods on the exinct orders would have a satisfactory Popperian corroborative effect, but it is evident that we would gain a good deal of additional confidence in the entire interlocking system of substantive and auxiliary conjectures if we found that a methodological claim (like an instrument's precision) is well corroborated for three of the methods used on extant forms for which the true values are known, namely, that these three methods do tend to give the right answer when it is directly known to us.

Agreement of the Four Methods

We now have the estimated number of extinct Carnivora species as inferred from the Discovery Asymptote Method, the Binomial Parameters Method, the Sandwich Method, and Dewar's Method of Extant Forms. Ditto for ungulates, insectivores, primates, etc. If the conjectured statistical model (reflecting the underlying historical *causal* model of the consecutive processes: fossilization + preservation + discovery) possesses moderate or good verisimilitude despite the idealizations and approximations involved, the four extinct species numbers estimated for a given order of, say, vertebrates should show a "reasonable agreement" with one another. I put the phrase "reasonable agreement" in double quotes to emphasize that this is a loose concept and that I do *not* intend to advocate performing a traditional statistical significance test. That would be inappropriate since it asks a question to which we already have an answer, namely, are the methods precisely equivalent in the unknown quantities they are estimating and have causally arisen from? The answer to this question is surely negative. The notion of "reasonable agreement" is not an exact notion even when stated in terms of tolerances, but biological and earth scientists are accustomed (more than social scientists) to dealing with that notion. It is scientifically a more useful and powerful notion by far than the mere trivial refutation of the null hypothesis that two population values are exactly equal—which, of course, in the life sciences they never are. (See Meehl 1967; Lykken 1968; Morrison and Henkel 1970; Meehl 1978.)

It may be objected by purists that, lacking a precise measure of agreement among the four numerical values (such as provided by the traditional statistical significance testing approach), one cannot say anything rational about the data. Philosophical replies to such perfectionism aside, one can only appeal to the history of the exact sciences, where for several centuries before the invention of statistical significance testing, or even "exact" procedures for setting up confidence intervals around point estimates (such as the physicists' and astronomers' probable error), great advances were made in these disciplines by studying graphs and noting reasonable numerical concordances in tables. Further, in evaluating a procedure for estimating unknown quantities one must evaluate the knowledge provided by the estimates, however crude, *in relation to our state of knowledge without them*. At the present time, almost nothing numerical has been said (given the neglect of Dewar's work) about the completeness of the fossil record. The standard comment is that it is surely very incomplete, a statement necessitated by the absence of the many thousands of transitional forms that neo-Darwinian theory requires to have existed but that are not found in the presently available fossil record. That being our current epistemic situation, even a ballpark estimate of so scientifically important a number as the "completeness index" is an advance over no quantitative estimate at all.

Nothing prevents us from treating the completeness index as a proportion or percentage and attaching an estimated standard error to it in the traditional way if we believe that is helpful. The reader may think it worthwhile to perform a significance test between completeness indices estimated by different methods. Because of my distaste for the widespread abuse of such procedures in the behavioral sciences, I would not myself prefer to do that (see Meehl 1978); but no doubt some others would. If it makes anybody feel better to calculate a significance test over the completeness indices obtained by the four methods, he may do that. At this stage, in this place, I wish primarily to emphasize that if we could set only the most extreme upper and lower bounds upon estimated completeness, we would have made a material advance on the present state of knowledge.

For example, the consensus among evolutionists would be that the fossil record is very considerably less than half complete. Now suppose that the completeness indices obtainable by the four methods for, say, order Carnivora were numbers like 10 percent, 12 percent, 20 percent, 31 percent. These four percentages are discouragingly dispersed from the standpoint of a statistician emphasizing a precise point value; but they are important nevertheless, because they tell us that four nonredundant methods of estimating the completeness of the record for carnivores are well under one-half. Imagine instead that we found four completeness indices that were, though varying considerably among themselves, all higher than 50 percent. (Thus, for instance, the completeness index for land mammals in Dewar's Method was found by him to be 60 percent—a proportion very much higher than anyone would expect, and one that obviously has grave implications for the received doctrine of evolutionary slowness, i.e., the number of transitional forms, and their species longevities, with which the evolutionary process occurs.) This would be an important finding despite the variance over four methods. In a nutshell, what I am saying is something that scientists working in such semi-speculative fields as cosmology have long been familiar with accepting as a part of their intellectual burden of uncertainty for studying a particular subject matter in which precision is difficult, to wit, sometimes it's worth estimating a quantity even within a few orders of magnitude, if the quantity is important enough, and if the orders of magnitude, from another point of view, differ only slightly in relation to the possible range we would assign in a state of complete Bayesian ignorance. A completeness index of .50 would be an order of magnitude larger than what paleontologists seem generally to consider "reasonable," presumably because of the sparsity of quasi-complete fossil phylogenies. Everyone interested in the theory of evolution would surely like to know whether the following statement can be considered more or less corroborated by reasonable estimates independent of each other: "The *majority* of species of animals with hard parts that have ever lived on the earth are known to us from the fossil record, at the present time." This would surely be a surprising generalization if corroborated by the proposed four methods. Given a fairly consistent set of completeness indices that were uniformly over 50 percent for different kinds of animals with hard parts—even if the percentages for a given order varied considerably, as between, say, 55 percent and 90 percent—we would possess a numerical result whose importance for evolutionary theory can hardly be exaggerated.

There is, however, a simple, straightforward statistical approach to evaluating the agreement, which I set forth briefly because it yields a rough quantitative measure of goodness of agreement. Most of us would find even a crude measure more reassuring than the mere verbal characterization "reasonably good agreement" among sets of four numbers. Let the reader visualize the state in which our data are now summarized, with four columns (corresponding to the four methods) of numbers, each row corresponding to an order of extinct animals. The numbers appearing in this matrix are not completeness percentages but estimated total extinct species frequencies N_a, N_b, N_c, . . . , based upon the completeness estimations for a = Carnivora, b = Primates, c = Insectivora, etc. We do it this way because the variation in number of species for different animal kinds is what we are studying when we ask whether the four methods agree to a closeness better than what one could plausibly attribute to "mere coincidence." The underlying reference base one tacitly presupposes in referring, however loosely, to "degree of agreement" among different conjectural ways of estimating one and the same underlying numerical value in nature, is always some prior information—however vague and common-sensical—about the antecedently expected range of numerical values for the kind of quantity being studied, as measured in the units that have been adopted. We need not be strict Bayesians in our theory of inductive inference (certainly not subscribers to the strong form of subjective/personalist probability theory) to insist upon the importance of background knowledge. We unavoidably rely on background knowledge in evaluating whether a convergence of two or more estimators of a theoretically conceived numerical quantity agree well enough that their agreement cannot plausibly be regarded as some kind of a happenstance or lucky accident. (From the Popperian standpoint, the accident is *un*lucky, since it misleads us by supplying a strong corroborator of a causally unrelated theory.)

I should emphasize that I do not here raise some novel philosophy of science of my own. I simply call attention to a fact that is common throughout the history of all quantitative empirical disciplines, to wit, the scientist develops conviction about the accuracy of his methods in the context of whatever theory (or even theory sketch) he is using to articulate and defend a method, by noting that numbers agree well. If one does not rely upon the traditional test of significance—it is inappropriate when one already knows that the theory is an idealization, that the instruments are fallible and may even contain slight systematic biases, along with random errors of measurement and sampling—then how "close" two or more numerical estimates are judged to be must obviously depend upon some antecedent knowledge, or at least expectation, as to their "ordinary range." The obvious example is the case of measuring a simple length, where whether a discrepancy of so many units in some physical measurement scale is counted for or against the accuracy of the measuring procedure in the context of the conjectured theory will depend entirely upon the range of "empirically plausible" lengths as expressed in those units. Example: If I tell you that my theory of genetic control of growth in the elephant predicts that the trunks of baby elephants born of the same mother tend to be within six inches of each other in length, you will be unimpressed. If I tell you that they are usually within two millimeters of each other, you either will be impressed or think I am faking the data. The only basis for this difference in attitude is your prior knowledge about how big, roughly, a baby elephant's trunk is and how organisms tend to show variation. In astrophysics and cosmology, it may be worthwhile to estimate a quantity within several orders of magnitude. In other branches of physics, it may be pointless to do an experiment whose accuracy of measurement is not at least 99.99 percent, and so on.

In the light of these considerations, a reasonable way to deal with our four columns of estimated species numbers for different orders of extinct animals is to express them not as percentages or proportions (the completeness index) but as estimated absolute frequencies of species per order. Thus we are taking advantage of the presumed wide range of differences in the number of species subsumed under a particular order to get a handle on the "more-than-coincidence agreement" question. Keep in mind that we are contrasting a claim of reasonable agreement with the counter-conjecture that the four methods are hardly any good at all, that they are nearly worthless for the purpose of estimating the mysterious number "C = completeness of the fossil record." If that pessimistic view is substantially correct, a set of calculations concerning the true unknown number of extinct species N_a for order a might as well be drawn from a hat, *and the hat to draw it from is the whole range of these species numbers, as generated by these invalid estimation procedures*. One way of looking at it is that in order to know whether estimates are in decent agreement on the species numbers for carnivores, insectivores, and rodents, we would like to know the basic range of species numbers for taxa at this level of classification. Thus an average disagreement of 20 species over the four methods would be discouraging if the average estimated number of species in an order ran around 40 and varied from that number down to 0; whereas if the number of species representing a particular order were characteristically tallied in the hundreds, an average disagreement of 20 species would be encouraging, given the present weak state of knowledge.

One simple and intuitively understandable expression of the degree of agreement among four measures "of the same thing" comes from classical psychometrics. It was invented three quarters of a century ago by the British engineer and psychologist Charles Spearman, who developed the so-called tetrad difference criterion for testing the psychological hypothesis that the pairwise correlations among mental tests could be attributed to the causal influence of one underlying common factor, Spearman's g. It was subsequently shown by L.L. Thurstone in the matrix development of multiple factor analysis (a generalization of Spearman's idea to the case of more than one latent causal variable determining the pattern of mental test

scores) that the tetrad difference criterion is the limiting case of evaluating
the rank of a matrix. The arrangement of correlation coefficients in
Spearman's tetrad difference criterion is seen, from the standpoint of
matrix algebra, to be the expansion of a second-order determinant in the
larger matrix of correlations among many tests. Hence the vanishing of the
tetrad differences tells us that the second-order determinants vanish (the
corresponding second order matrices being of unit rank); and that is what
Spearman's method requires if one factor suffices to account for all the
relationships.

Treating each species number in a column as a nonredundant estimator
of the true unknown species number, the correlations between columns in
our table summarize pairwise agreement of the methods. Given four
methods, there are six such inter-method correlation coefficients. As they
are all estimates of the same true quantity [N_t = Number of extinct species
in order t] that varies over different orders, then even if these estimates are
highly fallible, the six correlations are attributable to the influence of the
one latent common factor, to wit, the true unknown extinct species number
N_t, so the tetrad difference equations should be satisfied. They read:

$$[39] \quad \begin{aligned} r_{12}r_{34} - r_{13}r_{24} &= 0 \\ r_{12}r_{34} - r_{14}r_{23} &= 0 \\ r_{13}r_{24} - r_{14}r_{23} &= 0 \end{aligned}$$

There are statistical significance tests for the departure of these equations from the idealized value 0, but one does not expect them to be exactly
satisfied (any more than they were *exactly* satisfied in Spearman's psychometric model). One would like to know whether they are reasonably close
to 0. If they are nearly satisfied, we have reason to think that our conjecture
that each method is an informative way of estimating the unknown number
of species, has verisimilitude. Of course we would also expect the pairwise
correlations to be high, because these correlations represent the extent to
which the four methods agree. The average amount of their disagreement
is thus being compared (via the inner structure of a Pearson r) to the
variation of each method over the different orders of animals. If the tetrad
difference equations are almost satisfied, it is then possible to assign a
weight or loading to each method. One method may, so to speak, be "doing
the best job of whatever all four of them are doing," as represented by the
pattern of the six correlations of the methods taken pairwise. One can then
combine these weights to reach an improved estimate of the species

number for a given order by multiplying the standardized number for each order by its common factor weight and adding these four products. What that amounts to is relying more on one measure than another, while taking them all into account, each one being weighted as befits its apparent validity for the latent quantity. The loading or saturation of a given indicator among the four is obtained by one of several approximate (and nearly equivalent) formulas, the simplest of which is

$$[40] \quad a_{kg} = \frac{\Sigma r_{ks}}{\sqrt{\Sigma r_{ij}}}$$

where one divides the sum of indicator k's correlations with the other three indicators by the square root of the sum of all correlations in the table. (Thurstone 1947, pp. 153, 279; Gulliksen 1950, pp. 343-345; Harman 1960, pp. 337-360.)

I do not urge that this is the optimal way to analyze these data, but with four measures the tetrad method springs naturally to mind. Other procedures would be defensible, such as the so-called generalized reliability coefficient developed many years ago by the psychologist Paul Horst (1936). I suppose most contemporary statisticians not oriented to psychometrics would prefer some generalized breakdown of components in the analysis of variance where the total sum of squares of species numbers is assigned to methods, orders, and methods x orders interaction. (See, e.g., Hoyt 1941.) If the methods proposed have even a crudely measured degree of reliability and construct validity as shown by their agreement, there will then be plenty of time to select the optimal method of expressing disagreement, and of combining the statistics into one "best available estimate" of the species number for each order. In the present state of the art, most of us would be happy to contemplate even a table of simple "percent deviations" of the methods from each other and their unweighted means.

Summary

Given a neo-Popperian philosophy of science, when one estimates values of unobserved theoretical variables (latent, underlying, causative, or historical) in reliance on certain mediating assumptions, these assumptions are treated as *auxiliary conjectures*, more or less problematic. Being problematic, they should not be left in the status of "assumptions," in the strong sense of required postulates that we hope are true but have complete liberty to deny. It is argued that such idealizing auxiliary conjectures can motivate four distinct, nonredundant, observationally and mathematically independent methods for estimating the degree of completeness of the fossil record as known to paleontologists at a given time. We can ascertain the closeness of numerical agreement achieved by these four nonredundant methods when each is used to estimate the total number of species, known and unknown, of a given taxon at intermediate level in the taxonomic tree (e.g., a family, order, or class). This procedure subjects the auxiliary conjectures rationalizing each method to a relatively severe Popperian test, because if the methods are poor estimators, having little or no relation to the true unknown species number for each order, there is no reason why they should have any appreciable tendency to converge. On this kind of reasoning, it is urged that paleontologists need not — I would say, *should* not—content themselves with stating that the fossil record is "very incomplete." This posture has allowed philosophic critics, including Sir Karl Popper, to argue that the theory of organic evolution is not, as it is claimed to be by geologists and systematic zoologists, a scientific theory at all, but merely a fruitful metaphysical speculation. Four methods are described, three of which, The Discovery Asymptote Method, The Binomial Parameters Method, and the Sandwich Method, were devised by the present author, and the fourth had been invented by Douglas Dewar. It is suggested that the four methods could be applied to some classes of data already catalogued in paleontology, and that future research ought to be conducted in such a way as to make available the raw data necessary for applying the methods to the fossil record for animals of many different kinds. If the methods show reasonable agreement (as measured by any of several statistical procedures), each method can be assigned a weight in proportion to its validity as an indicator of the latent quantity: "Total extinct species of the order." A composite estimate obtained by summing these weighted estimates from the four methods can then be used to calculate a completeness index for each order studied. In addition to the validity argument from numerical convergence, the trustworthiness of the author's three indirect methods can also be checked more directly by applying them to extant forms, where the true species numbers are known. It is urged that crude approximate results from these methods would represent a material advance on the present state of paleontology, where only very broad and untested qualitative statements concerning the record's gross inadequacy are presently available. It is also argued that approximate measures of consistency are of much greater scientific value than traditional use of statistical significance tests, since we know that the auxiliary conjectures are idealizations, false if taken literally, so that showing this is pointless.

Appendix I

It would be surprising, given what we surely must presume about minuscule find-probabilities, to discover that the fossil specimen frequencies per species are distributed too symmetrically ($p \simeq q$) for the Poisson to provide a good empirical fit. Such a result, locating the modal fossil count clearly to the left of our empty column at $k = 0$, would in itself strongly discorroborate the received conjecture that the record is grossly incomplete. But if the Poisson did appear to give a rather poor fit, a somewhat more complicated Binomial Parameters procedure would be needed. I present it here briefly in two variants, without any conviction as to its optimality, and expecting it never to be needed.

The problem is to estimate the latent parameters of a binomial that characterizes the find-probability in relation to digs. The latent function conjectured to generate our observed distribution of fossil-specimens-per-species is simply $N(p + q)^n$ where p = Find-probability, n = Number of digs, N = Total number of extinct Carnivora species, found and unfound.

The unknown N cancels when we form ratios of successive terms, t_1/t_2, t_2/t_3, t_3/t_4, . . . , of the expansion of this binomial. If we then take ratios of these term-ratios, the latent parameters (p, q) also cancel, and the ratio-ratios (from the low-find end) run

$$[41] \quad \frac{1}{2} \frac{n-1}{n}, \frac{2}{3} \frac{n-2}{n-1}, \frac{3}{4} \frac{n-3}{n-2}, \quad \ldots$$

Which for large n (= number of digs, not empirically known but surely large enough) involves n-ratios of form $(n - r)/(n - r + 1) \simeq 1$. Hence the ratios of term ratios are closely approximated by simple fractions involving small integers (at the low find frequency end) and the last one, involving the missing term at $k = 0$ (no fossil found) is

$$[42] \quad R = (t_1/t_2)/(t_2/t_3) = \frac{1}{2} \left(\frac{n-1}{n} \right)$$

very near to $\frac{1}{2}$, surely an error of less than 1 percent. So we solve in [41] for the missing number of species at $k = 0$ finds.

Another possibility utilizes more distribution of information but I fear could be quite unstable, so the added information might not pay off. The ratios of successive terms, although the equations for them will be empirically inconsistent, could be solved for the parameters p, n. Then the missing term at $k = 0$ is calculated from these. Taking this value as a first approximation to the missing item, we compute the empirical skewness and kurtosis of our find-distribution. These numerical values characterizing the empirical distribution shape are then employed with the standard formulas for skewness and kurtosis of a binomial, two equations soluble for the two parameters (p, n). (Cramér 1946 p. 195; Kenney 1939 p. 15.) This gives a second approximation to them, and permits a revised estimate of the $k = 0$ probability. We iterate until the values settle down.

Appendix II.

Given the expression Equation [32] for the probability of a specified size sandwich arising from longevity L, we can get the probability P_{sL} of finding a sandwich (any size) from antecedent condition L. This is the sum over all sandwich thicknesses k of the terms p(k/L) as just obtained, i.e.,

$$[43] \quad P_{sL} = \sum_{k=1}^{L-2} (L - k - 1)\, p^2 q^{L-k-2}$$

which is like a geometric progression with common ratio q except that each term has an integer coefficient that undergoes unit increments from term to term. Thus for L = 6 the sum is

$$[44] \quad \Sigma = 4p^2 q^3 + 3p^2 q^2 + 2p^2 q^1 + 1p^2 q^0$$

$$[45] \quad \Sigma = p^2 (4q^3 + 3q^2 + 2q^1 + 1q^0).$$

Call the parenthesis S', multiply by q and subtract,

$$[46] \quad S'_4\, q - S'_4 = 4q^4 - (q^3 + q^2 + q) - 1$$

and now the parenthetical sum is a 3-term G.P. (call it S_3) with its ratio q equal to its initial term, that is, after dividing by $(q-1)$ to get S'_4 alone on the left, we have

$$[47] \quad S'_4 = \frac{4q^4 - S_3 - 1}{q - 1}.$$

The general expression for total sandwich probability is easily seen to be

[48] $P_{sL} = p^2 S' = p^2 \dfrac{S_{L-3} - (L-2)q^{L-2} + 1}{1-q}$

$= p[S_{L-3} - (L-2)\, q^{L-2} + 1].$

We have to substitute the usual formula for a G.P. for S_{L-3},

[49] $S_{L-3} = \dfrac{q(1-q^{L-3})}{1-q}$

in [48], obtaining

[50] $P_{sL} = p\left[\dfrac{q(1-q^{L-3})}{1-q} - (L-2)q^{L-2} + 1 \right]$

which after some grouping and cancelling gives us

[51] $P_{sL} = (L-2)\, q^{L-1} - (L-1)\, q^{L-2} + 1.$

This is a fairly simple expression for sandwich-probability (any size) as a function of the slice-*im*probability q_i for a given longevity L_i.

We note that this result can be obtained more directly (and easier intuitively) by subtraction, beginning with the probability of *not* finding a sandwich (of any size) from condition L. There are three non-sandwich cases:

	Probability
Not found at all	q^L
One bread slice	Lpq^{L-1}
Two adjacent bread slices (no "inner" tally available)	$(L-1)\, p^2 q^{L-2}$

The sandwich probability is the complement of the sum of these three nonsandwich probabilities,

[52] $P_{sL} = 1 - (q^L + Lpq^{L-1} + (L-1)\, p^2 q^{L-2}),$

which with a little manipulation readily yields the expression [51] reached positively.

These longevity-conditional probabilities cannot be quickly evaluated from our observed within-sandwich slice-probabilities because the empirical slice-p's do not correspond to the latent slice-p variable occurring in the conditional probability formula.

Note

1. Professor Malcolm Kottler has kindly called my attention to Popper's article "Natural selection and the emergence of mind." *Dialectica* 32, 1978, pp. 339-355, in which Sir Karl modifies his long-held view to this extent, that it is possible to reformulate Darwinism in such a strong fashion that it does become testable; in which strong form it is, he says, known to be false. Evolutionists will hardly be happier with this amended view than with his earlier one, and it will be clear to readers that the amendment does not nullify our original problem or the point of this paper.

References

Cramér, H. 1946. *Mathematical Methods of Statistics*. Princeton: Princeton University Press.

Cronbach, L.J. and Meehl, P.E. 1955. Construct Validity in Psychological Tests. *Psychological Bulletin* 52: 281-302.

Darwin, C. 1959. *On the Origin of Species by Means of Natural Selection*. London: John Murray.

Dewar, D. and Shelton, H.S. 1947. *Is Evolution Proved?* London: Hollis and Carter.

Dobzhansky, T. 1940. Catastrophism versus Evolutionism. *Science* 92: 356-358.

Eldredge, N. and Gould, S.J. 1972. Punctuated Equilibria: An Alternative to Phyletic Gradualism. In *Models in Paleobiology*, ed. T. J. M. Schopf. San Francisco: Freeman, Cooper and Company, 1972, pp. 82-115.

Feller, W. 1957. *An Introduction to Probability Theory and Its Applications* (2nd ed.). New York: Wiley.

Golden, R.R. and Meehl, P.E. 1978. Testing a Single Dominant Gene Theory without an Accepted Criterion Variable. *Annals of Human Genetics* 41: 507-514.

Golden, R.R. and Meehl, P.E. 1979. Detection of the Schizoid Taxon with MMPI Indicators. *Journal of Abnormal Psychology* 88: 217-233.

Golden, R.R. and Meehl, P.E. 1980. Detection of Biological Sex: An Empirical Test of Cluster Methods. *Multivariate Behavioral Research* 15: 475-496.

Goldschmidt, R.B.G. 1940. *The Material Basis of Evolution*. New Haven: Yale University Press.

Gulliksen, H. 1950. *Theory of Mental Tests*. New York: John Wiley and Sons.

Harman, H.H. 1960. *Modern Factor Analysis*. Chicago: University of Chicago Press.

Hays, W.L. 1973. *Statistics for the Social Sciences* (2nd ed.). New York: Holt, Rinehart and Winston.

Horst, P. 1936. Obtaining a Composite Measure from Different Measures of the Same Attribute. *Psychometrika* 1: 53-60.

Hoyt, C. 1941. Test Reliability Obtained by Analysis of Variance. *Psychometrika* 6: 153-160.

Hull, C.L. 1943. *Principles of Behavior*. New York: Appleton-Century.

Kenney, J.F. 1939. *Mathematics of Statistics*. New York: D. Van Nostrand Company.

Kyburg, H.G.E. 1969. *Probability Theory*. Englewood-Cliffs, NJ: Prentice Hall.

Lewis, D. 1960. *Quantitative Methods in Psychology*. New York: McGraw-Hill.

Lykken, D.T. 1968. Statistical Significance in Psychological Research. *Psychological Bulletin* 70: 151-159.

Meehl, P.E. 1965. Detecting Latent Clinical Taxa by Fallible Quantitative Indicators Lacking an Accepted Criterion. *Reports from the Research Laboratories of the Department of Psychiatry, University of Minnesota*, Report No. PR-65-2. Minneapolis.

Meehl, P.E. 1967. Theory-Testing in Psychology and Physics: A Methodological Paradox. *Philosophy of Science* 34:103-115.

Meehl, P.E. 1968. Detecting Latent Clinical Taxa, II: A Simplified Procedure, Some Additional Hitmax Cut Locators, a Single-Indicator Method, and Miscellaneous Theorems. *Reports from the Research Laboratories of the Department of Psychiatry, University of Minnesota*, Report No. PR-68-4. Minneapolis.

Meehl, P.E. 1970. Nuisance Variables and the Ex Post Facto Design. In Analyses of Theories and Methods in Physics and Psychology. ed. M. Radner and S. Winokur, *Minnesota Studies in the Philosophy of Science*, Volume IV. Minneapolis: University of Minnesota, pp. 373-402.

Meehl, P.E. 1978. Theoretical Risks and Tabular Asterisks: Sir Karl, Sir Ronald, and the Slow Progress of Soft Psychology. *Journal of Consulting and Clinical Psychology* 46: 806-834.

Meehl, P.E. 1979. A Funny Thing Happend to Us on the Way to the Latent Entities. *Journal of Personality Assessment* 43: 563-581.

Meehl, P.E. and Golden, R.R. 1982. Taxometric Methods. In *Handbook of Research Methods in Clinical Psychology*, ed. P. Kendall and J.N. Butcher. New York: Wiley.

Morrison, D.E. and Henkel, R.E., eds. 1970. *The Significance Test Controversy*. Chicago: Aldine.

Perrin, J. 1910. *Atoms*. New York: D. Van Nostrand Co.

Popper, K.R. 1962. *Conjectures and Refutations*. New York: Basic Books.

Popper, K.R. 1972. *Objective Knowledge: An Evolutionary Approach*. Oxford: Oxford University Press.

Popper, K.R. 1974. Intellectual Autobiography. In *The Philosophy of Karl Popper*, ed. P.A. Schilpp. LaSalle: Open Court. pp. 3-181.

Segré, Emilio. 1980. *From X-Rays to Quarks: Modern Physicists and Their Discoveries*. San Francisco: W.H. Freeman and Co.

Thurstone, L.L. 1947. *Multiple-Factor Analysis*. Chicago: University of Chicago Press.

Uspensky, J.V. 1937. *Introduction to Mathematical Probability*. New York: McGraw-Hill.

Yule, G.U. and Kendall, M.G. 1940. *An Introduction to the Theory of Statistics*. London: Charles Griffin.

INDEXES

Author Index

Subject Index

Analytic-synthetic distinction, 105
Astronomy: Aristotelian, 205, 273;
Babylonian, 202-203; Copernican, 69,
70, 75-76, 80-81, 84, 202, 214-217,
220-221, 223, 225-231, 233-258, 311;
Platonic, 203-204; Ptolemaic, 69, 70,
75-76, 80-81, 84, 206-214, 216-217,
220-221, 223, 225-258, 311; Tychonic,
220, 243, 258
Atomic sentences, 111-114
Auxiliary hypotheses, 43: relation to
hypothesis being tested, 45-53, 419-
420, 467-468
Average likelihood, 75-76, 79, 81-82, 89

Basic proposition: solution of, 28; truth
of, 28
Bayes' Theorem, 74-75, 85, 91, 133-
134, 218, 222, 376, 384-386, 454
Bayesian: confirmation theory, 16, 18,
33, 43, 53-55, 62-64, 66-97 *passim*,
99-100, 103-105, 133-134, 138, 300;
subjectivism, 82, 144-145, 218
Behaviorism, 360
Belief, 3, 39-40, 65, 101, 143-144, 302:
and acceptance, 168, 173, 175;
changes in, 106-108; degree of, 34,
62, 88, 100, 102-105, 108, 111, 113,
115, 118, 124, 143-144; functions of,
5, 20-21; logic of, 167; modeling of,
99; reasons for, 169; representation
of, 138-139, 141-142, 155
Bernoulli process, 159-160
Boolean derivative, 14-15, 24-26

Bootstrap condition, 32, 34, 45-46, 49-
50, 52, 56: modified, 47, 48
Bootstrap testing, 4-5, 24-26, 29-32, 35,
43-44, 48, 50-57, 59-61, 63-65, 69-72,
80-81, 96-97, 125-127, 133, 173-176,
180, 224-225, 230-231, 299, 309, 421-
422: and computation, 32, 309-310;
conditions for, 6-11, 16-17;
Hempelian version of, 43-44; and
Newton's argument for universal
gravitation, 180-196 *passim*
Brownian motion, 33

Causal relevance, 324, 332-333, 335,
340-341
Causation: in Freudian theory, 322,
332; and intentionality, 377
Celestial mechanics, 36
Coherence, 100, 104, 106-107, 109-110,
112-119, 136-137, 144
Conditionalization, 17, 100-101, 107,
119, 127, 133, 135, 142, 146-147,
152, 154, 157, 159
Confirmation, 3-5, 17-19, 40, 55, 64-65,
69, 72-74, 77, 83, 85-86, 92, 104,
124-125, 146, 165-166, 174, 176:
conjunction criterion, 55-56, 62, 64;
consequence condition, 8, 52, 57, 59-
61, 92, 96-97; consistency condition,
47, 60, 61; converse consequence
condition, 284; degree of, 63-64, 92,
292; equivalence condition, 10;
instance, 124-125, 311; necessary and
sufficient conditions for, 56, 58-60,
62; and novel predictions, 284-286;

481

John Earman received his doctorate in
philosophy at Princeton University in 1968. He
has taught at UCLA and the Rockefeller
University and is now a professor of philosophy
at the University of Minnesota, where he is also
a member of the Minnesota Center for
Philosophy of Science. He served as editor, with
Clark Glymour and John J. Stachel, of Volume
VIII in the series Minnesota Studies in the
Philosophy of Science, *Foundations of Space-Time
Theories.*